ATOMIC WEIGHTS OF THE ELEMENTS (Based on Carbon-12)

Values in parentheses are used for radioactive elements whose atomic weights cannot be quoted precisely without knowledge of the origin of the elements; the value given is the atomic mass number of the isotope of that element of longest known half-life.

Name	Symbol	Atomic Number	Atomic Weight	Footnotes
Actinium	Ac	89	227.0278	z
Aluminum	Al	13	26.98154	
Americium	Am	95	(243)	
Antimony (Stibium)	Sb	51	121.75	
Argon	Ar	18	39.948	w,x
Arsenic	As	33	74.9216	
Astatine	At	85	(210)	
Barium	Ba	56	137.33	x
Berkelium	Bk	97	(247)	
Beryllium	Be	4	9.01218	
Bismuth	Bi	83	208.9804	
Boron	B	5	10.81	w,y
Bromine	Br	35	79.904	
Cadmium	Cd	48	112.41	x
Calcium	Ca	20	40.08	x
Californium	Cf	98	(251)	
Carbon	C	6	12.011	w
Cerium	Ce	58	140.12	x
Cesium	Cs	55	132.9054	
Chlorine	Cl	17	35.453	
Chromium	Cr	24	51.996	
Cobalt	Co	27	58.9332	
Copper	Cu	29	63.546	w
Curium	Cm	96	(247)	
Dysprosium	Dy	66	162.50	
Einsteinium	Es	99	(252)	
Erbium	Er	68	167.26	
Europium	Eu	63	151.96	x
Fermium	Fm	100	(257)	
Fluorine	F	9	18.998403	
Francium	Fr	87	(223)	
Gadolinium	Gd	64	157.25	x
Gallium	Ga	31	69.72	
Germanium	Ge	32	72.59	
Gold	Au	79	196.9665	
Hafnium	Hf	72	178.49	
Hahnium	Ha	105	(262)	
Helium	He	2	4.00260	x
Holmium	Ho	67	164.9304	
Hydrogen	H	1	1.0079	w
Indium	In	49	114.82	x
Iodine	I	53	126.9045	
Iridium	Ir	77	192.22	
Iron	Fe	26	55.847	
Krypton	Kr	36	83.80	x,y
Lanthanum	La	57	138.9055	x
Lawrencium	Lr	103	(260)	
Lead	Pb	82	207.2	w,x
Lithium	Li	3	6.941	w,x,y
Lutetium	Lu	71	174.967	
Magnesium	Mg	12	24.305	x
Manganese	Mn	25	54.9380	
Mendelevium	Md	101	(258)	
Mercury	Hg	80	200.59	
Molybdenum	Mo	42	95.94	
Neodymium	Nd	60	144.24	x
Neon	Ne	10	20.179	y
Neptunium	Np	93	237.0482	z
Nickel	Ni	28	58.70	
Niobium	Nb	41	92.9064	
Nitrogen	N	7	14.0067	
Nobelium	No	102	(259)	
Osmium	Os	76	190.2	x
Oxygen	O	8	15.9994	w
Palladium	Pd	46	106.4	x
Phosphorus	P	15	30.97376	
Platinum	Pt	78	195.09	
Plutonium	Pu	94	(244)	
Polonium	Po	84	(209)	
Potassium (Kalium)	K	19	39.0983	
Praseodymium	Pr	59	140.9077	
Promethium	Pm	61	(145)	
Protactinium	Pa	91	231.0359	z
Radium	Ra	88	226.0254	x,z
Radon	Rn	86	(222)	
Rhenium	Re	75	186.207	
Rhodium	Rh	45	102.9055	
Rubidium	Rb	37	85.4678	x
Ruthenium	Ru	44	101.07	x
Rutherfordium	Rf	104	(261)	
Samarium	Sm	62	150.4	x
Scandium	Sc	21	44.9559	
Selenium	Se	34	78.96	
Silicon	Si	14	28.0855	
Silver	Ag	47	107.868	x
Sodium (Natrium)	Na	11	22.98977	
Strontium	Sr	38	87.62	x
Sulfur	S	16	32.06	w
Tantalum	Ta	73	180.9479	
Technetium	Tc	43	(98)	
Tellurium	Te	52	127.60	x
Terbium	Tb	65	158.9254	
Thallium	Tl	81	204.37	
Thorium	Th	90	232.0381	x,z
Thulium	Tm	69	168.9342	
Tin	Sn	50	118.69	
Titanium	Ti	22	47.90	
Tungsten (Wolfram)	W	74	183.85	
Uranium	U	92	238.029	x,y
Vanadium	V	23	50.9415	
Xenon	Xe	54	131.30	x,y
Ytterbium	Yb	70	173.04	
Yttrium	Y	39	88.9059	
Zinc	Zn	30	65.38	
Zirconium	Zr	40	91.22	x

w Element for which known variations in isotopic composition in normal terrestrial material prevent a more precise atomic weight being given; values should be applicable to any "normal" material.

x Element for which geological specimens are known in which the element has an anomalous isotopic composition, such that the difference between the atomic weight of the element in such specimens and that given in the table may exceed considerably the implied uncertainty.

y Element for which substantial variations from the value given can occur in commercially available material because of inadvertent or undisclosed change of isotopic composition.

z Element for which the value is that of the radioisotope of longest half-life.

PROBLEMS AND SOLUTIONS FOR

COLLEGE CHEMISTRY

AND

GENERAL CHEMISTRY

SEVENTH EDITIONS

BY HOLTZCLAW, ROBINSON, AND NEBERGALL

JOHN H. MEISER
Ball State University

FREDERICK K. AULT
Ball State University

HENRY F. HOLTZCLAW, JR.
University of Nebraska — Lincoln

WILLIAM R. ROBINSON
Purdue University

D. C. HEATH AND COMPANY
Lexington, Massachusetts   Toronto

Cover: This photograph of gas liquid chromatography columns was included in an exhibition entitled "The Arts and Science," which hung in the Louvre in Paris during the fall of 1979, © Chuck Chaput/Global Focus.

Published simultaneously in Canada.

Printed in the United States of America.

International Standard Book Number: 0-669-06340-1

PREFACE

Through chemistry we try to describe quantitatively the phenomena we observe in ourselves and our surroundings. Much of the work in chemistry thus involves measurement, and many of its concepts require the use of mathematics.

Problems and Solutions for College Chemistry and General Chemistry was developed because of requests by our students to have more problems, such as those presented in their textbook, worked out in detail. Many of the chapters in *College Chemistry* and *General Chemistry,* Seventh Editions, by Holtzclaw, Robinson, and Nebergall, involve discussion of concepts requiring specific use of mathematics. The problems solved in this manual are taken from the questions included at the ends of those chapters.

A major purpose of this manual is to provide students with helpful step-by-step analysis of solutions to specific types of problems. Each unit, keyed to a specific chapter of the Holtzclaw et al. texts, begins with a short introduction to the particular subject area. A detailed section on definitions and formulas pertinent to that subject area and the exercises with their solutions follow. A useful feature to note is that the presentation includes the statement of each problem selected, making the manual suitable for use as a supplement for problem-solving in general chemistry level courses. The exercises selected are indicated in the textbooks by the symbol [S]. At least one example of each type of problem in the chapter is chosen for a detailed solution. A full explanation is provided in addition to the required mathematical manipulations. The selected problems comprise about one-third to one-half the problems in each chapter. Where it seemed appropriate, we have also provided answers to non-numerical exercises that require mathematical insight.

Problems and Solutions for College Chemistry and General Chemistry also contains a self-instructional section (Part Two) that teaches the use of exponential numbers and logarithms. Self-evaluation exercises enable students to check on their progress at appropriate intervals.

We know that many thousands of students have successfully used the textbooks *College Chemistry* and *General Chemistry* through the past six editions in pursuing goals in chemistry and the allied health sciences. The use of this manual should help students become independent learners and gain a better understanding of the quantitative aspects of the sciences.

In producing this manual, we are indebted to our colleagues, our students, and our associates from other institutions for many helpful suggestions; our typist Joanne Beckham for the physical body of the manuscript; and our families, particularly for their understanding during the long hours spent in preparing this work.

J. H. Meiser
F. K. Ault
H. F. Holtzclaw, Jr.
W. R. Robinson

CONTENTS

Part One / Problems and Solutions

Unit 1 Chapter 1: Some Fundamental Concepts *1*

Unit 2 Chapter 2: Symbols, Formulas, and Equations; Elementary Stoichiometry *15*

Unit 3 Chapter 3: Applications of Chemical Stoichiometry *33*

Unit 4 Chapter 4: Structure of the Atom and the Periodic Law *53*

Unit 5 Chapter 5: Chemical Bonding — General Concepts *69*

Unit 6 Chapter 10: The Gaseous State and the Kinetic-Molecular Theory *83*

Unit 7 Chapter 11: The Liquid and Solid States *101*

Unit 8 Chapter 13: Solutions; Colloids *113*

Unit 9 Chapter 14: Acids and Bases *131*

Unit 10 Chapter 15: Chemical Kinetics and Chemical Equilibrium *139*

Unit 11 Chapter 16: Ionic Equilibria of Weak Electrolytes *157*

Unit 12 Chapter 17: The Solubility Product Principle *187*

Unit 13 Chapter 18: Chemical Thermodynamics *205*

Unit 14 Chapter 20: Electrochemistry and Oxidation-Reduction *223*

Unit 15 Chapter 28: Nuclear Chemistry *249*

Part Two / Exponential Notation and Logarithms *257*

Appendixes *279*

Appendix C: Units and Conversion Factors *279*

Appendix E: Solubility Products *280*

Appendix F: Formation Constants for Complex Ions *281*

Appendix G: Ionization Constants of Weak Acids *282*

Appendix H: Ionization Constants of Weak Bases *282*

Appendix I: Standard Electrode (Reduction) Potentials *283*

Appendix J: Standard Molar Enthalpies of Formation, Standard Molar Free Energies of Formation, and Absolute Standard Entropies [298.15 K (25°C), 1 atm] *284*

Appendix K: Composition of Commercial Acids and Bases *291*

Appendix L: Half-Life Times for Several Radioactive Isotopes *291*

Appendix M: Vapor Pressure of Ice and Water at Various Temperatures *292*

PART 1

PROBLEMS AND SOLUTIONS

UNIT 1

CHAPTER 1: SOME FUNDAMENTAL CONCEPTS

INTRODUCTION

This chapter of the textbook includes a brief introduction to the nature of matter and energy. It also details the units of measurement essential for quantifying and communicating observations of scientific phenomena.

The almost universally accepted system of measurement, the metric system, originally was adopted in France and has undergone continuous improvement since the early nineteenth century. In 1960 the Eleventh Conference on Weights and Measures, as a result of major revisions to the metric system, renamed the system the "International System of Units." The abbreviation of the original French title "le Système International d'Unités," SI, is now the preferred form for identifying the system.

The seven base units of the SI are listed in Table 1. Units needed for expressing quantities such as volume, density, velocity, energy, and force are

Table 1 SI Base Units

Quantity	Symbol for Quantity	SI Physical Unit	Symbol
Length	d	Meter	m
Mass	M	Kilogram	kg
Time	t	Second	s
Electric current	I	Ampere	A
Thermodynamic temperature	T	Kelvin	K
Amount of substance	n	Mole	mol
Luminous intensity	I_{ν}	Candela	cd

derived from these base units. This manual, as well as the major textbooks, emphasizes SI units, with the exception of several of the older units, such as the liter, the Celsius or centigrade temperature scale, and the mmHg for pressure, all of which remain very much a part of the language of science. These units, because of their convenience and because existing laboratory glassware and apparatus are mostly labeled in terms of these non-SI units, will remain part of the working language of scientists for some years to come. (Even some

SI units appear to be destined for a distinctively American spelling, such as "meter" rather than the official "metre.")

The problem-solving exercises included in this chapter stress development of skills for each of the following applications: expression of numbers in scientific notation; conversion among SI prefixes; expression of mass-to-volume ratios such as density and specific gravity; conversion of measurements expressed in English units to appropriate SI and derived units; conversion of temperatures from one system to another; and the transfer of heat.

Although scientific work is communicated in terms of SI language, U.S. industry is only gradually moving away from the English system of measurement. As a consequence scientific workers in the near future will have to tolerate working with two systems.

FORMULAS AND DEFINITIONS

The SI base units are frequently large with respect to the quantities to be measured. For example, the mass of a dime is approximately 0.0025 kg and the wavelength of red light is 0.0000007 m. To avoid the cumbersome use of decimals, a system of prefixes specifying a multiplication factor to use with the base units has been developed. The most commonly employed prefixes are shown in Table 2. The SI unit is multiplied by the prefix that permits the numerical value of the measurement to be kept as near to 1 as is practicable.

Table 2 Common Prefixes for Measurement Units

Prefix	Symbol	Factor Times Unit	
Kilo	k	1000	$= 10^{3}$ times
Deci	d	1/10	$= 10^{-1}$ times
Centi	c	1/100	$= 10^{-2}$ times
Milli	m	1/1000	$= 10^{-3}$ times
Micro	μ	1/1,000,000	$= 10^{-6}$ times
Nano	n	1/1,000,000,000	$= 10^{-9}$ times
Pico	p		$= 10^{-12}$ times

Length (meter, m) The meter is the standard unit of length and it is used in conjunction with the prefixes for most linear measurements in science. Another commonly employed non-SI unit is the angstrom ($1\ \text{Å} = 10^{-10}$ m), which appears in measurements and calculations in spectrographic work.

Mass (kilogram, kg) The kilogram is the standard unit of mass and is a measure of the quantity of matter a body possesses. The mass of a body is constant while at rest regardless of geographic position and always is considered constant for chemical studies. Weighings made in the chemical laboratory are balanced against the mass of an object used as a standard; values obtained in this manner are masses. Because the kilogram is a large unit of mass, the more familiar mass in laboratory practice is the gram (g), defined as 1/1000 kg. Stated differently, 1000 g = 1 kg.

Volume (cubic meter, m^3) The cubic meter, the standard unit of volume, is not widely used for measuring laboratory quantities. The volumes of laboratory glassware and other lab ware are more conventionally expressed in cubic centimeters, cm^3, or in terms of liters. Although the liter (L) is not an

SI unit, its value is defined to be one cubic decimeter (dm^3) or 0.001 m^3. The milliliter (mL) and the cubic centimeter (cm^3) now are defined as equal: 1 mL = 1 cm^3; and 1000 mL = 1000 cm^3 = 1.0 L.

Density (D) The density of a substance is the ratio of its mass M to its volume V. The usual expression for the density of solids and liquids is mass (grams) per cubic centimeter or per milliliter. Based on these units, the densities of elements range from a low of 8.99×10^{-5} g/cm^3 for hydrogen gas to a high of 22.57 g/cm^3 for osmium. Among compounds, water has a maximum density of 1.0 g/cm^3 at 3.98°C. The densities of gases — such as hydrogen or oxygen, for example — are typically reported in units of grams per liter to avoid the excessive use of decimals.

$$\text{Density} = \frac{\text{mass}}{\text{volume}} = \frac{M}{V} = \frac{\text{g}}{\text{mL or cm}^3}$$

Energy (joules, J) Energy is the capacity to do work, where work is the movement of a mass against an opposing force. The measurement of energy is expressed in units of joules.

Heat (q) Heat is a form of energy arising from the motion of particles composing a substance such as a solid or fluid. Like other forms of energy, heat is measured by using the derived unit for energy called the joule (J). The calorie is an older unit of energy and is still very much a part of our language. It is widely used for work in nutrition, physiology, and the health sciences in general. For purposes of conversion, 1 joule equals 0.239 calories or 1 calorie (cal) equals 4.184 J. The calorie is liberally defined as the amount of heat required to raise the temperature of 1.0 g of water 1.0°C. The nutrition calorie (Cal) equals 1.0 kcal or 4184 J (4.184 kJ).

Specific gravity (sp gr) The specific gravity of a substance is the ratio of the density of that substance to the density of a substance used as a reference standard. (Both densities are measured at the same temperature.) Water generally serves as the standard for both solids and liquids; air is the standard for gases. Specific gravity may be expressed alternatively as the ratio of the masses of equal volumes of two substances, where one of the substances is arbitrarily designated as the standard. (Again both substances must be at the same temperature.) Notice from the following expression that specific gravity values are unitless:

$$\text{sp gr} = \frac{\text{density of unknown}}{\text{density of known}} = \frac{\text{g/cm}^3}{\text{g/cm}^3} \quad \text{(Units divide out, leaving a dimensionless value.)}$$

In essence, the specific gravity of a body can be thought of as a measure of "heaviness." For example, when a person says that a cork stopper is much "lighter" than a lead stopper of the same size, what is implied is that lead has a much greater density than cork. Indeed, if the masses of both are compared to the mass of an equal volume of water, the specific gravities of lead and cork will be about 11.4 and 0.2, respectively; that is, lead is 11.4 times as dense as water and cork is 0.2 times as dense as water.

Specific heat capacity (C) Energy is intangible in that only changes in energy can be measured. Although the joule or calorie serves as the standard unit of energy, changes in heat must be determined by measuring temperature changes of specific substances. The specific heat capacity (formally known as the specific heat) of a substance is the amount of heat required to change the temperature of 1 g of that substance by 1°C; the specific heat capacity for a given substance is unique to that substance. An algebraic interpretation of this definition applied to any substance is

$$\begin{array}{ccccccc} \text{Heat} & = & \text{mass} & \times & \text{S.H.} & \times & \text{temp change} \\ Q & = & M & \times & C & \times & \Delta T \end{array}$$

The transfer of heat between two bodies (heat flow) always occurs from the hotter body to the colder body. In the process of heat flow both bodies eventually reach an equilibrium temperature. Because heat is conserved, the amount of heat lost by the hotter body equals the amount of heat gained by the colder body.

$$Q_{lost} = Q_{gained}$$

Significant figures In making measurements it is important to recognize the limitations of the measuring instruments. For example, a typical bathroom scale, although possibly accurate, is calibrated to register weights to the nearest pound and would not be useful for measuring the small quantities used in laboratory experiments; the uncertainty in each measurement would be too great.

Laboratory measurements generally require a higher degree of precision or less uncertainty than would be required for weighing oneself on a scale. Common laboratory balances used in general chemistry have a range of uncertainties from 0.01 g to 0.1 g for simple beam balances and from 0.0001 g to 0.001 g for analytical balances.

The reporting of measurements should be made in a way that conveys the maximum information about the measurements to the user. Numbers that are actually read from an instrument are considered to be ***significant figures***. As a general rule, answers to mathematical operations may not contain more significant figures (less uncertainty) than the operant with the smallest number of significant figures (greatest uncertainty). The solved problems in this manual follow the rules on significant figures as these are presented in the textbooks. Many of the solved problems have been analyzed into steps; and in these cases, intermediate answers are written with one more significant figure than is justified in the final answer. *Computations have been made with an electronic calculator. Generally, all the numbers that were generated in the calculator resulting from mathematical operations have been carried through to the final answer. The answer is rounded to the proper number of significant figures at the end of the computation.*

Use of prefixes A conversion scheme using some common prefixes and examples is shown below.

$$\text{Length} \quad \text{km} \underset{1000\,\div}{\overset{\times\,1000}{\rightleftharpoons}} \text{m} \underset{100\,\div}{\overset{\times\,100}{\rightleftharpoons}} \text{cm} \underset{10\,\div}{\overset{\times\,10}{\rightleftharpoons}} \text{mm} \underset{1000\,\div}{\overset{\times\,1000}{\rightleftharpoons}} \mu\text{m} \underset{1000\,\div}{\overset{\times\,1000}{\rightleftharpoons}} \text{nm}$$

$$\text{Mass} \quad \text{kg} \underset{1000\,\div}{\overset{\times\,1000}{\rightleftharpoons}} \text{g} \underset{1000\,\div}{\overset{\times\,1000}{\rightleftharpoons}} \text{mg} \underset{1000\,\div}{\overset{\times\,1000}{\rightleftharpoons}} \mu\text{g}$$

$$\text{Volume} \quad \text{m}^3 \underset{1000\,\div}{\overset{\times\,1000}{\rightleftharpoons}} \text{dm}^3 = \text{liter (L)} \underset{1000\,\div}{\overset{\times\,1000}{\rightleftharpoons}} \text{mL} \underset{1000\,\div}{\overset{\times\,1000}{\rightleftharpoons}} \mu\text{L}$$

USE OF CALCULATORS

Hand-held calculators are now available to most students. Calculators come in a wide variety of brands and designs, and so it would be impossible to attempt to give the details of operation for all of them. However, because Texas Instrument's TI* calculators are very popular among students and because most calculators operate according to the same basic principles, this chapter introduces a boxed section in some problems called *Keystrokes*, a feature also included in later chapters. *Keystrokes* sections will not appear in each problem, but only in those problems in which mathematical operations other than ×, ÷, +, and - occur. Thus, when exponentials are introduced, the student will have both the standard mathematical form to observe as well as the actual keystrokes to operate the calculator. Some *Keystrokes* sections will appear even with the four basic mathematical operations to demonstrate proper use of rounding procedures for expressing the proper number of significant figures.

EXERCISES

22. Perform the following calculations, and give each answer with the correct number of significant figures.

(b) $5.63 \times 10^2 \times 7.4 \times 10^3$

Solution

Because calculators provide answers to mathematical operations without regard to significant figures, it is important to be able to properly determine the number of significant figures in order to obtain a meaningful answer.

$$5.63 \times 10^2 \times 7.4 \times 10^3$$

Keystrokes on TI* Calculators: 5.63 [EE] 2 × 7.4 [EE] 3 = ans. 4.1662 06

The rule for multiplying or dividing is that the product or quotient should contain no more digits than the smaller number of significant figures in the computation. Applying this rule we find only two significant figures in 7.4×10^3, whereas 5.63×10^2 has three. Therefore, only two numbers may be expressed in the answer. Because there are five significant figures in 4.1662, this number must be rounded to only two places. The number 6 in the third place from the left is greater than 5 and, therefore, 1 is rounded to the next larger number. Thus, the answer is 4.2×10^6.

(c) $\frac{2734}{28.0}$

Solution

Keystrokes: 2734 ÷ 28.0 = ans. 97.64285719

---

* Trademark Texas Instruments.

Applying the same rule here, there are three significant figures in 28.0 and four in 2734. Only three figures may be maintained in the answer.. The answer is, therefore, 97.6 because the fourth number is less than 5.

(f) $\dfrac{42.7 + 0.259}{28.4445}$

Solution

Keystrokes: 42.7 + 0.259 = ÷ 28.4445 = ans. 1.510274394

Here the rule for addition and subtraction requires that the number of significant figures in the answer be the same as that number with the least number of decimal places. Both 42.7 and 0.259 have three significant figures. Only three significant figures result from the addition, the same as the lesser of the two numbers involved in division. Applying the rule for division stated earlier, only three significant figures are allowed in the answer. The answer is, therefore, 1.51.

29. Use exponential notation to express the following quantities in terms of SI base units (See Section 1.8):

(a) 0.13 g

Solution

Conversions among units must begin with defining the units as identities or equivalencies relative to each other. The definition can be rearranged into a fraction with the unit of the quantity to be converted in the denominator. Multiplication of this fraction times the quantity yields the desired unit. Prefixes used with SI units are exact definitions, and the number of significant figures included in computations depends only on the number of significant figures in the measurements.

$$1000 \text{ g} = 1 \text{ kg}$$

$$\text{Mass in g} = 0.13 \text{ g} \times \frac{1 \text{ kg}}{1000 \text{ g}} = 0.00013 \text{ kg} = 1.3 \times 10^{-4} \text{ kg}$$

Only two significant figures are allowed.

(b) 232 Gg

Solution

$1 \text{ Gg} = 10^9 \text{ g}$ (Note: base unit is 1 kg = 10^3 g.)

$$232 \text{ Gg} \times \frac{10^9 \text{ g}}{\text{Gg}} \times \frac{1 \text{ kg}}{10^3 \text{ g}} = 232 \times 10^6 \text{ kg} = 2.32 \times 10^8 \text{ kg}$$

33. Calculate the length of a football field (100.0 yd) in meters and kilometers.

Solution

The English measure, yard, is converted to meters. From Appendix C, 1 m = 1.094 yd. This unit conversion factor is used to convert from one system of units to another system of units. To convert yards to meters, yards must be multiplied by a factor that has yards in the denominator so that the unit yard will cancel. Thus,

$$100.0 \text{ yd} \times \frac{1.000 \text{ m}}{1.094 \text{ yd}} = 91.41 \text{ m}$$

To convert m to km use the unit conversion 1000 m = 1 km:

$$91.41 \text{ m} \times \frac{1 \text{ km}}{1000 \text{ m}} = 9.141 \times 10^{-2} \text{ km}$$

35. How many milliliters of a soft drink are contained in a 12-oz can?

Solution

Note that a liquid measure is required and thus the conversion oz = 28.35 g is not appropriate. From Appendix C, 1 liquid qt (U.S.) = 0.9463 L; there are 32 oz in 1 qt. Because 0.9463 L = 946.3 mL, then 946.3 mL/32 oz = 29.57 mL/oz. We now have the required unit conversion factor:

$$12 \text{ oz} \times 29.57 \text{ mL/oz} = 355 \text{ mL}$$

41. Gasoline is now often sold by the liter. How many liters are required to fill a 12.0-gal gas tank?

Solution

There are 4 qt per gallon and, from Appendix C, 1 liquid qt = 0.9463 L; the unit conversion factor is, therefore,

$$\frac{4 \text{ qt}}{\text{gal}} \times \frac{0.9463 \text{ L}}{1 \text{ qt}} = \frac{3.785 \text{ L}}{\text{gal}}$$

Then,

$$12.0 \text{ gal} \times 3.785 \text{ L/gal} = 45.4 \text{ L}$$

42. In a recent Belgian Grand Prix, the leader turned a lap with an average speed of 182.83 km/h. What was his speed in miles per hour, meters per second, and feet per second?

Solution

In the first part of this problem the time unit remains the same; as in previous problems only the distance need be converted. Because 1 mi = 1.609 km,

$$182.83 \frac{\text{km}}{\text{h}} \times \frac{1 \text{ mi}}{1.60934 \text{ km}} = 113.61 \text{ mi/h}$$

Both numerator and denominator change in this part. The conversions are familiar.

$$182.83\ \frac{km}{h} \times 1000\ \frac{m}{km} \times \frac{h}{60\ min} \times \frac{1\ min}{60\ s} = 50.786\ m/s$$

$$182.83\ \frac{km}{h} \times \frac{1\ mi}{1.60934\ km} \times \frac{5280\ ft}{1\ mi} \times \frac{1\ h}{3600\ s} = 166.62\ ft/s$$

45. Make the conversion indicated in each of the following:

(b) the height of Mt. Kilimanjaro, at 19,565 ft the highest mountain in Africa, to kilometers

Solution

Again, a unit conversion is required from Appendix C. The change to metric units is gradually occurring in the United States, so it is a good idea to memorize some of the more common conversions, such as 1 in = 2.54 cm. Because 12 in = 1 ft, 100 cm = 1 m, and 1000 m = 1 km, formation of factors to cancel all units except km gives

$$19{,}565\ ft \times \frac{12\ in}{1\ ft} \times \frac{2.54\ cm}{1\ in} \times \frac{1\ m}{100\ cm} \times \frac{1\ km}{1000\ m} = 5.9634\ km$$

Note that all of the unit conversions are exact numbers so that all significant figures in the original height should be kept.

(d) the area of the state of Oregon, 96,981 mi^2, to square kilometers

Solution

It is important to remember that if a squared quantity is to be converted, the unit conversion factors must also be squared. From Appendix C, 1 mi = 1.60934 km. Then,

$$96{,}981\ mi^2 \times \left(\frac{1.60934\ km}{1\ mi}\right)^2 = 96{,}981 \times 2.5900 = 2.5118 \times 10^5\ km^2$$

Keystrokes: 1.60934 [2nd] [x^2] × 96981 = ans. 251178.39

Note that five places are allowed in the final answer because the factor 1.60934 may be considered exact.

(f) the displacement of an automobile engine, 161 in^3, to liters

Solution

What was written in part (d) also applies to cubed quantities. Here the conversion is 1 in = 2.54 cm. The expression, therefore, is

$$161\ in^3 \times \frac{(2.54\ cm)^3}{(1\ in)^3} \times \frac{1\ L}{1000\ cm^3} = 2.64\ L$$

48. If a line of 1.0×10^8 water molecules is 1.00 in long, what is the average diameter of a water molecule in centimeters, angstrom units, and nanometers?

Solution

Assume that each molecule is placed in contact with the one adjacent to it and that each molecule is spherical. Then each molecule's diameter is $1/1.0 \times 10^8$ of the total length of 1 in = 2.54 cm. Consequently,

$$\text{Diameter} = 2.54 \text{ cm} \times \frac{1}{1.0 \times 10^8} = 2.5 \times 10^{-8} \text{ cm}$$

$$\text{Diameter} = 2.5 \times 10^{-8} \text{ cm} \times \frac{1 \text{ Å}}{10^{-8} \text{ cm}} = 2.5 \text{ Å}$$

$$\text{Diameter} = 2.5 \times 10^{-8} \text{ cm} \times \frac{1 \text{ m}}{100 \text{ cm}} \times \frac{1 \text{ nm}}{10^{-9} \text{ m}} = 0.25 \text{ nm}$$

49. The density of liquid bromine is 2.928 g/cm^3. What is the mass of 25 mL of bromine? What is the volume of 100.0 g of bromine?

Solution

$$1 \text{ mL} = 1 \text{ cm}^3 \qquad D = 2.928 \text{ g/mL}$$

$$M \text{ (Br)} = 25 \text{ mL} \times 2.928 \text{ g/mL} = 73 \text{ g}$$

Only two significant figures are allowed because there are only two in 25 mL.

$$V \text{ (Br)} = \frac{100.0 \text{ g}}{2.928 \text{ g/cm}^3} = 34.15 \text{ cm}^3$$

51. Osmium is the densest element known. What is its density if 2.72 g has a volume of 0.121 cm^3?

Solution

$$D = \frac{M}{V} = \frac{2.72 \text{ g}}{0.121 \text{ cm}^3} = 22.5 \text{ g/cm}^3$$

53. What is the volume of the following?

(b) 3.28 g of gaseous hydrogen, density = 0.089 g/L

Solution

Density is calculated in exactly the same way for each state of matter whether it be a gas, liquid, or solid.

$$D = \frac{M}{V}$$

$$V = \frac{M}{D} = \frac{3.28 \text{ g}}{0.089 \text{ g/L}} = 37 \text{ L}$$

54. What is the specific gravity of uranium at 25°C relative to water if 37.4 g of uranium has a volume of 2.00 cm^3?

Solution

Specific gravity is the comparison of one density to the density of a standard. The density of uranium is

$$D(\mathrm{U}) = \frac{37.4\ \mathrm{g}}{2.00\ \mathrm{cm}^3} = 18.7\ \mathrm{g/cm}^3$$

The density of water is taken as 0.99707 g/cm^3. Therefore,

$$\frac{D(\mathrm{U})}{D(\mathrm{H_2O})} = \frac{18.7\ \mathrm{g/cm}^3}{0.99707\ \mathrm{g/cm}^3} = 18.8$$

59. The *Voyager 1* flyby of Saturn revealed that the surface temperature of the moon Titan is 93 K. What is the surface temperature in degrees Celsius and degrees Fahrenheit?

Solution

$$\mathrm{K} = {}^\circ\mathrm{C} + 273.15$$

$${}^\circ\mathrm{C} = \mathrm{K} - 273.15 = 93 - 273.15 = -180^\circ\mathrm{C}$$

$${}^\circ\mathrm{F} = \frac{9}{5}{}^\circ\mathrm{C} + 32 = \frac{9}{5}(-180) + 32 = -292^\circ\mathrm{F}$$

Keystrokes: 9 × 180 [+/-] ÷ 5 + 32 = ans. -292

61. Make the conversion indicated in each of the following:

(a) temperature of the interior of a rare roast, 125°F, to degrees Celsius.

Solution

$${}^\circ\mathrm{C} = \frac{5}{9}({}^\circ\mathrm{F} - 32) = \frac{5}{9}(125 - 32) = 52^\circ\mathrm{C}$$

(d) temperature of dry ice, -77°C, to degrees Fahrenheit and Kelvin.

Solution

$${}^\circ\mathrm{F} = \frac{9}{5}{}^\circ\mathrm{C} + 32 = \frac{9}{5}(-77) + 32 = -107^\circ\mathrm{F}$$

$$\mathrm{K} = {}^\circ\mathrm{C} + 273.15 = -77 + 273.15 = 196\ \mathrm{K}$$

62. If 1.506 kJ of heat are added to 30.0 g of water at 26.5°C, what is the resulting temperature of the water?

Solution

The formula relating the quantities in the problem is

$$\text{Heat added} = \text{mass} \times \text{specific heat} \times \text{temperature change}$$

Substitution gives

$$1506 \text{ J} = 30.0 \text{ g} \times 4.184 \text{ J/g °C} \times (\Delta T)$$

$$\Delta T = \frac{1506 \text{ J}}{30.0 \text{ g} \times 4.184 \text{ J/g °C}} = 12.0\text{°C}$$

Because the starting temperature is 26.5°C and the temperature increases 12.0°C, the final temperature is 26.5°C + 12.0°C = 38.5°C.

There is, however, another way to calculate the new temperature directly. Because ΔT is $T_f - T_i$, where T_f is the final temperature and T_i is the initial temperature, then

$$1506 \text{ J} = 30.0 \text{ g} \times 4.184 \text{ J/g °C} \times (T_f \text{ °C} - 26.5\text{°C})$$

$$1506 \text{ J} = (125.52 \text{ J})T_f - 3326.28 \text{ J}$$

$$T_f = \frac{1506 \text{ J} + 3326.28 \text{ J}}{125.52 \text{ J}} = 38.5\text{°C}$$

The answer is, of course, the same in the two methods.

64. The specific heat of ice is 1.95 J/g °C. How much heat in joules and calories is required to heat a 35.5-g ice cube from -23.0°C to -1.0°C?

Solution

Heat required = mass x specific heat x temperature change

$$\text{Heat required} = 35.5 \text{ g} \times 1.95 \text{ J/g °C} \times (-1.0 - -23.0)\text{°C}$$
$$= 1.52 \times 10^3 \text{ J}$$

From Appendix C, 1 cal = 4.184 J.

$$\text{Heat} = 1.5229 \times 10^3 \text{ J} = 1.5229 \times 10^3 \text{ J} \times \frac{1 \text{ cal}}{4.184 \text{ J}} = 364 \text{ cal}$$

65. Calculate the heat capacity of each of the following:

(a) 22.6 g of water (specific heat = 4.184 J/g °C).

Solution

The heat capacity of 22.6 g of water is

$$22.6 \text{ g} \times 4.184 \text{ J/g °C} = 94.6 \text{ J/°C}$$

67. How many milliliters of water at 25°C must be mixed with 150 mL (about 5 oz) of coffee at 100°C so that the resulting combination will have a temperature of 70°C? Assume that coffee and water have the same density and specific heat.

Solution

This is a heat balance problem.

Heat lost by coffee = heat gained by water

Because the densities are assumed equal and appear on both sides of the equation, the actual value does not matter. For convenience, assume that the density is 1.00 g/mL. This assumption makes 150 mL = 150 g. The set-up is as follows:

$$150 \text{ g (coffee)} \times \text{S.H.} \times (100 - 70) = X \text{ g (water)} \times \text{S.H.} \times (70 - 25)$$

Note that no value appears for S.H. because it will cancel on both sides, and ΔT is written as a difference of temperature ($T_f - T_i$) in °C. Solving

$$150 \text{ g} \times 30 = x \times 45$$

$$x = 100 \text{ g}$$

Dividing by the assumed density 1.00 g/mL, gives the volume 100 mL.

73. Given that 1 bushel (bu) is defined as 32 dry qt (1 dry qt = 1.1012 L), what is the mass in kilograms of a liter of the following types of grain?

(c) corn, 56.0 lb/bu

Solution

From Appendix C, 1 qt (dry) = 1.1012 L, and 1 lb = 0.45359237 kg. We use unit conversions to convert lb/bu to kg/L.

$$56.0 \frac{\text{lb}}{\text{bu}} \times \frac{1 \text{ bu}}{32 \text{ qt}} \times \frac{1 \text{ qt (dry)}}{1.1012 \text{ L}} \times \frac{0.454 \text{ kg}}{1 \text{ lb}} = 0.721 \text{ kg/L}$$

Notice that in solving this problem, dry measure was used for the quart quantity. This confusing difference with liquid measures again emphasizes the need to complete the change to the metric system.

75. Moon rock is estimated to contain about 0.1% water. What mass of moon rock (in pounds) would a moon base have to process to recover 1 L of water (density = 1.0 g/cm^3)?

Solution

First convert the required volume of water to mass in pounds. Then determine the mass of rock required.

$$1 \text{ L } (H_2O) = 1000 \text{ mL} \times 1.0 \text{ g/mL} = 1000 \text{ g}$$

$$1 \text{ lb} = 453.59237 \text{ g}$$

so that

$$\frac{1000 \text{ g}}{453.59237 \text{ g/lb}} = 2 \text{ lb}$$

Only one significant figure is justified because the 1 L is known only to one place. Because 2 lb is 0.1% of the total,

$$\frac{2 \text{ lb}}{0.001} = 2000 \text{ lb}$$

76. A solid will float in a liquid that has a density greater than its own. A pound of butter forms a block about 6.5 × 6.5 × 12.0 cm. Will this butter float in water (density = 1.0 g/cm^3)?

Solution

First determine the density of butter.

$$\text{Volume} = 6.5 \text{ cm} \times 6.5 \text{ cm} \times 12.0 \text{ cm} = 507 \text{ cm}^3$$

$$\text{Mass} = 1 \text{ lb} = 454 \text{ g}$$

$$\text{Density} = \frac{M}{V} = \frac{454 \text{ g}}{507 \text{ cm}^3} = 0.89 \text{ g/cm}^3$$

Yes, it will float; its density is 0.89 g/cm^3, which is less than 1.0 g/cm^3.

82. What mass, in kilograms, of concentrated hydrochloric acid is contained in a standard 5.0-pt container? The specific gravity of concentrated hydrochloric acid is 1.21.

Solution

$$1 \text{ pt is } \frac{1}{2} \text{ qt}; \quad 1 \text{ qt} = 0.9463 \text{ L}$$

$$\text{Volume of acid} = 5 \text{ pt} \times \frac{1 \text{ qt}}{2 \text{ pt}} \times \frac{0.9463 \text{ L}}{1 \text{ qt}} = 2.366 \text{ L}$$

Compared to water, which would weigh 2.366 kg (because the density of water is 1.00 g/cm^3), hydrochloric acid is 1.21 times as dense. Therefore,

$$2.366 \text{ kg} \times 1.21 = 2.9 \text{ kg}$$

RELATED EXERCISES

1. Recently a sign in front of a fast-food restaurant indicated that 40 billion hamburgers had been sold. If the buns used for the sandwiches have a diameter of 4 in., what length (in miles) of a chain of buns could be constructed from this number of buns? in meters?
   Answer: 2.5×10^6 mi; 4.0×10^9 m

2. Assuming you walk on the chain of buns described above, and you can walk 8 km/h, how long would it take you to walk the chain?
   Answer: 5.0×10^5 h; 2.1×10^4 d; 58 yr

3. Meteorologists usually indicate the amount of rainfall in inches and centimeters. Let's assume that a square city block, 0.10 mile per edge, is drenched by 3.0 cm of rain. Calculate the volume of water (in gallons) resulting from the rainfall in this area. What is the volume in liters?
   Answer: 2.1×10^5 gal; 7.8×10^5 L

4. Which one of the following numbers has the greatest number of significant figures?

(a) 1.0001 g (b) 2.2000 mL (c) 0.000001 g
(d) 4.01010 mL (e) 1.26×10^8 kg

Answer: d

5. Given the following data, calculate the density of ethanol (C_2H_5OH).

weighing bottle plus ethanol	162.65 g
weighing bottle	126.95 g
volume of ethanol	45.25 mL

Answer: 0.789 g/ml

6. The food energy content of white sugar is 3.85 kcal/g. Given that 1 J of energy is sufficient to lift a 1.0-kg mass to a height of 4.0 in., calculate the amount of white sugar that would be required at 50% utilization efficiency for a 160-lb gymnast to climb a 5.0-m rope.

Answer: 0.44 g

UNIT 2

CHAPTER 2: SYMBOLS, FORMULAS, AND EQUATIONS; ELEMENTARY STOICHIOMETRY

INTRODUCTION

It is important in chemistry to understand the interactions of matter that occur both in nature and in the laboratory. These interactions are described in the language of chemistry, which includes symbols, formulas, equations, and the interpretations used to provide qualitative and quantitative information about the substances involved.

Symbols consisting of a single capital letter or a capital letter and a small letter, such as O, N, Cl, Na, and S, are used as shorthand abbreviations for elements. Combinations of symbols are used to write formulas of compounds, for example, methane, CH_4; vitamin C (ascorbic acid), $C_6H_8O_6$; and table sugar (sucrose) $C_{12}H_{22}O_{11}$. Elements and compounds enter into chemical reactions that are described in the form of an equation using symbols and formulas.

Chemical equations are similar to household recipes in that both provide information for preparing products from particular quantities of reactants. Unlike household recipes, however, chemical equations usually do not detail reaction conditions or laboratory procedures other than noting that energy or a specific catalyst must be used in conjunction with the reactants. The purpose of a chemical equation is to convey quickly to the reader pertinent information about the reaction that includes the substances involved and their quantitative ratio.

In this chapter you have the opportunity to study quantitative aspects of reactions. By working with the symbols, formulas, and equations introduced here, you soon will become comfortable with the basic language tools of chemistry. The problems given here involve using formulas and equations to calculate the amounts of substances involved in reactions, determining the formulas of substances from experimental data, and using solutions in the laboratory.

FORMULAS AND DEFINITIONS

Atomic weight The atomic weight of an element is the average mass of all atoms composing a normal sample of the element. For example, consider a sample of naturally occurring carbon atoms. Most of the carbon atoms

(98.89%) would contain six protons and six neutrons, ^{12}C; some (1.11%) would contain six protons and seven neutrons, ^{13}C; and a trace (1 atom in 10^{12} atoms) would contain six protons and eight neutrons, ^{14}C. Carbon-12, -13, and -14 are called isotopes of carbon and differ only in the number of neutrons in their nuclei.

Carbon-12 has been arbitrarily assigned a mass of exactly 12.00 and is used as the reference standard for determining the masses of other atoms. Based on this standard the atomic weight of carbon is 12.011 and represents the average mass of the atoms composing a statistical sample of carbon atoms.

Empirical formula A formula that gives the simplest whole number ratio of atoms.

Formula weight (FW) The formula weight of a substance is the sum of the atomic weights of all of the atoms appearing in a formula. For example, the formula weight of copper(II) sulfate, $CuSO_4$, is calculated in the following way:

No. atoms: 1 Cu 1 S 4 O

$$FW = 1(63.546) + 1(32.06) + 4(15.9994) = 159.60$$

Formula weights are conventionally written as unitless entities, with atomic mass units implied or understood.

Molecular weight (mol. wt) The formula weight of a substance that exists as discrete molecules.

Gram-atomic weight (GAW) The atomic weight of an element expressed in grams.

Gram-formula weight (GFW) The formula weight of a substance expressed in grams.

Gram-molecular weight (GMW) The molecular weight of a substance expressed in grams.

Mole (mol) The mole is defined as a number equal to the number of atoms in exactly 12 grams of carbon-12. This number, called Avogadro's number, 6.022×10^{23}, has wide application in chemistry. For the purposes of this chapter, 1 mole (mol) can be interpreted in the following ways:

The number of atoms in a gram-atomic weight of an elemental substance. For example, the number of atoms in 22.9898 g of sodium (Na) is 6.022×10^{23}.

The number of formula units in a gram-formula weight of an ionic substance. For example, the number of formula units in 58.443 g of sodium chloride (NaCl) is 6.022×10^{23}.

The number of molecules in a gram-molecular weight of a molecular substance. For example, the number of molecules in 44.010 g of carbon dioxide (CO_2) is 6.022×10^{23}.

Identities used for computations involving the mole and quantities of substances should be interpreted as equivalencies and not equalities as in mathematics. As cited above, 1 mol of sodium atoms (GAW) has a mass of 22.9898 g, and the equivalence is written

$$1 \text{ mol Na} \triangleq \text{GAW Na} = 22.9898 \text{ g} \triangleq 6.022 \times 10^{23} \text{ atoms}$$

where the symbol $\hat{=}$ is read "is equivalent to" or "corresponds to." This equivalence can be used to calculate the amount of a substance (in moles) contained in a specified amount of that substance in the following ways:

$$\text{Moles} = \text{mass of substance} \times \frac{1 \text{ mol}}{\text{GAW, GFW, or GMW of substance}}$$

or $= \text{no. of particles (atoms, formula units, or molecules)}$

$$\times \frac{1 \text{ mol}}{6.022 \times 10^{23} \text{ particles}}$$

EXERCISES

9. A sample of TlCl contains 0.85217 g of Tl and 0.14783 of Cl. From this information calculate the atomic weight of Tl using 35.453 as the atomic weight of Cl.

Solution

From the formula, 1 atom of thallium (Tl) combines with 1 atom of chlorine (Cl). The mass of each element is in direct proportion to its respective atomic weight and to its mass in the compound. Consequently, the mass ratio of Tl to Cl is

$$\frac{\text{Mass of Tl}}{\text{Mass of Cl}} = \frac{\text{at. wt Tl}}{35.453} = \frac{0.85217 \text{ g}}{0.14783 \text{ g}}$$

$$\text{At. wt Tl} = 5.7645 \times 35.453 = 204.37$$

11. Calculate the molecular weight of each of the following compounds:

(d) silver sulfate, Ag_2SO_4

Solution

Mol. wt = molecular weight

= sum of all atomic weights of the elements that appear in the formula

At. wts: Ag = 107.870, S = 32.06, O = 15.9994

Mol. wt = 2(107.868) + 32.06 + 4(15.9994) = 311.79

Keystrokes: 2 × 107.868 + 32.06 + 4 × 15.9994 = ans. 311.7936

The rule for addition of significant figures applies here. Only two numbers after the decimal point may be kept since the atomic weight of S is known only to two decimal places.

(e) acetic acid, CH_3CO_2H

Solution

At. wts: C = 12.011, H = 1.0079, O = 15.9994

Mol. wt = 2(12.011) + 4(1.0079) + 2(15.9994) = 60.052

13. Determine the number of moles in each of the following:

(a) 2.12 g of potassium bromide, KBr.

Solution

In order to determine the number of moles in a given number of grams of KBr, the molecular weight of KBr must first be known. Proceed as in Exercise 11.

$$\text{At. wts: } K = 39.098, \quad Br = 79.904$$

$$1 \text{ mol KBr} = 39.098 \text{ g} + 79.904 \text{ g} = 119.002 \text{ g}$$

$$\text{Amount of KBr} = 2.12 \text{ g KBr} \times \frac{1 \text{ mol}}{119.002 \text{ g}}$$

$$= 0.0178 \text{ mol} = 1.78 \times 10^{-2} \text{ mol}$$

(b) 235 g of calcium carbonate, $CaCO_3$

Solution

$$1 \text{ mol } CaCO_3 = 40.08 \text{ g} + 12.001 \text{ g} + 3(15.9994 \text{ g}) = 100.08 \text{ g}$$

$$\text{Amount of } CaCO_3 = 235 \text{ g } CaCO_3 \times \frac{1 \text{ mol}}{100.08 \text{ g}} = 2.35 \text{ mol}$$

15. Up to 0.0100% by mass of copper(I) iodide may be added to table salt as a dietary source of iodine. How many moles of CuI would be contained in 1 lb (454 g) of such supplemented table salt?

Solution

This is a two-part problem. The permitted amount of CuI must be found and then the number of moles in that amount must be found. The maximum amount permitted is 0.0100%. This corresponds to

$$0.000100 \times 454 \text{ g} = 4.54 \times 10^{-2} \text{ g}$$

$$1 \text{ mol CuI} = 63.546 \text{ g} + 126.9046 \text{ g} = 190.451 \text{ g}$$

$$\text{Amount of CuI} = 4.54 \times 10^{-2} \text{ g} \times \frac{1 \text{ mol}}{190.451 \text{ g}} = 2.38 \times 10^{-4} \text{ mol}$$

17. Determine the mass, in grams, of each of the following:

(a) 0.250 mol of ammonia, NH_3

Solution

$$1 \text{ mol } NH_3 = 14.0067 \text{ g} + 3(1.0079 \text{ g}) = 17.0304 \text{ g}$$

$$\text{Mass } NH_3 = 0.250 \text{ mol} \times 17.0304 \frac{\text{g}}{\text{mol}} = 4.26 \text{ g}$$

22. Determine the mass in grams of zirconium and of oxygen in 0.3384 mol of the mineral zircon, $ZrSiO_4$.

Solution

The sample cannot contain more than 0.3384 mol of each element present times the number of times the element appears in the formula. There are, therefore, 0.3384 mol of Zr present and 4(0.3384) mol of O present.

$$\text{Mass Zr} = 0.3384 \text{ mol} \times \frac{91.22 \text{ g}}{\text{mol}} = 30.87 \text{ g Zr}$$

$$\text{Mass O} = 4(0.3384) \text{ mol} \times \frac{15.9994 \text{ g}}{\text{mol}} = 21.66 \text{ g O}$$

23. Determine each of the following:

(b) grams of magnesium in 7.52 mol MgO

Solution

Because 1 mol Mg is present in 1 mol MgO,

$$\text{Mass Mg} = 7.52 \text{ mol Mg} \times \frac{24.305 \text{ g}}{1 \text{ mol Mg}} = 183 \text{ g}$$

24. Determine which of the following contains the greatest mass of hydrogen: 1 mol of CH_4, 0.6 mol of C_6H_6, 0.4 mol of C_3H_8.

Solution

In this problem the moles of hydrogen present in each compound may be determined. The substance with the greatest number of moles of hydrogen will also contain the most mass of hydrogen.

For CH_4: $$1 \text{ mol } CH_4 \times \frac{4 \text{ mol H}}{\text{mol } CH_4} = 4 \text{ mol H}$$

For C_6H_6: $$0.6 \text{ mol } C_6H_6 \times \frac{6 \text{ mol H}}{\text{mol } C_6H_6} = 3.6 \text{ mol H}$$

For C_3H_8: $$0.4 \text{ mol } C_3H_8 \times \frac{8 \text{ mol H}}{\text{mol } C_3H_8} = 3.2 \text{ mol H}$$

Therefore, 1 mol of CH_4 has the most mass of H:

$$4 \text{ mol H} \times \frac{1.0079 \text{ g H}}{\text{mol H}} = 4.0316 \text{ g H}$$

28. Determine which of the following contains the greatest total number of atoms: 10.0 g of potassium metal; 1.02×10^{23} molecules of nitrogen, N_2; 0.10 mol of chloroform, $CHCl_3$; 5 g of carbon dioxide, CO_2.

Solution

This problem involves converting the grams of each substance to the moles of that substance. Once the number of moles is known, multiplication by the number of atoms present in each molecule and by the Avogadro constant will give the total number of atoms.

$$10.0 \text{ g K} \times \frac{1 \text{ mol}}{39.102 \text{ g}} \times \frac{6.022 \times 10^{23} \text{ atoms}}{\text{mol}} = 1.54 \times 10^{23} \text{ atoms}$$

$$1.02 \times 10^{23} \text{ molecules } N_2 \times \frac{2 \text{ atoms}}{\text{molecule}} = 2.04 \times 10^{23} \text{ atoms}$$

$$0.10 \text{ mol } CHCl_3 \times \frac{5 \text{ atoms}}{\text{molecule}} \times \frac{6.022 \times 10^{23} \text{ molecules}}{\text{mol}} = 3.0 \times 10^{23} \text{ atoms}$$

$$5 \text{ g } CO_2 \times \frac{1 \text{ mol}}{44.010 \text{ g}} \times \frac{3 \text{ atoms}}{\text{molecule}} \times \frac{6.022 \times 10^{23} \text{ molecules}}{\text{mol}} = 2 \times 10^{23} \text{ atoms}$$

The substance with the largest number of atoms is chloroform.

29. Calculate the percent composition of each of the following compounds to three significant figures:

(a) potassium bromide, KBr

Solution

The percent composition is found by dividing the atomic weight of each element (multiplied by its stoichiometric coefficient) by the molecular weight of the compound and multiplying the resulting fraction by 100%.

For K: $$\frac{\text{At. wt K}}{\text{mol. wt KBr}} \times 100\% = \frac{39.098}{39.098 + 79.904} \times 100\% = 0.3285491 \times 100\%$$
$$= 32.9\% \text{ K}$$

By subtraction from 100.0% the percent Br is 67.1%, or the same answer may be obtained with a procedure similar to the one just used.

For Br: $$\frac{\text{At. wt Br}}{\text{mol. wt KBr}} \times 100\% = \frac{79.904}{119.002} \times 100\% = 0.671 \times 100\% = 6.71\% \text{ Br}$$

(e) cryolite, Na_3AlF_6

Solution

We proceed in the same manner as in (a).

For Na: $$\frac{3(\text{at. wt Na})}{\text{mol. wt } Na_3AlF_6} \times 100\% = \frac{3(22.9898) \times 100\%}{3(22.9898) + 26.9815 + 6(18.9984)}$$
$$= \frac{68.9695 \times 100\%}{209.941} = 32.9\% \text{ Na}$$

For Al: $$\frac{26.9815 \times 100\%}{209.941} = 12.8\% \text{ Al}$$

For F: $$\frac{113.9904 \times 100\%}{209.941} = 54.3\% \text{ F}$$

30. Determine the percent composition of each of the following to three significant figures:

(b) vitamin C, ascorbic acid, $C_6H_8O_6$.

Solution

$$\%C = \frac{6(\text{at. wt C}) \times 100\%}{\text{mol. wt } C_6H_8O_6} = \frac{6(12.011) \times 100\%}{6(12.011) + 8(1.0079) + 6(15.9994)}$$

$$= \frac{72.066 \times 100\%}{72.066 + 8.0632 + 95.9964} = \frac{7206.6\%}{176.1256} = 40.9\%$$

$$\%H = \frac{8(\text{at. wt H}) \times 100\%}{\text{mol. wt } C_6H_8O_6} = \frac{8.0632 \times 100\%}{176.1256} = 4.58\%$$

$$\%O = \frac{6(\text{at. wt O}) \times 100\%}{\text{mol. wt } C_6H_8O_6} = \frac{95.9965 \times 100\%}{176.1256} = 54.5\%$$

34. The light-emitting diode used in some electronic calculator displays contains a compound composed of 69.24% Ga and 30.76% P. What is the empirical formula of this compound?

Solution

An empirical formula shows the relative numbers of moles of atoms in a compound. It is the simplest whole-number atom ratio of the elements in the compound. Normally, either the masses of the elements are given or the percent by mass of the compound is known. Percent composition data can be used to determine the atomic ratio of elements in a compound because the percent composition represents the mass fraction of each element. If the percent by mass is given, use a 100.0 g sample of compound for convenience and find the mass of each element present in it. Convert the masses to moles, and then reduce the mole ratios to the simplest whole numbers. In this problem there are 69.24 g Ga (100.00 g × 0.6924) and 30.76 g P. The next step is to find the number of moles.

$$\text{No. of mol Ga} = 69.24 \text{ g} \times \frac{1 \text{ mol}}{69.72 \text{ g}} = 0.9931 \text{ mol}$$

$$\text{No. of mol P} = 30.76 \text{ g} \times \frac{1 \text{ mol}}{30.97376 \text{ g}} = 0.9931 \text{ mol}$$

The ratio of atoms of Ga to atoms of P, $Ga_{0.9931} : P_{0.9931}$, can be converted to a whole-number ratio by dividing each term by the smallest value of subscript.

$$\text{Ga} : \frac{0.9931}{0.9931} = 1.000 \qquad \text{P} : \frac{0.9931}{0.9931} = 1.000$$

Thus, the ratio is 1 : 1 and the formula is GaP.

38. Most polymers are very large molecules composed of simple units repeated many times. Thus, they often have relatively simple empirical formulas. Determine the empirical formula of the following polymer:

(e) Lucite (Plexiglas), 71.40% C, 9.59% H, 19.02% O

Solution

Assume 100.00 g of Lucite. In the sample there are then 71.40 g C, 9.59 g H, and 19.02 g O. Next, determine the amount of each element in moles.

$$\text{Amount of C} = 71.40 \text{ g} \times \frac{1 \text{ mol}}{12.011 \text{ g}} = 5.94 \text{ mol}$$

$$\text{Amount of H} = 9.59 \text{ g} \times \frac{1 \text{ mol}}{1.0079 \text{ g}} = 9.51 \text{ mol}$$

$$\text{Amount of O} = 19.02 \text{ g} \times \frac{1 \text{ mol}}{15.9994 \text{ g}} = 1.19 \text{ mol}$$

The ratio of atoms is

$$C_{5.94} : H_{9.51} : O_{1.19}$$

and is converted to a whole-number ratio by dividing each term by the smallest subscript:

$$\text{C: } \frac{5.94}{1.19} = 4.99 \qquad \text{H: } \frac{9.51}{1.19} = 7.99 \qquad \text{O: } \frac{1.19}{1.19} = 1$$

The numbers 4.99 and 7.99 are close enough to the whole numbers 5 and 8 that no further reduction is necessary. Hence, the formula is C_5H_8O.

39. Determine the empirical and molecular formulas of a compound with a molecular weight of 32 that contains 87.5% N and 12.5% H.

Solution

The empirical formula is determined as in the previous problem:

Element	Mass of Element in 100 g of Sample	Relative Number of Moles	Divide by the Smaller Number	Smallest Integral Number of Moles
N	87.5 g	$87.5 \text{ g N} \times \frac{1 \text{ mol N}}{14.0067 \text{ g C}}$ = 6.25 mol	$\frac{6.25}{6.25} = 1.00$	1
H	12.5 g	$12.5 \text{ g H} \times \frac{1 \text{ mol H}}{1.0079 \text{ g H}}$ = 12.40 mol	$\frac{12.40}{6.25} = 1.98$	2

The molecular weight is 32 and the smallest unit is NH_2, which has a mass of 14 + 2(1) = 16. This is the empirical formula. Divide the mol. wt 32 by the weight of the smallest unit, that is, 16. The result, 2, is the number of formula units of NH_2 in the compound. The molecular formula is, therefore, $(NH_2)_2$ or N_2H_4.

42. Plaster of paris contains 6.2% water and 93.8% calcium sulfate, $CaSO_4$. What is the empirical formula of plaster of paris?

Solution

This exercise involves finding the mol ratio of water and $CaSO_4$ rather than the ratio of individual elements. Exactly the same procedure is used here as in Exercise 39 except that gram-formula weights rather than atomic weights are used.

Formula Unit	Mass of Unit in 100 g of Sample	Relative Number of Moles	Divide by the Smaller Number	Smallest Integral Number of Moles
$CaSO_4$	93.8 g	$93.8 \text{ g } CaSO_4 \times \frac{1 \text{ mol } CaSO_4}{136.14 \text{ g}} = 0.689 \text{ mol}$	$\frac{0.689}{0.344} = 2.00$	2
H_2O	6.2 g	$6.2 \text{ g } H_2O \times \frac{1 \text{ mol } H_2O}{18.0152 \text{ g}} = 0.344 \text{ mol}$	$\frac{0.344}{0.344} = 1.00$	1

The formula is, then, $(CaSO_4)_2 \cdot H_2O$ or $Ca_2(SO_4)_2 \cdot H_2O$.

45. Determine the empirical formula of gold chloride if a 126-mg sample contains 81.8 mg of gold.

Solution

The mass of gold is given and the mass of Cl is obtained by subtraction: 126.0 - 81.8 = 44.2 mg. It is convenient to multiply both mass quantities by 1000. This does not change the ratio of the two elements but, in effect, converts the masses to gram quantities to be consistent with the units of gram-atom weights. The number of moles is then found.

Relative Number of Moles	Divide by the Smaller Number	Smallest Integral Number of Moles
$81.8 \text{ g Au} \times \frac{1 \text{ mol Au}}{196.9665 \text{ g}} = 0.415 \text{ mol}$	$\frac{0.415}{0.415} = 1$	1
$44.2 \text{ g Cl} \times \frac{1 \text{ mol Cl}}{35.453 \text{ g}} = 1.247 \text{ g}$	$\frac{1.247}{35.453} = 3.00$	3

The formula, therefore, is $AuCl_3$.

48. A 0.709-g sample of iron reacts with HCl(g) to give hydrogen gas, H_2, and 1.610 g of a compound containing iron and chlorine. Identify the product and write the balanced equation for its formation.

Solution

In the 1.610-g sample of the iron and chlorine compound, 0.709 g is Fe and 1.610 g - 0.709 g = 0.901 g is Cl. The empirical formula of the compound is found as in Exercise 45.

Sample Size	Relative Number of Moles	Divide by the Smaller Number	Smallest Integral Number of Moles
Fe 0.709 g	$0.709 \text{ g Fe} \times \frac{1 \text{ mol Fe}}{55.847 \text{ g}} = 0.0127 \text{ mol}$	$\frac{0.0127}{0.0127} = 1$	1
Cl 0.901 g	$0.901 \text{ g Cl} \times \frac{1 \text{ mol Cl}}{35.453 \text{ g}} = 0.0254 \text{ mol}$	$\frac{0.0254}{0.0127} = 2$	2

The compound is $FeCl_2$ and the reaction is $Fe + 2HCl \longrightarrow FeCl_2 + H_2$.

53. How many molecules of I_2 are produced by the reaction of 0.3600 mol of $CuCl_2$ according to the following equation?

$$2CuCl_2 + 4KI \longrightarrow 2CuI + 4KCl + I_2$$

Solution

First calculate the amount of I_2 in moles. The balanced equation shows that 1 mol of I_2 is produced per 2 mol of $CuCl_2$ consumed (1 mol I_2/2 mol $CuCl_2$). Use this relationship to convert the number of moles of reacting $CuCl_2$ to the number of moles of I_2 produced.

$$0.3600 \text{ mol } CuCl_2 \times \frac{1 \text{ mol } I_2}{2 \text{ mol } CuCl_2} = 0.1800 \text{ mol } I_2$$

The number of molecules of I_2 is found by multiplying Avogadro's constant by the moles of I_2.

$$\text{No. of molecules } I_2 = 0.1800 \text{ mol } I_2 \times 6.022 \times 10^{23} \text{ mol}^{-1}$$
$$= 1.084 \times 10^{23}$$

(b) What mass of I_2 is produced?

Solution

Because 0.1800 mol I_2 is produced,

$$0.1800 \text{ mol } I_2 \times \frac{2(126.9045) \text{ g } I_2}{1 \text{ mol } I_2} = 45.69 \text{ g } I_2$$

If the number of molecules is used in the calculation,

$$\frac{1.084 \times 10^{23}}{6.022 \times 10^{23} \text{ mol}^{-1} \; I_2} \times \frac{2(126.9045) \text{ g } I_2}{1 \text{ mol } I_2} = 45.69 \text{ g } I_2$$

57. Tooth enamel consists of hydroxyapatite, $Ca_5(PO_4)_3(OH)$. This substance is converted to the more decay-resistant fluorapatite, $Ca_5(PO_4)_3F$, by treatment with tin(II) fluoride, SnF_2 (commonly referred to as stannous fluoride). Products of this reaction are SnO and water. What mass of hydroxyapatite can be converted to fluorapatite by reaction with 0.100 g of SnF_2?

Solution

The balanced equation required is

$$2Ca_5(PO_4)_3(OH) + SnF_2 \longrightarrow 2Ca_5(PO_4)_3F + SnO + H_2O$$

The equation indicates that 2 mol hydroxyapatite reacts with 1 mol SnF_2 [2 mol $Ca_5(PO_4)_3(OH)$/1 mol SnF_2]. Included are factors to convert g SnF_2 to mol SnF_2 and mol hydroxyapatite (OHA) to g OHA. Thus,

$$0.100 \text{ g } SnF_4 \times \frac{2 \text{ mol OHA}}{1 \text{ mol } SnF_2} \times \frac{502.32 \text{ g OHA}}{1 \text{ mol OHA}} \times \frac{1 \text{ mol } SnF_2}{156.69 \text{ g } SnF_2}$$

$$= 0.641 \text{ g } Ca_5(PO_4)_3(OH)$$

60. Which is the limiting reagent when 0.5 mol of P_4 and 0.5 mol of S_8 react according to the following equation?

$$4P_4 + 5S_8 \longrightarrow 4P_4S_{10}$$

Solution

The term *limiting reagent* implies that the amount of one reactant necessary for a reaction is in excess relative to another reactant. That is, one reactant will be completely consumed during the reaction while an amount of the other reactant remains unreacted. The maximum amount of product that can be produced by a reaction is determined from the amount of the reactant that is totally consumed. To find the limiting reagent, assume that each reactant is the limiting factor and compute the potential quantities of product based on each reactant. It follows that the reactant quantity yielding the smallest amount of product is the limiting factor, the smallest amount of product is the maximum amount that can be produced. For the reaction,

$$4P_4 + 5S_8 \rightarrow 4P_4S_{10}$$

calculate the amount of P_4S_{10} that could be produced from 0.5 mol of P_4 and the amount of P_4S_{10} that could be produced from 0.5 mol of S_8.

$$\begin{array}{ccccc} 0.5 \text{ mol} & & 0.5 \text{ mol} & & \text{No. mol?} \\ 4P_4 & + & 5S_8 & \rightarrow & 4P_4S_{10} \\ 4 \text{ mol} & & 5 \text{ mol} & & 4 \text{ mol} \end{array}$$

No. mol P_4S_{10} based on P_4:

$$\text{Mol } P_4S_{10} = 0.5 \text{ mol } P_4 \times \frac{4 \text{ mol } P_4S_{10}}{4 \text{ mol } P_4} = 0.5 \text{ mol } P_4S_{10}$$

No. mol P_4S_{10} based on S_8:

$$\text{Mol } P_4S_{10} = 0.5 \text{ mol } S_8 \times \frac{4 \text{ mol } P_4S_{10}}{5 \text{ mol } S_8} = 0.4 \text{ mol } P_4S_{10}$$

A comparison of the results shows that the limiting reagent is S_8 because 0.4 mol of P_4S_{10} is the maximum amount that can be produced by the given reagents.

63. Which is the limiting reagent when 42.0 g of propane, C_3H_8, is burned with 115 g of oxygen?

$$C_3H_8(g) + 5O_2(g) \rightarrow 3CO_2(g) + 4H_2O(g)$$

Solution

A slightly different approach will be taken to solve this problem than that used in Exercise 60. Either C_3H_8 or O_2 could be in excess. Which one? With 42.0 g of propane to react, how much oxygen will be required? Is this amount more or less than that available — 115 g? To find out, calculate the amount of oxygen required to react with 42.0 g of propane by using the balanced equation and the data given.

$$\begin{array}{ccccccc} 42.0 \text{ g} & & \text{Mass?} & & & & \\ C_3H_8 & + & 5O_2 & \rightarrow & 3CO_2 & + & 4H_2O \\ 1 \text{ mol} & + & 5 \text{ mol} & & & & \end{array}$$

Molar mass $C_3H_8 = 44.0962$ g mol^{-1}; Molar mass $O_2 = 31.9988$ g mol^{-1}
Mass of O_2 required for 42.0 g of C_3H_8:

$$\text{Mass } O_2 = 42.0 \text{ g} \times \frac{1 \text{ mol}}{44.0962 \text{ g}} \times \frac{5 \text{ mol } O_2}{1 \text{ mol } C_3H_8} \times \frac{31.9988 \text{ g}}{\text{mol}} = 152 \text{ g}$$

The complete reaction of 42.0 g of propane requires 152 g of oxygen. Only 115 g of oxygen are available; insufficient oxygen is available to react with 42.0 g of propane and, therefore, oxygen is the limiting reagent.

65. A student spilled some of an ethanol preparation and consequently isolated only 15 g rather than the 47 g that was theoretically possible. What was the percent yield?

Solution

Percent yield represents the fraction of product actually produced or recovered from a reaction, divided by the theoretical or expected yield. In this reaction the expected yield of ethanol was 47 g, but only 15 g was recovered. The percent yield is

$$\%\text{ yield} = \frac{\text{actual yield}}{\text{theoretical yield}} \times 100\%$$

$$\%\text{ yield} = \frac{15 \text{ g}}{47 \text{ g}} \times 100\% = 32\%$$

For further information on percent yield, see FORMULAS AND DEFINITIONS, page 34.

67. Toluene, $C_6H_5CH_3$, is oxidized by air under carefully controlled conditions to benzoic acid, $C_6H_5CO_2H$, which is used to prepare the food preservative sodium benzoate, $C_6H_5CO_2Na$. What is the yield of a reaction in which 1.000 kg of toluene is converted to 1.21 kg of benzoic acid?

$$2C_6H_5CH_3 + 3O_2 \rightarrow 2C_6H_5CO_2H + 2H_2O$$

Solution

The theoretical yield of a reaction is the amount of product expected from a reaction if all of the reagents are converted to product. The calculation of theoretical yield is based on the balanced equation and the amount of starting reagents. In this case how much benzoic acid can be produced from the complete reaction of 1.000 kg of toluene? Perform this calculation. Then take the ratio of the amount actually produced to the amount so found in order to compute the yield of the reaction. Theoretical yield of benzoic acid: 1.000 kg = 1000 g.

$$2C_6H_5CH_3 + 3O_2 \rightarrow 2C_6H_5CO_2H + 2H_2O$$

$$2 \text{ mol} \quad\quad \rightarrow \quad\quad 2 \text{ mol}$$

Molar mass $C_6H_5CH_3 = 92.1402\ g\ mol^{-1}$; Molar mass $C_6H_5CO_2H = 122.1232\ g\ mol^{-1}$

$$\text{Theoretical mass} = 1000\ g\ C_6H_5CH_3 \times \frac{1\ mol}{92.1402\ g} \times \frac{2\ mol\ C_6H_5CO_2H}{2\ mol\ C_6H_5CH_3}$$

$$\times \frac{122.1232\ g}{mol} = 1325\ g$$

Percent yield of $C_6H_5CO_2H$:

$$\%\ yield = \frac{1210\ g}{1325\ g} \times 100\% = 91.3\%$$

70. Citric acid, $C_6H_8O_7$, a component of jams, jellies, and "fruity" soft drinks, is prepared industrially by fermentation of sucrose by the mold *Aspergillus niger*. The overall reaction is

$$C_{12}H_{22}O_{11} + H_2O + 3O_2 \rightarrow 2C_6H_8O_7 + 4H_2O$$

What is the amount, in kilograms, of citric acid produced from 1 metric ton (1.000×10^3 kg) of sucrose if the percent yield is 85.4%?

Solution

The actual yield from the reaction process is 85.4% of the theoretical yield. From the balanced equation the theoretical yield is

$$\underset{1\ mol}{\overset{1.000 \times 10^3\ kg}{C_{12}H_{22}O_{11}}} + H_2O + 3O_2 \rightarrow \underset{2\ mol}{\overset{Mass?}{2C_6H_8O_7}} + 4H_2O$$

Molar mass $C_{12}H_{22}O_{11} = 342.2992\ g\ mol^{-1}$; Molar mass $C_6H_8O_7 = 192.125\ g\ mol^{-1}$

$$\text{Theoretical yield} = 1.000 \times 10^6\ g \times \frac{1\ mol\ C_{12}H_{22}O_{11}}{342.2992\ g} \times \frac{2\ mol\ C_6H_8O_7}{1\ mol\ C_{12}H_{22}O_{11}}$$

$$\times \frac{192.125\ g}{mol} = 1.123 \times 10^6\ g$$

The yield expected from the process is

$$Yield = \frac{85.4\%}{100\%} \times 1.123 \times 10^6\ g = 9.59 \times 10^5\ g = 959\ kg$$

72. The reaction of 3.0 mol of H_2 with 2.0 mol of I_2 produces 1.0 mol of HI.

$$H_2 + I_2 \rightarrow 2HI$$

Determine the limiting reagent, the theoretical yield in moles, and the percent yield for this reaction.

Solution

H_2 and I_2 react on a 1 : 1 basis; I_2, therefore, being in lesser amount, is the limiting reagent. The theoretical yield is calculated from the limiting reagent, I_2. From the balanced equation, 1 mol I_2 produces 2 mol HI. Thus, for 2 mol I_2, 4 mol HI can be produced. Because only 1 mol HI is produced, the percent yield is

$$\frac{\text{Amount produced}}{\text{Theoretical yield}} \times 100\% = \frac{1 \text{ mol}}{4 \text{ mol}} \times 100\% = 25\%$$

75. What is the percent yield of $NaClO_2$ if 106 g is isolated from the reaction of 202.3 g of ClO_2 with a solution containing 3.229 mol of NaOH?

$$2ClO_2(g) + 2NaOH(aq) \longrightarrow NaClO_2(aq) + NaClO_3(aq) + H_2O(\ell)$$

Solution

The grams of $NaClO_2$ theoretically possible from the reaction of 202.3 g ClO_2 and NaOH are determined. This requires that the limiting agent be determined. The percent yield then is computed comparing 106 g to the theoretical yield.

$$\text{Mol } ClO_2 = 202.3 \text{ g } ClO_2 \times \frac{1 \text{ mol}}{67.452 \text{ g } ClO_2} = 2.999 \text{ mol}$$

The ratio of reactants is 1 : 1; therefore, NaOH with 3.229 mol is in excess and ClO_2 is the limiting reagent. The following calculation is thus based on ClO_2:

$$\text{g } NaClO_2 = 2.999 \text{ mol } ClO_2 \times \frac{1 \text{ mol } NaClO_2}{2 \text{ mol } ClO_2} \times \frac{90.442 \text{ g}}{1 \text{ mol } NaClO_2} = 135.6 \text{ g}$$

$$\% \text{ Yield} = \frac{106 \text{ g}}{135.6 \text{ g}} \times 100\% = 78.2\%$$

77. Naphthalene, a compound with a molecular weight of about 130 and containing only carbon and hydrogen, is the principal component of mothballs. A 3.000-mg sample of naphthalene burns to give 10.3 mg of CO_2. Determine its empirical and molecular formula.

Solution

All of the C in naphthalene must appear as C in CO_2 when burned. The amount of C in the original sample is

$$\text{mg C} = 10.3 \text{ mg} \times \frac{12.011 \text{ g C}}{44.0098 \text{ g } CO_2} = 2.81 \text{ mg C}$$

The composition of this sample of naphthalene is

$$3.00 - 2.81 = 0.19 \text{ mg H} \quad \text{and} \quad 2.81 \text{ mg C}$$

The determination of the empirical formula proceeds as in earlier exercises (1 mmol = 0.001 mol).

$$\text{C:} \quad 2.81 \text{ mg} \times \frac{1 \text{ mmol C}}{12.011 \text{ mg C}} = 0.000234$$

$$\text{H:} \quad 0.19 \text{ mg} \times \frac{1 \text{ mmol C}}{1.0079 \text{ mg H}} = 0.000189$$

Division of both by the smallest amount gives

$$\text{C:} \quad 1.238 \qquad \text{H:} \quad 1.000$$

These are not whole numbers; so the smallest number that can multiply 1.238 and give an integer or nearly integer number must be found. The number 4 will give a ratio of 4.95 to 1, which is sufficiently close to 5 to 1 to allow C_5H_4 as the empirical formula. This has a mass of approximately 64, which is only one-half of the known molecular weight of 130. The molecular formula, therefore, is $C_{10}H_8$.

82. Sodium bicarbonate, baking soda, $NaHCO_3$, can be purified by dissolving in hot water (60°C), filtering to remove insoluble impurities, cooling to 0°C to precipitate solid $NaHCO_3$, and then filtering to remove the solid $NaHCO_3$, leaving soluble impurities in solution. Any $NaHCO_3$ that remains in solution is not recovered. The solubility of $NaHCO_3$ in hot water at 60°C is 164 g/L. The solubility in cold water at 0°C is 69 g/L. What is the percent yield of $NaHCO_3$ when purified by this method?

Solution

The mass of $NaHCO_3$ precipitated and hence recovered is

$$164 \text{ g/L} - 69 \text{ g/L} = 95 \text{ g/L}$$

$$\% \text{ yield} = \frac{95 \text{ g/L}}{164 \text{ g/L}} \times 100\% = 58\%$$

RELATED EXERCISES

1. Calculate the formula weight of each of the following compounds:
   (a) a common soap, sodium stearate, $C_{17}H_{35}COONa$
   (b) MSG, monosodium glutamate, $HOOCCH_2CH_2CHNH_2COONa$

(c) the acid responsible for the odor in rancid butter, butyric acid or butanoic acid, $CH_3CH_2CH_2COOH$

(d) a food preservative, sodium benzoate, C_6H_5COONa

Answer: (a) 306; (b) 169; (c) 88; (d) 144

2. The composition of regular gasoline is approximately 70% heptane, C_7H_{16}, and 30% octane, C_8H_{18}, by volume. The densities of heptane and octane are 683.76 g/L and 702.5 g/L, respectively, at 20°C. Calculate the mass of water that would be produced by the combustion of 1.00 gallon of gasoline according to the following equations.

$$C_7H_{16} + 11O_2 \rightarrow 7CO_2 + 8H_2O$$

$$2C_8H_{18} + 25O_2 \rightarrow 16CO_2 + 18H_2O$$

Answer: 3740 g

3. The analysis of a sample of brass shows the composition to be 72.0% Cu and 28.0% Zn by weight. Calculate the mass of brass that can be made from 5.5 kg of copper assuming adequate zinc is available. What mass of zinc would be required?

Answer: 7.64 kg of brass; 2.14 kg of zinc

4. Much of the world reserve of mercury is located in Spain and Italy in the form of cinnabar, HgS. Assuming that ore is pure HgS, how much ore would be required to produce 8.0 million kg of mercury, equal to about a one-year supply?

Answer: 9.3×10^6 kg HgS

5. Calculate the mass of oxalic acid dihydrate, $H_2C_2O_4 \cdot 2H_2O$, required to prepare 5.00 L of 0.250 *M* solution.

Answer: 158 g

UNIT 3

CHAPTER 3: APPLICATIONS OF CHEMICAL STOICHIOMETRY

INTRODUCTION

This chapter extends the presentation of Chapter 2 to include broader classes of reactions and equations and to bring out their utility in the study of chemistry. Various examples are used to emphasize systematic procedures for analyzing a problem into its essential components.

Although the simplicity of most chemical equations tends to convey the idea that reactions occur in a single step, most reactions require completion of several steps before the desired product is formed. For example, some complex biochemical compounds may require a hundred individual steps in which intermediate products result, yet the equation representing the formation of such a compound may be a simple one-line expression.

To illustrate the complexity of what appears to be a simple reaction, consider the production of table salt, NaCl(*s*), from its elements according to the equation

$$Na(s) + \frac{1}{2}Cl_2(g) \rightarrow NaCl(s)$$

The production of NaCl(*s*) is exothermic (that is, it proceeds with the liberation of heat) and occurs spontaneously in the presence of sunlight or heat. Without going into great detail, the overall reaction process can be broken down into the following five reaction steps:

1. $Na(s) + 108.8\ kJ \rightarrow Na(g)$ (endothermic, requires heat to proceed)
2. $Na(g) + 496.0\ kJ \rightarrow Na^+ + e^-$ (endothermic)
3. $\frac{1}{2}Cl_2(g) + 119.7\ kJ \rightarrow Cl(g)$ (endothermic)
4. $Cl(g) + e^- \rightarrow Cl^-(g) + 364.8\ kJ$ (exothermic, heat is given off)
5. $Na^+(g) + Cl^-(g) \rightarrow NaCl(g) \rightarrow NaCl(s) + 770.7\ kJ$ (exothermic)

The total heat released in producing 1 mol of NaCl(*s*) is 411.0 kJ, obtained by subtracting the sum of the endothermic values from the sum of the exothermic values. Much of the work in chemical research lies in the elucidation of similar reaction processes and their attendant energy changes.

Most of the reactions studied in general chemistry involve substances in solution. The solution phase either facilitates and/or provides the medium for reactions to occur. When substances are in solution, the solution must be described in terms of how much substance (solute) is contained in a specified quantity (volume or mass) of solvent. Although several units of concentration are commonly used in chemistry, molarity, *M*, is presented in this chapter to enable you to begin work with solutions in the laboratory as soon as possible. Other units of concentration are presented in detail in Chapter 13 of the text.

A variety of problems that have industrial and/or physiological importance appear in this chapter. The solutions to many of these problems involve several steps that can generally be simplified into a convenient format for a systematic solution. To develop such a format, read each problem carefully in order to identify what is known and unknown. Then write the equation for the reaction and balance it as necessary. The next step is to develop an algebraic equation that relates the known quantities to those that are unknown, and then solve for the unknown quantity(ies) accordingly. After studying examples and solving a few problems, you will begin to develop a personal method for organizing data and solving problems. Do not become concerned that your method may be slightly different from the way the text approaches the problem — after all, solving problems is a human endeavor, and probably no two people solve problems in exactly the same way.

There are many possible methods in chemistry for solving problems that involve reactants and products. The problem-solving format used in the texts is basically the same as that used in this manual, but you occasionally will notice slightly different organizational practices that should enable you to see still other ways of approaching problems.

FORMULAS AND DEFINITIONS

Limiting reagent The reagent (or reagents) completely consumed by a reaction is called the limiting reagent(s). In a balanced chemical equation, which represents a reaction, the proper mole ratio for reactants is given. If, however, the reaction mixture contains one reactant in excess of the proper mole ratio, the other reactant(s) will be completely consumed and the excess amount of the reactant in excess will remain unreacted. The amount of product produced by the reaction, therefore, is limited by the amount of reagent available for reaction. As an analogy, a chemical reaction may be compared to the assembly of bicycles begun with three bicycle frames and eight tires. Three complete bicycles can obviously be built, with two tires in excess. The available bicycle frames *limit* (like the limiting reagent) the number of bicycles that can be built when tires are in excess.

Theoretical yield The potential amount of product that can be produced in a reaction determined by the mole ratio in the balanced equation and calculated from the amounts of available reactant(s).

Yield The actual amount of product produced in a chemical reaction.

Percent yield The percent yield for a reaction is the yield fraction, actual yield divided by theoretical yield, times 100.

$$\%\ \text{yield} = \frac{\text{actual yield}}{\text{theoretical yield}} \times 100\%$$

Molarity (M) Molarity is a unit of concentration used to describe a solution in terms of the moles of solute present in 1 L of solution.

$$\text{Molarity} = \frac{\text{mol of solute}}{\text{vol of solution in liters}} = \frac{\text{mol}}{\text{L}}$$

This definition can be rearranged to express the amount of solute (in moles) contained in a given volume of solution:

$$\text{Amount of solute} = \text{molarity}\left(\frac{\text{mol}}{\text{L}}\right) \times \text{vol (L)} = \text{mol}$$

and to express the volume of a solution in terms of molarity and the amount of solute:

$$\text{Vol in L} = \frac{\text{mol of solute}}{\text{molarity}}$$

The definition of molarity is readily extended to calculate the amount of solute required to prepare a specified volume of solution.

$$\text{Molarity} = \frac{\text{mol}}{\text{vol in L}} = \frac{\text{mass in g} \times \frac{1\ \text{mol}}{\text{GFW}}}{\text{vol in L}}$$

or

$$\text{Mass} = \text{molarity} \times \text{vol in L} \times \text{GFW} = \frac{\text{mol}}{\text{L}} \times \text{L} \times \frac{\text{g}}{\text{mol}} = \text{g}$$

EXERCISES

1. Calculate the molarity of each of the following solutions:

(a) 98.1 g of sulfuric acid, H_2SO_4, in 1.00 L of solution

Solution

The molarity is defined as moles of solute divided by liters of solution. The number of moles of sulfuric acid present is given by its mass divided by the molar mass (g-molecular weight) 98.07 g mol^{-1}. In equation form,

$$\text{Molarity} = M = \frac{\text{mol of solute}}{\text{L of soln}} = \frac{\text{g/molar mass}}{\text{L}}$$

$$= \frac{98.1\ \text{g}/98.07\ \text{g mol}^{-1}}{1.00\ \text{L}} = \frac{98.1\ \text{g mol}}{98.07\ \text{g} \times 1.00\ \text{L}} = 1.00\ M$$

(d) 2.12 g of potassium bromide, KBr, in 458 mL of solution

Solution

The gram-formula weight of KBr is 119.002 g. The volume must be expressed in liters. Then,

$$M = \frac{\text{g/molar mass}}{\text{L}} = \frac{2.12\ \text{g}/119.002\ \text{g mol}^{-1}}{0.458\ \text{L}}$$

$$= 3.89 \times 10^{-2}\ M$$

2. Determine the moles of solute present in each of the following solutions:

(b) 2.0 L of 0.480 *M* $MgCl_2$ solution

Solution

From the definition of molarity, *M* = moles/liters, rearrangement and substitution gives

$$\text{mol } MgCl_2 = M \times L = 0.480\ M \times 2.0\ L = 0.96 \text{ mol}$$

(d) 2.50 mL of 0.1812 *M* $KMnO_4$ solution

Solution

The volume must be expressed in liters: 2.50 mL = 0.00250 L. Substitution gives

$$\text{mol } KMnO_4 = M \times L = 0.1812\ M \times 0.00250\ L = 0.000453 = 4.53 \times 10^{-4} \text{ mol}$$

3. Determine the mass of each of the following solute that is required to make the indicated amount of solution:

(b) 4.2 L of 2.45 *M* C_2H_5OH solution

Solution

The definition of molarity in this case is

$$M = \frac{\text{mol } C_2H_5OH}{\text{L of soln}} = \frac{\text{g } C_2H_5OH/\text{molar mass } C_2H_5OH}{\text{L of soln}}$$

Rearrangement gives

$$\text{g } C_2H_5OH = M \times L \times \text{molar mass } C_2H_5OH$$

The molar mass of C_2H_5OH is 46.0688 g mol^{-1}. Substitution gives

$$\text{g } C_2H_5OH = 2.45\ M \times 4.2\ L \times 46.0688 \text{ g mol}^{-1} = 4.7 \times 10^2 \text{ g } C_2H_5OH$$

4. The lowest limit of $MgSO_4$ that can be detected by taste in drinking water is about 0.400 g/L. What is this molar concentration of $MgSO_4$?

Solution

Use of the definition of molarity is required along with the molar mass of $MgSO_4$ (120.3606 g mol^{-1}).

$$M = \frac{\text{mol of } MgSO_4}{\text{L of soln}} = \frac{\text{g/molar mass}}{\text{L of soln}} = \frac{0.400 \text{ g}}{\text{L}} \times \frac{1}{120.3606 \text{ g/mol}}$$

$$= 3.32 \times 10^{-3}\ M\ MgSO_4$$

7. An iron content of 0.1 mg/L in drinking water can usually be detected by taste. Will the iron in a water sample with an iron(II) carbonate, $FeCO_3$, concentration of 5.25×10^{-7} *M* be detectable by taste?

Solution

This problem requires the determination of the mg of Fe present in a 5.25×10^{-7} *M* solution of $FeCO_3$ and comparison with the standard value. First convert the molarity of the $FeCO_3$ solution to mg/L ($FeCO_3$). Then the percent composition of Fe in $FeCO_3$ can be used to determine the mg of Fe present. The molar mass of $FeCO_3$ is 115.856 g mol^{-1}.

$$\text{Molarity} = M = \frac{\text{mol}}{\text{L}} = \frac{\text{g/molar mass}}{\text{L}}$$

Rearrangement gives

$$\text{g/L} = M \times \text{molar mass} = 5.25 \times 10^{-7}\ M \times 115.856\ \text{g mol}^{-1}$$
$$= 6.08 \times 10^{-5}\ \text{g FeCO}_3\text{/L}$$

$$\%\ \text{Fe in FeCO}_3 = \frac{55.847}{115.856} \times 100 = 48.20\%$$

Then, $$6.08 \times 10^{-5}\ \text{g/L} \times 0.4820 = 2.93 \times 10^{-5}\ \text{g Fe/L}$$
$$= 2.93 \times 10^{-2}\ \text{mg Fe/L}$$

This is less than 0.1 mg/L and cannot be detected.

9. How many moles of silver nitrate, $AgNO_3$, are required to react with the calcium chloride in 14.96 mL of a 2.244 *M* solution of $CaCl_2$?

$$2AgNO_3(aq) + CaCl_2(aq) \longrightarrow 2AgCl(s) + Ca(NO_3)_2(aq)$$

Solution

From the definition *M* = mol/L, the equation is solved for moles of $CaCl_2$ and values are substituted.

$$\text{mol}(CaCl_2) = M(CaCl_2) \times L(CaCl_2) = 2.244 \times 0.01496$$
$$= 0.03357 = 3.357 \times 10^{-2}\ \text{mol}$$

From the equation, 2 mol ($AgNO_3$) are required to react with 1 mol ($CaCl_2$). Thus,

$$2 \times 3.357 \times 10^{-2}\ \text{mol}\ (AgNO_3) = 6.714 \times 10^{-2}\ \text{mol}$$

are required.

14. A 25.00-mL sample of sulfuric acid solution from an automobile battery reacts exactly with 87.42 mL of a 1.95 *M* solution of sodium hydroxide, NaOH. What is the molar concentration of the battery acid?

$$H_2SO_4(aq) + 2NaOH(aq) \longrightarrow Na_2SO_4(aq) + 2H_2O(\ell)$$

Solution

At the equivalence point the amount of NaOH present is calculated from the molarity.

$$M\ (NaOH) = \frac{mol(NaOH)}{L(NaOH)}$$

$$mol = M \times L = 1.95\ M \times 0.08742\ L = 0.170\ mol\ NaOH$$

From the balanced equation, 2 mol NaOH react with 1 mol H_2SO_4 so that there is 0.170/2 = 0.0852 mol H_2SO_4 present. Then,

$$M\ (H_2SO_4) = \frac{mol}{L} = \frac{0.0852\ mol}{0.02500\ L} = 3.41\ M$$

17. What is the molar concentration of $AgNO_3$ in a solution if titration of 25.00 mL of the solution with 0.300 *M* NaCl requires 37.05 mL of the NaCl solution to reach the end point?

$$AgNO_3(aq) + NaCl(aq) \longrightarrow AgCl(s) + NaNO_3(aq)$$

Solution

In this reaction 1 mol of $AgNO_3$ reacts with 1 mol of NaCl. The number of moles of $AgNO_3$ in 25.00 mL, therefore, must equal the number of moles of NaCl in 37.05 mL of 0.300 *M* solution. The molarity of the $AgNO_3$ solution can then be computed from the amount of $AgNO_3$ present and the volume of the $AgNO_3$ solution.

$$Mol\ AgNO_3\ in\ 25.00\ mL = mol\ NaCl$$

$$Mol\ NaCl = 0.03705\ L \times 0.300\ \frac{mol}{L} = 0.011115\ mol$$

$$Mol\ AgNO_3 = mol\ NaCl = 0.011115\ mol$$

$$M\ (AgNO_3) = \frac{mol\ AgNO_3}{L} = \frac{0.011115\ mol}{0.02500\ L} = 0.445\ M$$

19. What is the molar concentration of H_2SO_4 in a solution if 31.91 mL of it is required to titrate 2.474 g of Na_2CO_3?

$$H_2SO_4(aq) + Na_2CO_3(s) \longrightarrow Na_2SO_4(aq) + H_2O(\ell) + CO_2(g)$$

Solution

One mol of H_2SO_4 reacts with 1 mol of Na_2CO_3. The number of moles of H_2SO_4 in 31.91 mL of solution, therefore, equals the number of moles of Na_2CO_3 in 2.474 g of Na_2CO_3.

Moles of Na_2CO_3 in 2.474 g:

$$\text{Molar mass } (Na_2CO_3) = 105.989 \text{ g mol}^{-1}$$

$$\text{Mol } Na_2CO_3 = 2.474 \text{ g} \times \frac{1 \text{ mol}}{105.989 \text{ g}} = 0.02334 \text{ mol}$$

Molarity of H_2SO_4 solution:

$$M \ (H_2SO_4) = \frac{0.02334 \text{ mol}}{0.03191 \text{ L}} = 0.7315 \ M$$

24. Crystalline potassium hydrogen phthalate, $KHC_8H_4O_4$, is often used as a "standard" acid for standardizing basic solutions because it is easy to purify and to weigh. If 1.5428 g of this salt is titrated with a solution of $Ca(OH)_2$, the reaction is complete when 42.37 mL of the solution has been added. What is the concentration of the $Ca(OH)_2$ solution?

$$2KHC_8H_4O_4 + Ca(OH)_2 \longrightarrow Ca(KC_8H_4O_4)_2 + 2H_2O$$

Solution

In this reaction 1 mol of $Ca(OH)_2$ reacts with 2 mol of $KHC_8H_4O_4$; therefore, the number of moles of $Ca(OH)_2$ contained in the 42.37-mL sample equals one-half the number of moles of $KHC_8H_4O_4$ in the reaction sample.

$$\text{Molar mass } Ca(OH)_2 = 40.08 + 2(15.9994) + 2(1.0079) = 74.09 \text{ g mol}^{-1}$$

$$\begin{aligned}\text{Molar mass } KHC_8H_4O_4 &= 39.0983 + 1.0079 + 8(12.011) + 4(1.0079) + 4(15.9994) \\ &= 204.223 \text{ g mol}^{-1}\end{aligned}$$

$$\text{Mol } KHC_8H_4O_4 = 1.5428 \text{ g} \times \frac{1 \text{ mol}}{204.223 \text{ g}} = 0.0075545 \text{ mol}$$

$$\text{Mol } Ca(OH)_2 \text{ reacted} = \frac{0.0075545 \text{ mol}}{2} = 0.0037772 \text{ mol}$$

Use the definition of molarity to obtain the value M:

$$M \ [Ca(OH)_2] = \frac{\text{mol } Ca(OH)_2}{\text{L}} = \frac{0.0037772 \text{ mol}}{0.04237 \text{ L}} = 8.915 \times 10^{-2} \ M$$

27. Begin with the expression for each of the following conversions, and derive the expression for the reverse conversion:

(a) moles A to grams A

Solution

Conversion of moles of A to grams of A requires an identity statement relating moles to mass. One mole of any substance has a mass equal to its molar mass (molecular weight) in grams. The required expression is written with units in parentheses as

$$\text{Amount of A(mol)} = \text{amount of A(mol)} \times \text{molar mass of A}\left(\frac{\text{g}}{\text{mol}}\right) = \text{mass A(g)}$$

The reverse conversion of grams of A to moles of A is

$$\text{Mass A(g)} = \text{mass A(g)} \times \frac{1}{\text{molar mass of A}}\left(\frac{1\ \text{mol}}{\text{g}}\right) = \text{amount of A(mol)}$$

(c) moles of solute to molar concentration (given volume of solution)

Solution

Concentration units, such as molarity, are defined as a ratio of the amount of solute to amount of solvent or to the volume of solution. Hence, conversions among concentration units require a definition or identity relating the amount of solute to the amount of solvent or solution.

Molarity is defined as

$$\text{Molarity } (M) = \frac{\text{moles of solute}}{\text{volume of solution in liters}}$$

The conversion is

$$\frac{\text{Amount of solute (mol)}}{\text{Volume of solution (L)}} = M\left(\frac{\text{mol}}{\text{L}}\right)$$

The reverse conversion is

$$\text{Molarity}\left(\frac{\text{mol}}{\text{L}}\right) \times \text{volume of solution (L)} = \text{amount of solute (mol)}$$

29. Determine the mass and the moles of calcium phosphate, $Ca_3(PO_4)_2$, in a 374-g sample of phosphate rock containing 75.9% $Ca_3(PO_4)_2$.

Solution

The $Ca_3(PO_4)_2$ in this sample comprises 75.9% of the sample mass. To calculate the mass of $Ca_3(PO_4)_2$ in the sample, multiply the sample mass by the percentage of $Ca_3(PO_4)_2$.

$$\text{Mass } Ca_3(PO_4)_2 = 374\ \text{g} \times \frac{75.9\%}{100\%} = 284\ \text{g}$$

The number of moles of $Ca_3(PO_4)_2$ is calculated from the identity.

$$1\ \text{mol } Ca_3(PO_4)_2 = \text{molar mass of } Ca_3(PO_4)_2 \text{ in grams} = 310.18\ \text{g mol}^{-1}$$

$$\text{Mol } Ca_3(PO_4)_2 = 284\ \text{g} \times \frac{1\ \text{mol}}{310.18\ \text{g}} = 0.916\ \text{mol}$$

31. What is the molar concentration of isotonic saline solution for injection, if it has a density of 1.007 g/mL and contains 0.95% NaCl by mass?

Solution

The molarity of this solution is calculated from the definition of molarity after converting the mass of NaCl per milliliter to mass per liter and then converting mass per liter to mole per liter. The solution is 0.95% NaCl per unit mass of solution. The mass of NaCl per milliliter of solution is

$$\text{Mass NaCl} = \frac{0.95\%}{100\%} \times 1.007 \frac{\text{g}}{\text{mL}} = 0.00957 \frac{\text{g}}{\text{mL}}$$

$$\text{Mass NaCl per liter} = 0.00957 \frac{\text{g}}{\text{mL}} \times 1000 \frac{\text{mL}}{\text{L}} = 9.57 \frac{\text{g}}{\text{L}}$$

$$1 \text{ mol NaCl} = 58.443 \text{ g}$$

Then, $$\text{Mol NaCl} = 9.57 \text{ g} \times \frac{1 \text{ mol}}{58.443 \text{ g}} = 0.164 \text{ mol}$$

$$M = \frac{0.164 \text{ mol}}{1.00 \text{ L}} = 0.16\ M \text{ (two significant figures)}$$

33. Sulfuric acid for laboratory use is supplied in 2.5-L bottles, which contain 98.0% of H_2SO_4 by mass with a density of 1.92 g/mL. What mass of pure H_2SO_4 in kilograms is contained in such a bottle?

Solution

To determine the mass of H_2SO_4 in 2.5 L concentrate, first calculate from the percentage purity and density the mass of H_2SO_4 per milliliter of concentrate. Next convert from grams of H_2SO_4 per milliliter to kilograms in 2.5 L of solution. Identities required are:

$$2.5 \text{ L} = 2500 \text{ mL} \qquad 1 \text{ kg} = 1000 \text{ g}$$

Then, $$\text{Mass } H_2SO_4 \text{ per mL} = 1.92 \frac{\text{g}}{\text{mL}} \times \frac{98.0\%}{100\%} = 1.88 \frac{\text{g}}{\text{mL}}$$

$$\text{Mass } H_2SO_4 \text{ in 2.5 L} = 1.88 \frac{\text{g}}{\text{mL}} \times 2500 \text{ mL} = 4704 \text{ g}$$

$$\text{Mass (kg)} = 4704 \text{ g} \times \frac{1 \text{ kg}}{1000 \text{ g}} = 4.7 \text{ kg (two significant figures)}$$

35. What volume of 2.00 *M* acetic acid solution can be prepared from 1 pint (0.473 L) of 99.7% (by mass) acetic acid, CH_3CO_2H, with a density of 0.960 g/cm^3?

Solution

To calculate the volume of 2.00 *M* solution that can be prepared from 1 pint (0.473 L) of concentrate, first determine the number of moles of acetic acid in 0.473 L of concentrate. The volume of the diluted solution can then be calculated from the definition of molarity.

$$\text{Mass acetic acid per mL} = \frac{99.7\%}{100\%} \times 0.960 \frac{g}{cm^3} = 0.9571 \frac{g}{cm^3} = 0.9571 \frac{g}{mL}$$

Then, since 1 mol acetic acid = 60.052 g,

$$\text{Mass acetic acid in 0.473 L} = 0.9571 \frac{g}{mL} \times 1000 \frac{mL}{L} \times 0.473\ L = 452.7\ g$$

$$\text{Mol acetic acid} = 452.7\ g \times \frac{1\ mol}{60.052\ g} = 7.538\ mol$$

Because 0.473 L contains 7.545 mol of acetic acid, the volume of 2.00 *M* solution that can be prepared from this amount of acid is calculated from the definition of molarity.

$$M = \frac{\text{Mol of solute}}{\text{L of soln}}, \text{ or } \quad \text{L of soln} = \frac{mol}{M}$$

$$V\ (2.00\ M \text{ solution}) = \frac{mol}{M} = \frac{7.538\ mol}{2.00\ \frac{mol}{L}} = 3.77\ L$$

39. What volume of concentrated nitric acid, a 69.5% solution of HNO_3 with a density of 1.42 g/cm^3, is required to prepare 275 mL of a 0.150 *M* solution of HNO_3?

Solution

First calculate the mass of HNO_3 required in the dilute solution. The volume of concentrate needed is then calculated from the stated density and purity. Mass of HNO_3 in the dilute solution is calculated from the molar mass and the molarity.

$$M = \frac{\text{Mol of solute}}{\text{L of soln}}, \text{ or Mol} = M\ (\text{L of soln})$$

$$\text{Mol } HNO_3 = 0.150 \frac{mol}{L} \times 0.275\ L = 0.04125\ mol$$

$$\text{Mass} = \text{mol} \times \text{molar mass} = \text{mol} \times \frac{g}{mol}$$

$$\text{Molar mass } HNO_3 = 63.0134\ g\ mol^{-1}$$

$$\text{Mass } HNO_3 = 0.04125\ mol \times 63.0134 \frac{g}{mol} = 2.599\ g$$

Mass HNO_3 per mL of concentrate:

$$\frac{\text{Mass}}{\text{Volume}} = \frac{69.5\%}{100\%} \times 1.42 \frac{g}{mL} = 0.987 \frac{g}{mL}$$

The volume of concentrate required equals the mass of HNO_3 needed for the solution divided by the mass of HNO_3 per mL of concentrate.

$$V = \frac{2.599\ g}{0.987 \frac{g}{mL}} = 2.63\ mL$$

42. The limit for occupational exposure to ammonia, NH_3, in the air is set at $2.8 \times 10^{-3}\%$ by mass. What mass of ammonia, in grams, per cubic meter of air (density = 1.20 g/L) is required to reach this limit?

Solution

Given that the upper limit for exposure to ammonia is $2.8 \times 10^{-3}\%$ of the total mass of air, the allowable mass of NH_3 per cubic meter is calculated from the mass of air found from the density. First convert g/L to g/m^3. The mass of NH_3 is the percentage factor times the mass of air per cubic meter.
Mass of air per cubic meter: $1\ m^3 = 1000\ L$.

$$\text{Mass (air)} = 1.20\ \frac{g}{L} \times \frac{1000\ L}{m^3} = 1.20 \times 10^3\ \frac{g}{m^3}$$

The allowable mass of NH_3 is

$$\text{Mass } NH_3 = 1.20 \times 10^3\ \frac{g}{m^3} \times \frac{2.8 \times 10^{-3}\%}{100\%} = 0.034\ g$$

44. How many grams of a solution of insect repellent containing 17.0% by mass of N,N-dimethyl-*meta*toluamide, $C_{10}H_{13}NO$, can be prepared from 1.50 mol of $C_{10}H_{13}NO$?

Solution

Of the total mass of the solution, 17.0% is derived from 1.50 mol of $C_{10}H_{13}NO$. To calculate the solution mass, first the mass of 1.5 mol of $C_{10}H_{13}NO$ needs to be determined. The mass of $C_{10}H_{13}NO$ can be equated to the solution mass. Mass of 1.5 mol of $C_{10}H_{13}NO$:

$$\text{Molar mass } C_{10}H_{13}NO = 163.22\ g\ mol^{-1}$$

$$\text{Mass } C_{10}H_{13}NO = 1.50\ mol \times \frac{163.22\ g}{mol} = 244.8\ g$$

Mass of solution: 17.0% of the solution mass equals 244.8 g. This fact is stated algebraically as

$$0.170 \times \text{solution mass} = 244.8\ g$$

$$\text{Solution mass} = \frac{244.8\ g}{0.170} = 1440\ g$$

45. How many grams of CaO are required for reaction with the HCl in 275 mL of a 0.523 *M* HCl solution? The equation for the reaction is

$$CaO + 2HCl \longrightarrow CaCl_2 + H_2O$$

Solution

Two mol of HCl react with 1 mol of CaO. The HCl, however, is in solution, and the number of moles of HCl available must be computed from the solution data before the mass of CaO needed can be calculated. Mol HCl available is calculated from the volume times molarity.

$$\text{Mol HCl} = 275 \text{ mL} \times \frac{1 \text{ L}}{1000 \text{ mL}} \times 0.523 \frac{\text{mol}}{\text{L}} = 0.1438 \text{ mol}$$

$$\text{Mol CaO required} = \frac{1}{2} \text{ mol HCl} = \frac{0.1438 \text{ mol}}{2} = 0.0719 \text{ mol}$$

$$\text{Mass CaO needed} = 0.0719 \text{ mol} \times 56.08 \frac{\text{g}}{\text{mol}} = 4.03 \text{ g}$$

47. Aspirin, $C_6H_4(CO_2H)(CO_2CH_3)$, can be prepared in a chemistry laboratory by the reaction of salicylic acid, $C_6H_4(CO_2H)(OH)$, with acetic anhydride, $(CH_3CO)_2O$, a corrosive liquid.

$$2C_6H_4(CO_2H)(OH) + (CH_3CO)_2O \longrightarrow 2C_6H_4(CO_2H)(CO_2CH_3) + H_2O$$

What mass of aspirin can be prepared by the reaction of 5.00 mL of acetic anhydride (density = 1.0820 g/cm^3), assuming a 100% yield?

Solution

According to the balanced equation, 1 mol of acetic anhydride reacts to produce 2 mol of aspirin. The mass of acetic anhydride available is calculated from the density and volume of the liquid. The theoretical yield of aspirin can be calculated from the known mass of acetic anhydride. Mass of acetic anhydride available:

$$\text{Mass } (CH_3CO)_2O = 5.00 \text{ mL} \times 1.0820 \frac{\text{g}}{\text{mL}} = 5.41 \text{ g}$$

Mass of aspirin if all of the acetic anhydride is converted to aspirin:

$$\begin{array}{ccc} & 5.41 \text{ g} & \text{Mass?} \\ 2C_6H_4(CO_2H)OH + & (CH_3CO)_2O \longrightarrow & 2C_6H_4(CO_2H)(CO_2CH_3) + H_2O \\ & 1 \text{ mol} \longrightarrow & 2 \text{ mol} \end{array}$$

Molar mass $(CH_3CO)_2O$ = 102.10 g mol^{-1}; molar mass (aspirin) = 180.17 g mol^{-1}

$$\text{Mass (aspirin)} = 5.41 \text{ g } (CH_3CO)_2O \times \frac{1 \text{ mol}}{102.10 \text{ g}} \times \frac{2 \text{ mol aspirin}}{1 \text{ mol } (CH_3CO)_2O} \times \frac{180.17 \text{ g}}{\text{mol}} = 19.1 \text{ g}$$

49. Automotive "air bags" inflate when a sample of sodium azide, NaN_3, is very rapidly decomposed.

$$2NaN_3(s) \longrightarrow 2Na(s) + 3N_2(g)$$

What mass of NaN_3 is required to produce 13.0 ft^3 (368 L) of nitrogen gas with a density of 1.25 g/L?

Solution

The decomposition of NaN_3 produces 3 mol of nitrogen gas when 2 mol of NaN_3 decompose. The mass of nitrogen produced is found by multiplying the volume of 368 L by its density of 1.25 g/L.

$$\text{Mass } N_2 = 368 \text{ L} \times 1.25 \frac{\text{g}}{\text{L}} = 460 \text{ g}$$

Mass of NaN_3 is found from the balanced equation:

$$\begin{array}{ccc} \text{Mass?} & & 460 \text{ g} \\ 2NaN_3(s) & \longrightarrow 2Na(s) + & 3N_2(g) \\ 2 \text{ mol} & \longrightarrow & 3 \text{ mol} \end{array}$$

Molar mass NaN_3 = 65.0099 g mol^{-1}; molar mass N_2 = 28.0134 g mol^{-1}

$$\text{Mass } NaN_3 = 460 \text{ g } N_2 \times \frac{1 \text{ mol}}{28.0134 \text{ g}} \times \frac{2 \text{ mol } NaN_3}{3 \text{ mol } N_2} \times \frac{65.0099 \text{ g}}{\text{mol}} = 712 \text{ g}$$

51. Reaction of rhenium metal with rhenium heptaoxide, Re_2O_7, gives a solid of metallic appearance that conducts electricity almost as well as copper. A 0.788-g sample of this material, which contains only rhenium and oxygen, was oxidized in an acidic solution of hydrogen peroxide. Addition of an excess of KOH gave 0.973 g of $KReO_4$. Determine the empirical formula of the metallic solid and write the equation for its formation.

Solution

The formula of the unknown solid must be known before the equation for the reaction can be written. The empirical formula of the unknown compound containing only Re and O can be determined after calculating the total amount of Re in 0.973 g of $KReO_4$; the Re in $KReO_4$ originated from the oxide of rhenium.

$$\text{Molar mass } KReO_4 = 39.0983 + 186.207 + (4 \times 15.9994) = 289.3 \text{ g mol}^{-1}$$

$$\text{Mass Re in } KReO_4 = \frac{186.207 \text{ g Re}}{289.303 \text{ g } KReO_2} \times 0.973 \text{ g } KReO_4 = 0.626 \text{ g}$$

$$\text{Mass O in unknown sample} = 0.788 \text{ g} - 0.626 \text{ g} = 0.162 \text{ g}$$

The empirical formula of the unknown is found from the amount (mol) of the two elements in the compound.

$$\text{Mol Re} = 0.626 \text{ g} \times \frac{1 \text{ mol}}{186.207 \text{ g}} = 0.00336 \text{ mol}$$

$$\text{Mol O} = 0.162 \text{ g} \times \frac{1 \text{ mol}}{15.9994 \text{ g}} = 0.0101 \text{ mol}$$

Determine the smallest ratio:

$$\text{Re: } \frac{0.00336}{0.00336} = 1 \qquad \text{O: } \frac{0.0101}{0.00336} = 3 \text{ or } ReO_3$$

The equation for the reaction of Re with Re_2O_7, therefore, is

$$Re + 3Re_2O_7 \longrightarrow 7ReO_3$$

54. Glauber's salt, $Na_2SO_4 \cdot 10H_2O$, is an important industrial chemical that is isolated from naturally occurring brines in New Mexico. A 0.3440-g sample of this material was allowed to react with an excess of barium nitrate, $Ba(NO_3)_2$, and 0.2398 g of barium sulfate, $BaSO_4$, was isolated. What is the percent of $Na_2SO_4 \cdot 10H_2O$ in the sample?

$$Na_2SO_4 \cdot 10H_2O(aq) + Ba(NO_3)_2(aq) \longrightarrow BaSO_4(s) + 2NaNO_3(aq) + 10H_2O$$

Solution

The balanced equation indicates that 1 mol of $Na_2SO_4 \cdot 10H_2O$ reacts to produce 1 mol of $BaSO_4$. To determine the purity of the sample, first calculate the mass of $Na_2SO_4 \cdot 10H_2O$ necessary to produce 0.2398 g of $BaSO_4$, then calculate the percentage composition. Mass of $Na_2SO_4 \cdot H_2O$ required to produce 0.2398 g of $BaSO_4$:

$$\begin{array}{ccc} \text{Mass?} & & 0.2398\text{ g} \\ Na_2SO_4 \cdot 10H_2O + Ba(NO_3)_2 & \longrightarrow & BaSO_4 + 2NaNO_3 + 10H_2O \\ 1\text{ mol} & \longrightarrow & 1\text{ mol} \end{array}$$

Molar mass $Na_2SO_4 \cdot 10H_2O$ = 322.1891 g mol^{-1}; molar mass $BaSO_4$ = 233.39 g mol^{-1}

$$\text{Mass } Na_2SO_4 \cdot 10H_2O = 0.2398\text{ g } BaSO_4 \times \frac{1\text{ mol}}{233.39\text{ g}} \times \frac{1\text{ mol } Na_2SO_4 \cdot 10H_2O}{1\text{ mol } BaSO_4} \times \frac{322.1891}{\text{mol}}\text{ g}$$

$$= 0.3310\text{ g}$$

Percentage $Na_2SO_4 \cdot 10H_2O$ in sample:

The 0.3440-g sample actually contains 0.3310 g of $Na_2SO_4 \cdot 10H_2O$. The percentage is:

$$\%\ Na_2SO_4 \cdot 10H_2O = \frac{0.3310\text{ g}}{0.3440\text{ g}} \times 100\% = 96.23\%$$

56. Potassium perchlorate, $KClO_4$, may be prepared from KOH and Cl_2 by the following series of reactions:

$$2KOH + Cl_2 \longrightarrow KCl + KClO + H_2O$$

$$3KClO \longrightarrow 2KCl + KClO_3$$

$$4KClO_3 \longrightarrow 3KClO_4 + KCl$$

What mass of $KClO_4$ can be prepared from 25.0 g of KOH?

Solution

The production of $KClO_4$ from KOH and Cl_2 involves three reactions. The last two depend on the first reaction. The mole ratio of KOH to $KClO_4$ for the overall reaction can be determined by combining the ratios of pertinent components in the three reactions.

Reaction	(1)	*2* mol KOH	: *1* mol KClO
	(2)	*3* mol KClO	: *1* mol $KClO_3$
	(3)	*4* mol $KClO_3$	: *3* mol $KClO_4$

The combined ratio of reactants to products is the product of the individual ratios.

Reactants		Products
2 x 3 x 4 = 24	:	1 x 1 x 3 = 3
or 8	:	1

That is, 8.0 mol of KOH will produce 1.0 mol of $KClO_4$. The mass of $KClO_4$ that can be produced by the reaction of 25.0 g of KOH is

$$\begin{array}{ccc} 25.0\text{ g} & & \text{Mass?} \\ \text{KOH} & \longrightarrow & KClO_4 \\ 8\text{ mol} & \longrightarrow & 1\text{ mol} \end{array}$$

Molar mass KOH = 56.1056 g mol^{-1}; Molar mass $KClO_4$ = 138.549 g mol^{-1}

$$\text{Mass } KClO_4 = 25.0\text{ g KOH} \times \frac{1\text{ mol}}{56.1056\text{ g}} \times \frac{1\text{ mol } KClO_4}{8\text{ mol KOH}} \times \frac{138.549\text{ g}}{\text{mol}}$$

$$= 7.72\text{ g}$$

60. The concentrations of sodium ion, potassium ion, chloride ion, and dihydrogen phosphate ion in Gatorade are Na^+, 21.0 m*M*; K^+, 2.5 m*M*; Cl^-, 17.0 m*M*; $H_2PO_4^-$, 6.8 m*M*. These, incidentally, match the relative concentration of these ions lost through perspiration. What masses of Na^+, K^+, Cl^-, and $H_2PO_4^-$ are contained in an 8.00-oz glass of Gatorade?

Solution

Several conversions are required: m*M* to *M* to g, and oz to L. Since 1 qt = 0.9463 L = 32 oz, then

$$\frac{8.00\ \text{oz}}{32.00\ \text{oz}} = \frac{1}{4}\ \text{qt} = \frac{1}{4}\ (0.9463\ \text{L}) = 0.2366\ \text{L}$$

For Na^+: 21.0 m*M* = 0.0210 *M*. Since *M* = (g/molar mass)/L,

$$\text{g } Na^+ = M \times \text{L} \times \text{molar mass}$$

$$= 0.0210\ \frac{\text{mol}}{\text{L}} \times 0.2366\ \text{L} \times 22.98977\ \frac{\text{g}}{\text{mol}}$$

$$= 0.114\ \text{g}$$

For K^+: 2.5 m*M* = 0.0025 *M*

$$\text{g } K^+ = 0.0025\ \frac{\text{mol}}{\text{L}} \times 0.2366\ \text{L} \times 39.0983\ \frac{\text{g}}{\text{mol}}$$

$$= 0.023\ \text{g}$$

For Cl^-: 17.0 m*M* = 0.0170 *M*

$$\text{g } Cl^- = 0.0170\ \frac{\text{mol}}{\text{L}} \times 0.2366\ \text{L} \times 35.453\ \frac{\text{g}}{\text{mol}}$$

$$= 0.143\ \text{g}$$

For $H_2PO_4^-$: 6.8 m*M* = 0.0068 *M*

$$\text{g } H_2PO_4^- = 0.0068\ \frac{\text{mol}}{\text{L}} \times 0.2366\ \text{L} \times 96.9872\ \frac{\text{g}}{\text{mol}}$$

$$= 0.16\ \text{g}$$

63. What mass of a sample that is 98.0% sulfur would be required in the production of 75.0 kg of H_2SO_4 by the following reaction sequence?

$$S_8 + 8O_2 \longrightarrow 8SO_2$$

$$2SO_2 + O_2 \longrightarrow 2SO_3$$

$$SO_3 + H_2O \longrightarrow H_2SO_4$$

Solution

Although the production of sulfuric acid requires three reactions, these reactions can be summarized as follows: 1 mol of S atoms combines with 2 mol of H atoms and 4 mol of O atoms to produce 1 mol of H_2SO_4. Calculate the mass of S needed to produce 75.0 kg of H_2SO_4 at 100% reaction efficiency, then account for the 98.0% purity factor.

$$\text{Molar mass } H_2SO_4 = 2(1.0079) + 32.06 + 4(15.9994)$$

$$= 98.07\ \text{g mol}^{-1}$$

In kilograms, 98.07 kg of H_2SO_4 contains 32.06 kg of S.

$$32.06 \text{ kg S} \longrightarrow 98.07 \text{ kg } H_2SO_4$$

To prepare 75.0 kg of H_2SO_4 at 100% efficiency, the mass of S needed is

$$\text{Mass S} = \frac{32.06 \text{ kg S}}{98.07 \text{ kg } H_2SO_4} \times 75.0 \text{ kg} = 24.5 \text{ kg}$$

At 98.0% S purity, the mass of the sample needed is

$$98.0\% \text{ (sample mass)} = 24.5 \text{ kg}$$

$$\text{Sample mass} = \frac{24.5 \text{ kg}}{0.980} = 25.0 \text{ kg}$$

64. For many years the standard medical laboratory technique for determination of the concentration of calcium (as Ca^{2+}) in blood serum used the following reaction sequence involving ammonium oxalate, $(NH_4)_2C_2O_4$, and potassium permanganate, $KMnO_4$:

$$Ca^{2+}(aq) + (NH_4)_2C_2O_4(s) \longrightarrow CaC_2O_4(s) + 2NH_4^+(aq)$$

$$CaC_2O_4(s) + H_2SO_4(aq) \longrightarrow H_2C_2O_4(aq) + CaSO_4(aq)$$

$$2KMnO_4(aq) + 5H_2C_2O_4(aq) + 3H_2SO_4(aq) \longrightarrow$$

$$K_2SO_4(aq) + 2MnSO_4(aq) + 10CO_2(g) + 8H_2O(\ell)$$

What is the molar concentration of calcium in a 2.0-mL serum sample if 0.68 mL of 2.44×10^{-3} *M* $KMnO_4$ is required for the final reaction? How many milligrams of calcium are contained in 1.00×10^2 mL of the serum?

Solution

The amount of Ca^{2+} (mol) is required to find the molarity. From the series of reactions, the amount of $KMnO_4$ used can be related to the moles of Ca^{2+}. From the equations, 1 Ca^{2+} reacts to form 1 CaC_2O_4, which in turn reacts to form 1 $H_2C_2O_4$. Thus, the ratio is 1 Ca^{2+} to 1 $H_2C_2O_4$. Then, $5H_2C_2O_4$ react with $2KMnO_4$, so that the ratio is $5Ca^{2+}$ to $2KMnO_4$.

Mol $KMnO_4$ equals:

$$0.0068 \text{ L} \times 2.44 \times 10^{-3}\ M = 1.659 \times 10^{-5} \text{ mol}$$

$$\text{Mol } Ca^{2+} = 1.659 \times 10^{-5} \text{ mol } KMnO_4 \times \frac{5 \text{ mol } Ca^{2+}}{2 \text{ mol } KMnO_4}$$

$$= 4.148 \times 10^{-5} \text{ mol}$$

Then, $M = \dfrac{\text{amount of solute}}{\text{volume of solution (L)}} = \dfrac{4.148 \times 10^{-5} \text{ mol}}{0.0020 \text{ V}} = 0.0021\ M$

The milligrams of calcium ion are calculated from the molarity:

$$\text{mg Ca}^{+} = 0.0021\ M \times 1.00 \times 10^2\ \text{mL} \times \frac{1\ \text{L}}{1000\ \text{mL}} \times 40.08\ \frac{\text{g}}{\text{mol}} \times \frac{1000\ \text{mg}}{\text{g}} = 8.4\ \text{mg}$$

67. Calculate the mass of sodium nitrate required to produce 5.00 L of O_2 (density = 1.43 g/L) according to the reaction

$$2NaNO_3 \xrightarrow{\Delta} 2NaNO_2 + O_2$$

if the percent yield of the reaction is 78.4%.

Solution

Solution of this problem requires the following steps: Determination of the mass of O_2 in 5.00 L, conversion of mass of O_2 to amount of O_2 (mol), then conversion to amount of $NaNO_3$, and mol $NaNO_3$ to mass (mg). Allowance for the 78.4% efficiency of the reaction can then be made.

$$\text{Mol } O_2 = 5.00\ \text{L} \times 1.43\ \frac{\text{g}}{\text{L}} \times \frac{1\ \text{mol}}{31.9988\ \text{g}} = 0.2234\ \text{mol}$$

$$\text{g } NaNO_3 = 0.2234\ \text{mol } O_2 \times \frac{2\ \text{mol } NaNO_3}{1\ \text{mol } O_2} \times \frac{84.9947\ \text{g}}{1\ \text{mol } NaNO_3}$$

$$= 37.98\ \text{g } NaNO_3$$

At 78.4% efficiency

$$37.98\ \text{g} \times \frac{1}{.784} = 48.4\ \text{g}$$

70. Uranium may be isolated from the mineral pitchblende, which contains U_3O_8. The pitchblende in a 4.835-g sample of an ore containing uranium was subjected to the following sequence of reactions:

$$2U_3O_8 + O_2 + 12HNO_3 \longrightarrow 6(UO_2)(NO_3)_2 + 6H_2O$$

$$(UO_2)(NO_3)_2 + 4H_2O + H_3PO_4 \longrightarrow (UO_2)HPO_4 \cdot 4H_2O + 2HNO_3$$

$$2[(UO_2)HPO_4 \cdot 4H_2O] \xrightarrow{\Delta} (UO_2)_2P_2O_7 + 9H_2O$$

What is the percent by mass of U_3O_8 in the ore if 1.432 g of $(UO_2)_2P_2O_7$ is isolated?

Solution

The three reactions take U_3O_8 to $(UO_2)_2P_2O_7$. The ratios of major reactants containing uranium are

$$2U_3O_8 : 6(UO_2)(NO_3)_2$$

$$(UO_2)(NO_3)_2 : (UO_2)HPO_4 \cdot 4H_2O$$

$$2[(UO_2)HPO_4 \cdot 4H_2O] : (UO_2)_2P_2O_7$$

The ratio $U_3O_8 \longrightarrow (UO_2)_2P_2O_7$ is found by multiplying together the coefficients on the left and the coefficients on the right and reducing to the smallest whole number ratio: $4 : 6 \longrightarrow 2 : 3$, that is, $2U_3O_8 \longrightarrow 3(UO_2)_2P_2O_7$. The amount of U_3O_8 producing 1.432 g of $(UO_2)_2P_2O_7$ is found in the normal way.

$$\text{g } U_3O_8 = 1.432 \text{ g } (UO_2)_2P_2O_7 \times \frac{2 \text{ mol } U_3O_8}{3 \text{ mol } (UO_2)_2P_2O_7} \times \frac{\text{mol } (UO_2)_2P_2O_7}{713.999 \text{ g } (UO_2)_2P_2O_7}$$

$$\times \frac{842.082 \text{ g } U_3O_8}{\text{mol } U_3O_8} = 1.126 \text{ g}$$

The percentage is

$$\frac{1.126 \text{ g}}{4.835 \text{ g}} \times 100 = 23.29\%$$

72. What is the limiting reagent when 5.0×10^{-2} mol of nitric acid, HNO_3, reacts with 225 mL of 0.10 *M* calcium chloride, $Ca(OH)_2$, solution?

$$Ca(OH)_2(aq) + 2HNO_3(aq) \longrightarrow Ca(NO_3)_2(aq) + 2H_2O(\ell)$$

Solution

Two mol HNO_3 react with 1 mol $Ca(OH)_2$. Calculate the mol of $Ca(OH)_2$ and compare with one-half the mol HNO_3. Since M = mol/L,

$$\text{mol } Ca(OH)_2 = 0.10\ M \times 0.225 \text{ L} = 0.0225 \text{ mol} = 2.25 \times 10^{-2} \text{ mol}$$

From the equation, 2.25×10^{-2} mol $Ca(OH)_2$ requires $2 \times 2.25 \times 10^{-2}$ mol of HNO_3. There is an excess of HNO_3, so $Ca(OH)_2$ is the limiting reagent.

RELATED EXERCISES

1. Which one of the following solutions has the greatest molarity?

   (a) 60.0 mg of NaOH in 250 ml of solution
   (b) 200. g of glucose, $C_6H_{12}O_6$, in 5.0 L of solution
   (c) 5.00 mg of $Al_2(SO_4)_3$ in 200 ml of solution

   Answer: (b) 0.22 M

2. Calculate the volume of 0.125 *M* HNO_3 required to neutralize 550 mL of 2.0 *M* NaOH.

$$HNO_3(aq) + NaOH(aq) \longrightarrow NaNO_3(aq) + H_2O(\ell)$$

   Answer: 8.8×10^3 *mL*

UNIT 4

CHAPTER 4: STRUCTURE OF THE ATOM AND THE PERIODIC LAW

INTRODUCTION

Ancient records point to the fact that human beings have long pondered the basic structure of matter in the hope of better understanding themselves and their surroundings. The early writings of the physician and philosopher Empedocles of Agrigentum in Sicily (ca. 490-435 B.C.) suggested that all visible objects are composed of four unchanging elements — air, earth, water, and fire. The Greek philosophers Leucippos and his pupil, Demokritos of Abdera (ca. 460-370 B.C.), postulated that matter is composed of small particles of these four elements and that these particles are in motion. This was the first evidence of an atomic and kinetic theory. Building upon these ideas, Aristotle moved away from the atomistic concept and popularized a theory to explain the transformation of one "element" to another. His ideas, including that of the void, or "ether," dominated Western scientific thought for centuries. Unfortunately, since Aristotle's theory did not lend itself to experimental verification, no one tried to test his theories. Consequently, very little experimental evidence existed to support the theory until about 1800.

Most recently, the development of sophisticated instruments has enabled scientists who study atoms to probe into the atoms' structure. We know now that the nature and composition of atoms depend on the interaction of the many smaller particles that provide the building blocks of all atoms. Several theoretical atomic models were proposed during the early 1900s to explain atomic phenomena and to predict atomic and molecular behavior.

This chapter treats the discovery of atomic particles, the unparalleled experimentation during the evolution of modern atomic theory, the particulars of the quantum mechanical model of the atom, and the relationship of atomic structure to the Periodic Table. Part of the chapter explains the nature of electromagnetic radiation and its relationship to electronic transition within atoms, and, hence, to the entire field of modern chemistry.

The exercises are directed to the various atomic models and especially to the current quantum mechanical model. Both numerical exercises and exercises requiring only verbal answers have been selected so that their solution will provide a useful framework to help you organize the important information of this chapter. Remember as you proceed that the development of the models and theories represents the attempt by scientists to advance our knowledge of matter and that the latest models generally will require further refinement in the future.

FORMULAS AND DEFINITIONS

Valence electrons Valence electrons are the electrons that occupy the highest energy level of an atom. The electrons in the outermost sublevel when that sublevel is not in the highest energy level (for example, $3d$ electrons when the $4s$ level is occupied) are also valence electrons. The valence electrons are largely responsible for the chemical behavior of the atom. All elements within a periodic group or family have the same outer electron configuration, so they also have the same number of valence electrons. For the representative elements, the number of valence electrons in an atom equals its periodic group number.

Representative elements The representative elements are the A-group elements, IA through VIIA and Group IIB, as listed in the Periodic Table. These elements have valence electrons either in s or s and p sublevels. Their valence electrons are in the same primary level (n level) as their period location. For example, the three valence electrons in gallium, $Z = 31$, are in the fourth n level, that is, $4s^2$ and $4p^1$ electrons.

Transition elements The transition elements are metals and are located in the B groups of the Periodic Table (except Group IIB). These elements have outer electrons that include the $(n - 1)d$ sublevels, where n is the period location of the elements. For example, titanium, Ti, $Z = 22$, is in period 4, but its outermost electrons are in the $3d$ orbitals. The period 4 transition metals are relatively abundant and are extremely important to the world steel industry. With a few exceptions the transition elements in periods 5 and 6 are in limited supply as far as commercially exploitable deposits are concerned. Several are highly valued for use in jewelry, coinage, and electronic systems.

Pauli Exclusion Principle The rigorous solution of the wave equation representing electrons orbiting an atomic nucleus involves four components. Four quantum numbers describe the behavior of these components and represent the effective volume of space in which an electron moves, the shape of the orbital in space, the orientation of the electronic charge cloud in space, and the direction of the spin of each electron on its own axis. The Pauli Exclusion Principle states that each electron has a unique set of four quantum numbers; that is, no two electrons in the same atom can have the same four quantum numbers.

Hund's Rule Atomic energy levels are filled by electrons in such a way as to achieve the lowest possible total energy state for the atom. Hund's Rule states that each orbital in a subshell or sublevel must be singly filled before any one orbital is completely (doubly) filled. In addition, all electrons in singly filled orbitals within a sublevel must have parallel spins; that is, their spin orientations are the same (aligned).

Aufbau Process The Aufbau Process is the way in which electronic structures of complex atoms are formed by assigning one electron at a time to a sublevel based on Hund's Rule and the Pauli Exclusion Principle.

Ionization potential The ionization potential (IP) of an atom, frequently called *ionization energy*, is the amount of energy required to completely remove the most loosely bound electron from a gaseous atom in its ground state. Values for ionization potentials are usually listed in terms of mole quantities; some examples are

$$\text{General case:} \quad X(g) + \text{energy} \longrightarrow X^+(g) + e^-$$

$$Li(g) \longrightarrow Li^+(g) + e^- \qquad IP = 5.19 \times 10^5 \text{ J/mol}$$

$$N(g) \longrightarrow N^+(g) + e^- \qquad IP = 1.40 \times 10^6 \text{ J/mol}$$

$$He(g) \longrightarrow He^+(g) + e^- \qquad IP = 2.37 \times 10^6 \text{ J/mol}$$

Electron affinity The electron affinity (EA) of an atom is a measure of the attraction a neutral gaseous atom in its ground state has for an additional electron. In general, atoms that have half-filled orbitals or incomplete valence levels will attract additional electrons and release energy upon the ionization. Some examples defining the electron affinity for several elements along with their numerical values are

$$\text{General case:} \quad X(g) + e^- \longrightarrow X^-(g) + \text{energy}$$

$$Cl(g) + e^- \longrightarrow Cl^-(g) + 3.48 \times 10^5 \text{ J/mol} \qquad EA = 3.48 \times 10^5 \text{ J/mol}$$

$$O(g) + e^- \longrightarrow O^-(g) + 2.25 \times 10^5 \text{ J/mol} \qquad EA = 2.25 \times 10^5 \text{ J/mol}$$

$$Mg(g) + e^- + 2.89 \times 10^4 \text{ J/mol} \longrightarrow Mg^-(g) \qquad EA = -2.89 \times 10^4 \text{ J/mol}$$

Constants and equations

$$\text{Planck's constant} \quad (h) = 6.626 \times 10^{-34} \text{ J s}$$
$$= 6.626 \times 10^{-27} \text{ erg s}$$
$$= 1.584 \times 10^{-37} \text{ kcal s}$$

$$\text{Speed of light} \quad (c) = 2.998 \times 10^8 \text{ m sec}^{-1} \text{ (vacuum)}$$

$$\text{Rydberg equation} \quad E = 2.179 \times 10^{-18} \left(\frac{1}{n_1^2} - \frac{1}{n_2^2}\right) \text{ J, where } n_1 < n_2$$

The Rydberg equation is empirically derived from spectral data resulting from studies on hydrogen. The energy value calculated from the equation is the energy associated with a photon produced by an electron transition from a higher energy level (n_2) to a lower energy level (n_1) in a hydrogen atom.

Bohr model energies

$$E_n = -2.179 \times 10^{-18} \frac{Z^2}{n^2} \text{ J} \quad \text{or} \quad -2.179 \times 10^{-18}/n^2 \text{ J}$$

where Z is the atomic number for a hydrogen atom

The energy calculated from this equation is the energy associated with a specific energy level in a hydrogenlike atom or ion. By convention all the values calculated using this equation have a negative sign, and when $n = 1$ the value is -2.179×10^{-18} J. As n increases (higher energy levels), the values of E increase and approach zero at $n = \infty$. The energy required to raise an electron from $n = 1$ to $n = \infty$ is the ionization energy for a hydrogen-like atom or ion in its ground state.

Electromagnetic wave characteristics Visible light and other types of radiant energy described in this chapter are propagated as waves in the form of electromagnetic radiation consisting of two components: one with an electric field, the other with a magnetic field. These components are coupled at right angles to each other and travel as a sine wave as shown in the following figure, where E is the electric field and H is the magnetic field.

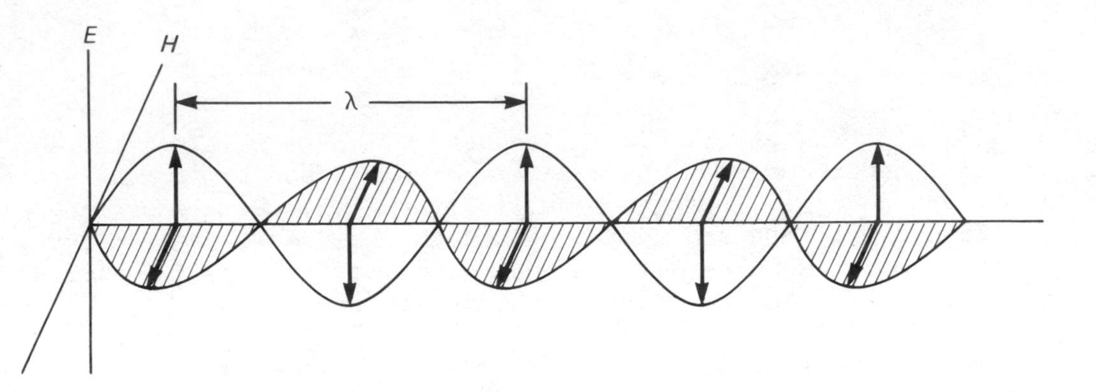

Wavelength (λ, lambda) Wavelength is the distance between equivalent positions on adjacent waves, for example, from one peak to the next.

Frequency (ν, nu) Frequency is the number of wavelengths passing a reference point in unit time. The unit for frequency is per second and is usually written as 1/sec or sec^{-1}. The words *cycle* or *wave* are implied but not written.

Velocity (c) The velocity of a wave is the distance traveled by the wave per unit time. For electromagnetic radiation in a vacuum, $c = 2.998 \times 10^8$ m sec^{-1}

Energy-wavelength-frequency relations Velocity, frequency, and wavelength are related through the equation $\lambda\nu = c$. The velocity of all electromagnetic radiation passing through a continuous medium (air, water, etc.) is the same and varies only with the medium. Here, the assumption will be made that the velocity of radiation is constant and that the medium of transmission is a vacuum in which the velocity = 2.998×10^8 m/sec. Because the value of c is constant, λ and ν are inversely related to each other; that is,

$$\lambda(\text{short})\nu(\text{high}) = \text{constant}$$

$$\lambda(\text{long})\nu(\text{low}) = \text{constant}$$

The energy of a photon is related to frequency as

$$E = h\nu \quad (h = \text{Planck's constant})$$

in which energy is directly proportional to frequency:

$$\text{High energy} \longleftrightarrow \text{high frequency}$$

$$\text{Low energy} \longleftrightarrow \text{low frequency}$$

where $\longleftrightarrow$ may be read as "is associated with." The relationship between energy and wavelength is derived by solving for ν in the equation $c = \lambda\nu$ and substituting $E = h\nu$:

$$\lambda\nu = c \quad \text{and} \quad \nu = c/\lambda$$

$$E = h\nu = hc/\lambda$$

This is an inverse relationship of energy to wavelength; that is,

$$\text{High energy} \longleftrightarrow \text{short wavelength}$$

$$\text{Low energy} \longleftrightarrow \text{long wavelength}$$

These relationships can be extended to the frequency of light in the following way:

$$\text{High energy} \longleftrightarrow \text{high frequency} \longleftrightarrow \text{short wavelength}$$

$$\text{Low energy} \longleftrightarrow \text{low frequency} \longleftrightarrow \text{long wavelength}$$

The proper units for use with energy relationships are

$$E \text{ in joules} = h\nu = (6.626 \times 10^{-34}\text{ J s})(\nu \text{ in 1/s or s}^{-1})$$

$$E \text{ in joules} = \frac{hc}{\lambda} = \frac{(6.626 \times 10^{-34}\text{ J s})(c \text{ in m/s})}{\lambda \text{ in m}}$$

EXERCISES

8. Using the data in Table 4-2, calculate the atomic weight of naturally occurring boron.

Solution

The composition of naturally occurring boron consists of ${}^{10}_{5}B$, mass = 10.0129 daltons (amu), % = 19.6; for ${}^{11}_{5}B$, mass = 11.0042 daltons, % = 80.4. The atomic weight of an element is the average mass of all atoms composing a normal sample of the element. For boron, 19.6% of the atomic weight is attributable to ${}^{10}B$ and 80.4% to ${}^{11}B$. The atomic weight is calculated as follows:

$$\text{At. wt} = \frac{19.6\ {}^{10}\text{B atoms}}{100\ {}^{10}\text{B atoms}}(10.0129\text{ amu}) + \frac{80.4\ {}^{11}\text{B atoms}}{100\ {}^{11}\text{B atoms}}(11.0093\text{ amu})$$

$$= 10.8\text{ amu}$$

14. Figure 4-10 gives the energies of an electron in a hydrogen atom for values of n from 1 to 5. What is the energy, in electron-volts, of an electron with $n = 8$ in a hydrogen atom?

Solution

The energy of the electron is given by the equation

$$E = -\frac{kZ^2}{n^2}$$

For hydrogen, $Z = 1$ and when $n = 1$, $E = -13.595$ eV. Since k is a constant, the value of k must be 13.595 eV. Thus, for hydrogen with $n = 8$,

$$E = -13.595\text{ eV}\ \frac{1}{(8)^2} = -0.2125\text{ eV}$$

16. Recently, hydrogen atoms with electrons in very high-energy orbitals have been isolated. What is the radius (in centimeters) of a hydrogen atom with an electron characterized by an n value of 110? How many times larger is this than the radius of the normal hydrogen atom?

Solution

The radius of the orbit is given by the equation

$$R = \frac{n^2 a_0}{Z}$$

where $a_0 = 0.529$ Å $= 5.29 \times 10^{-9}$ cm. Substitution of quantities into the equation for R gives

$$R = \frac{110^2(5.29 \times 10^{-9}\text{ cm})}{1} = 6.40 \times 10^{-5}\text{ cm}$$

In order to find the relative increase in size of the radius, divide the new radius by that of hydrogen in the ground state.

$$\frac{6.40 \times 10^{-5}\text{ cm}}{5.29 \times 10^{-9}\text{ cm}} = 1.21 \times 10^4$$

or 12,100 times larger.

18. According to the Bohr model, which is larger: a hydrogen atom with an electron in an orbit with $n = 4$; or a He^+ ion with an electron in an orbit with $n = 5$?

Solution

From the Bohr theory the radius of the orbit is given by the equation

$$R = \frac{n^2 a_0}{Z}$$

where $a_0 = 0.529$ Å. For hydrogen, $Z = 1$; and for helium, $Z = 2$. Thus,

$$R_H = \frac{4^2(0.529\text{ Å})}{1} = 8.46$$

$$R_{He} = \frac{5^2(0.529\text{ Å})}{2} = 6.61$$

Consequently, the H radius is larger.

19. From the data given in Fig. 4-12, calculate the energy in electron volts, of the photon emitted when an electron in an excited hydrogen atom moves from an orbit with $n = 4$ to one with $n = 2$.

Solution

The difference in energies of the two states is calculated from

$$E = \frac{-kZ^2}{n^2}$$

where $k = 13.595$ eV and for hydrogen, $Z = 1$.

$$E_{n_4} - E_{n_2} = \frac{-kZ^2}{n_4^2} - \left(\frac{-kZ^2}{n_2^2}\right) = -13.595\left(\frac{1}{4^2} - \frac{1}{2^2}\right) \text{ eV} = 2.549 \text{ eV}$$

21. When lithium metal is heated, lithium atoms emit photons of red light with a wavelength of 6708 Å. What is the energy in joules and the frequency of this light?

Solution

The wavelength of light is related to the frequency by the equation

$$\lambda\nu = c$$

where $c = 2.998 \times 10^8$ m/sec.

The energy is related to the frequency by the expression

$$E = h\nu$$

where h is Planck's constant and is equal to 6.626×10^{-34} J s.

Our first step is to convert 6708 Å to m.

$$6708 \text{ Å} \times \frac{10^{-10} \text{ m}}{\text{Å}} = 6.708 \times 10^{-7} \text{ m}$$

Rearrangement of $\lambda\nu = c$ and substitution of c and λ gives

$$\nu = \frac{c}{\lambda} = \frac{2.998 \times 10^8 \text{ m/s}}{6.708 \times 10^{-7} \text{ m}} = 4.469 \times 10^{14} \text{ s}^{-1}$$

Substitution into $E = h\nu$ gives

$$E = h\nu = 6.626 \times 10^{-34} \text{ J s} \times 4.469 \times 10^{14} \text{ s}^{-1}$$
$$= 2.961 \times 10^{-19} \text{ J}$$

24. Consider a collection of hydrogen atoms with electrons randomly distributed in the $n = 1,2,3,4$, and 5 shells. How many different wavelengths of light will be emitted by these atoms as the electrons fall into the lower-energy orbitals?

Solution

Imagine an electron being placed in each of the five n levels. Each possible transition from a higher level to a lower level releases energy in the form of radiation. The number of possible transitions beginning with level 5 through level 2 is

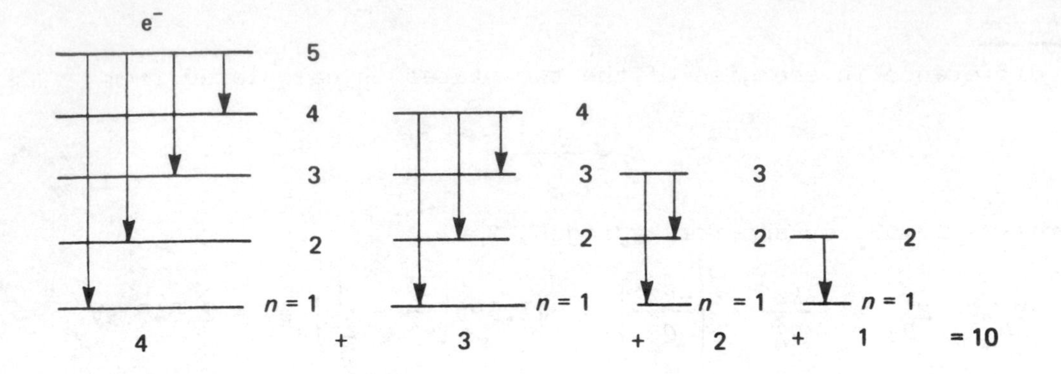

28. Does a photon of the green light described in Exercise 27 have enough energy to excite the electron in a hydrogen atom from the $n = 1$ to the $n = 2$ shell?

Solution

The required energy for the transition can be calculated by using the Rydberg equation to determine the difference in energy for the transition $n = 1$ to $n = 2$.

$$E_{n_1 \to n_2} = 2.179 \times 10^{-18} \left(\frac{1}{n_1^2} - \frac{1}{n_2^2}\right) \text{ J} = 2.179 \times 10^{-18} \left(\frac{1}{1^2} - \frac{1}{2^2}\right)$$

$$= 2.179 \times 10^{-18} \left(\frac{3}{4}\right) \text{ J} = 1.634 \times 10^{-18} \text{ J}$$

The required energy is greater than 3×10^{-19} J, the energy of a photon of green light; hence the photon will not excite the electron from the $n = 1$ to the $n = 2$ shell.

29. X rays are produced when the electron stream in an X-ray tube knocks an electron out of a low-energy shell of an atom in the target, and an electron in a higher shell falls into the low-energy shell. The X ray is the photon given off as the electron falls into the low-energy shell. The most intense X rays produced by an X-ray tube with a copper target have wavelengths of 1.542 Å and 1.392 Å. These X rays are produced by an electron from the L or M shell falling into the K shell of a copper atom. Calculate the energy separation in electron-volts of the K, L, and M shells in copper.

Solution

The two X rays are produced from electron shifts between the K and L shells and the K and M shells. The transition from M to K emits the X ray with the greatest energy and the shortest wavelength. The energy difference between the L and M shell is the difference between the energies of the two X rays.

1.392 Å (M → K) ; 1.542 Å (L → K)

$$E_{M \to L} = E_{M \to K} - E_{L \to K}$$

$$E_{M \to K} = \frac{hc}{\lambda} = \frac{(6.626 \times 10^{-34} \text{ J sec})(2.998 \times 10^{10} \text{ cm/sec})}{1.392 \times 10^{-8} \text{ cm}}$$

$$E_{M\to L} \cong E_{M\to K} - E_{L\to K}$$
$$= (1.427 \times 10^{-15} - 1.288 \times 10^{-15})\ \text{J} = 1.39 \times 10^{-16}\ \text{J}$$
$$= 1.427 \times 10^{-15}\ \text{J}$$

$$E_{L\to K} = \frac{hc}{\lambda} = \frac{(6.626 \times 10^{-34}\ \text{J sec})(2.998 \times 10^{10}\ \text{cm/sec})}{1.542 \times 10^{-8}\ \text{cm}} = 1.288 \times 10^{-15}\ \text{J}$$

These values in electron-volts are calculated from the relation

$$1\ \text{eV} = 1.602189 \times 10^{-19}\ \text{J}$$

$$E_{M\to K} = 1.427 \times 10^{-15}\ \text{J} \times \frac{1\ \text{eV}}{1.602189 \times 10^{-19}\ \text{J}} = 8.907 \times 10^{3}\ \text{eV}$$

$$E_{M\to L} = 1.39 \times 10^{-16}\ \text{J} \times \frac{1\ \text{eV}}{1.602189 \times 10^{-19}\ \text{J}} = 8.67 \times 10^{2}\ \text{eV}$$

$$E_{L\to K} = 1.288 \times 10^{-15}\ \text{J} \times \frac{1\ \text{eV}}{1.602189 \times 10^{-19}\ \text{J}} = 8.040 \times 10^{3}\ \text{eV}$$

36. Consider three atomic orbitals (a), (b), and (c), whose outlines are shown below.

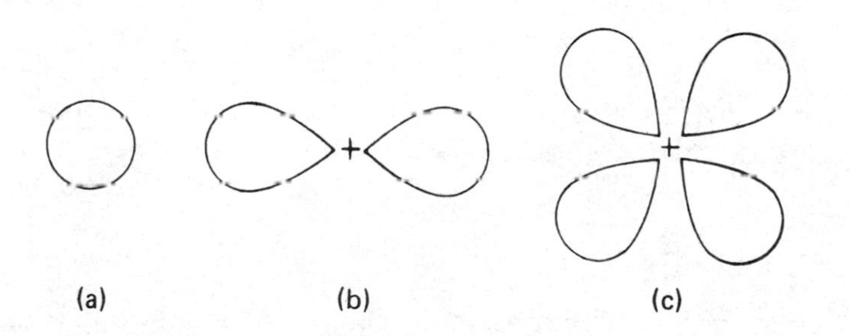

(a) What is the maximum number of electrons that can be contained in orbital (c)?
(b) How many orbitals with the same value of ℓ as orbital (a) can be found in the shell with $n = 4$? How many with the same value as orbital (b)? How many with the same value as orbital (c)?
(c) What is the smallest n value possible for an electron in an orbital of type (c)? of type (b)? of type (a)?
(d) What are the ℓ values that characterize each of these three orbitals?
(e) Arrange these orbitals in order of increasing energy in the M shell. Is this order different in other shells?

Solution

(a) The maximum number of electrons that can occupy an orbital is two, regardless of its shape or volume. Hence the orbitals represented in (a), (b), and (c) above can contain a maximum of two electrons.
(b) The orbital represented in (a) is an s or $\ell = 0$ type. Each primary level, n level, has a corresponding s or $\ell = 0$ type orbital; only the volume of the s orbital space varies. The orbital represented in (b) is a p or $\ell = 1$ type, and as a requirement of the quantum model, all n levels except $n = 1$ contain three p orbitals. The orbital represented in (c) is a d or $\ell = 2$ type. According to the model, all n levels of 3 and greater have five d or $\ell = 2$ type orbitals.

(c) Type c, $n = 3$; type b, $n = 2$; type a, $n = 1$.
(d) Type c, $\ell = 2$; type b, $\ell = 1$; type a, $\ell = 0$.
(e) The M shell is the $n = 3$ energy level and it has three possible ℓ values: 0, 1, and 2. These ℓ levels are the 3*s*, 3*p*, and 3*d* orbitals with the orbital energies increasing as written. Note, however, that conclusions derived from experimental evidence indicate that the $n = 4$, $\ell = 0$ level (4*s*) has a lower energy than the $n = 3$, $\ell = 2$ level (3*d*) and is filled before the 3*d* level. The *s* and *p* orbitals for all values of *n* fill in order, but the *d* orbitals fill after the $\ell = 0$ level of the next higher *n* level. Thus, for the orbitals shown, $a < b < c$. This order of orbital energies is the same in other shells.

37. Identify the subshell in which electrons with the following quantum numbers are found:

(a) $n = 1$, $\ell = 0$ (c) $n = 3$, $\ell = 2$ (e) $n = 4$, $\ell = 3$

(b) $n = 3$, $\ell = 1$ (d) $n = 7$, $\ell = 4$

Solution

The *s* orbital is always associated with an $\ell = 0$ value, the *p* orbital with $\ell = 1$, the *d* orbital with $\ell = 2$, the *f* orbital with $\ell = 3$, and the *g* orbital with $\ell = 4$. Therefore, the *n* value that is stated first is associated with the following orbitals: (a) 1*s*; (b) 3*p*; (c) 3*d*; (d) 7*g*; (e) 4*f*

39. Write the quantum numbers for each electron found in an oxygen atom. For example, the quantum numbers for one of the 2*s* electrons are $n = 2$, $\ell = 0$, $m = 0$, $s = +1/2$.

Solution

Oxygen ($Z = 8$) has eight electrons and its electron configuration is $1s^2 2s^2 2p^4$. The quantum numbers for the eight electrons in oxygen are

e^-	n	ℓ	m	s
1	1	0	0	$+\frac{1}{2}$
2	1	0	0	$-\frac{1}{2}$
3	2	0	0	$+\frac{1}{2}$
4	2	0	0	$-\frac{1}{2}$
5	2	1	-1	$+\frac{1}{2}$
6	2	1	0	$+\frac{1}{2}$
7	2	1	+1	$+\frac{1}{2}$
8	2	1	-1	$-\frac{1}{2}$

To avoid confusion when assigning quantum numbers, it is important to adopt and consistently use a convention for assigning m and s quantum numbers. The convention used here is to begin by assigning negative values and for m proceed through positive values, that is $-\ell$ through $+\ell$ and $+1/2$ and then $-1/2$ for s.

40. The quantum numbers that describe the electron in the lowest energy level of a hydrogen atom (the ground state) are $n = 1$, $\ell = 0$, $m = 0$, $s = +1/2$. Excitation of the electron can promote it to energy levels described by other sets of quantum numbers. Which of the following sets of quantum numbers cannot exist in a hydrogen atom (or any other atom)?

(a) $n = 2$, $\ell = 1$, $m = -1$, $s = +\frac{3}{2}$

(b) $n = 3$, $\ell = 2$, $m = 0$, $s = +\frac{1}{2}$

(c) $n = 3$, $\ell = 3$, $m = -2$, $s = -\frac{1}{2}$

(d) $n = 4$, $\ell = 1$, $m = -2$, $s = +\frac{1}{2}$

(e) $n = 27$, $\ell = 14$, $m = -8$, $s = -\frac{1}{2}$

Solution

In theory there are an infinite number of possible quantum levels in an atom. But for each level a unique set of four numbers exists to describe an electron occupying it, and the quantum numbers must be allowable as indicated by the model. The hydrogen electron may occupy any one of the n levels cited in (b) and (e) above. In the case of (a), s may not have a value of $+3/2$; the values of s may only be $+1/2$ or $-1/2$.

In (c) allowed values of ℓ include all values of 0 through $n - 1$ for a specific atom. Thus, a value of $\ell = 3$ is not allowed when $n = 3$.

In (d) the value of ℓ limits the values that m may have. The integer values of m may range from $-\ell$ through 0 to $+\ell$. In (d) the value of $m(-2)$ exceeds the limiting value of $-\ell(-1)$ and therefore cannot be an acceptable quantum number.

In (e), in spite of the rather unusual quantum numbers, the values are allowed by the model and consequently (e) is a suitable set.

45. Using subshell notation ($1s^2 2s^2 2p^6$, etc.), predict the electron configurations of the following ions:

(a) Ti^{3+}

(b) Ni^{2+}

(c) Cr^{3+}

(d) Ru^{3+}

(e) Ag^{+}

(Read Exercise 44 before attempting this exercise.)

Solution

(a) Ti^{3+}: The subshell notation for Ti is

$$\text{Ti:}\quad 1s^2 2s^2 2p^6 3s^2 3p^6 3d^2 4s^2$$

When ionization occurs the electrons are removed from the outermost orbital first. Thus, the $4s^2$ electrons are removed first; then one of the $3d$ electrons are removed. The final configuration is

$$\text{Ti}^{3+}\text{:}\quad 1s^2 2s^2 2p^6 3s^2 3p^6 3d$$

(d) Ru^{3+}: In this ion the inner *d*'s are filled and the electrons are removed from the outer 4*d* level after the single 5*s* electron is removed. For Ru,

$$\text{Ru} \;:\; 1s^2 2s^2 2p^6 3s^2 3p^6 3d^{10} 4s^2 4p^6 4d^7 5s$$

$$\text{Ru}^{3+}:\; 1s^2 2s^2 2p^6 3s^2 3p^6 3d^{10} 4s^2 4p^6 4d^5$$

This latter arrangement is particularly stable because of the half-filled *d* orbital set.

47. Which of the following sets of quantum numbers describes the most easily removed electron in an aluminum atom in its ground state? Which of the electrons described is most difficult to remove?

(a) $n = 1,\ \ell = 0,\ m = 0,\ s = -\frac{1}{2}$

(b) $n = 2,\ \ell = 1,\ m = 0,\ s = -\frac{1}{2}$

(c) $n = 3,\ \ell = 0,\ m = 0,\ s = \frac{1}{2}$

(d) $n = 3,\ \ell = 1,\ m = 1,\ s = -\frac{1}{2}$

(e) $n = 4,\ \ell = 1,\ m = 1,\ s = \frac{1}{2}$

Solution

The most easily removed electron in any atom is the one occupying the highest energy level. In an aluminum atom in its ground state the quantum numbers for the outermost electron ($3p^1$) are

n	ℓ	m	s		n	ℓ	m	s		n	ℓ	m	s
3	1	−1	$-\frac{1}{2}$	or	3	1	1	$-\frac{1}{2}$	or	3	1	0	$-\frac{1}{2}$

These three sets are all equivalent because they have the same energy and cannot be distinguished except in a magnetic field. The set in answer (d) corresponds to the center set as well as being equivalent to the other two sets. The most difficult electron to remove from any atom is the one occupying the lowest energy level.

n	ℓ	m	s
1	0	0	$-\frac{1}{2}$

This set is given in answer (a).

52. Identify the groups that have the following electron configurations in their valence shells (*n* represents the principal quantum number):

(a) ns^1

(b) ns^2np^3

(c) ns^2np^5

(d) $ns^2(n-1)d^1$

(e) $ns^2(n-1)d^6$

Solution

The elements represented by (a), (b), and (c) are representative elements, the A-group classification. Elements with electron configurations ending in *ns* or *np* are located in A groups whose numbers equal the sum of the electrons occupying the *s* and *p* orbitals in a specific *n* level.

(a) ns^1 - (1 e^-, Group IA)

(b) ns^2np^3 - (5 e^-, Group VA)

(c) ns^2np^5 - (7 e^-, Group VIIA)

Group numbers for the transition elements cannot be assigned on the basis of the total number of electrons in the $ns(n-1)d$ levels without knowledge of the order of placement of electrons in the orbitals. To identify the periodic location of the elements in (d) and (e), sum the electrons in the s and d levels, then count from left to right in a given period (n) the number of groups equaling the number of $ns(n-1)d$ electrons.

(d) $ns^2(n-1)d^1 = 3\ e^-$ - Group IIIB

(e) $ns^2(n-1)d^6 = 8\ e^-$ - Group VIIIB

57. Why is the radius of a positive ion smaller than the radius of its parent atom?

Solution

The formation of a positive ion from a neutral atom involves the removal of one or more electrons from the valence level of the atom. Consider the formation of a Mg^{2+} ion from a Mg atom:

$$Mg(g) + \text{energy} \longrightarrow Mg^{2+}(g) + 2e^-$$

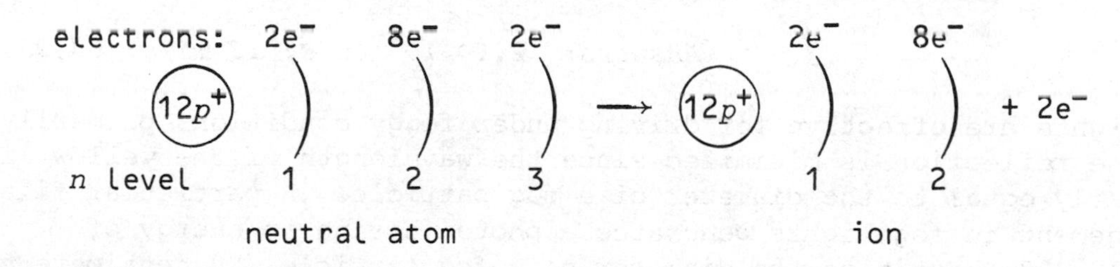

The ionization of the neutral atom leaves the $n = 3$ level empty. As a consequence the outer electrons in the ion are in a filled level that is one level closer to the nucleus than in the neutral atom. With the electrons closer to the nucleus, the nuclear-charge-to-electronic-charge ratio is increased, which further decreases the radius of the ion from that of the neutral atom.

64. The isotope of uranium used in nuclear fission is $^{235}_{92}U$.

(a) How many protons, neutrons, and electrons are contained in this atom?

Solution

The number of protons is written to the lower left of the symbol representing the element. There are, therefore, 92 protons. The mass number, the sum of the number of protons and neutrons, is written to the upper left of the symbol. Thus, 235 (protons + neutrons) - 92 (protons) = 143 protons. The number of electrons in the neutral atom is the same as the number of protons; there are 92 protons, consequently, there are 92 electrons.

(b) How many protons, neutrons, and electrons are contained in a ${}^{235}_{92}U$ ion with a charge of +3?

Solution

In the ion ${}^{235}_{92}U^{3+}$, the number of protons and neutrons are the same as in (a), 92 protons and 143 neutrons. However, there must be 3 fewer electrons than protons for the charge to be +3. Thus, 92 - 3 = 89 electrons.

RELATED EXERCISES

1. If nuclear physicists are successful in synthesizing element 114, what would be the predicted n and ℓ values for its outermost electron?

   Answer: $n = 7$, $\ell = 1$

2. An accelerated particle, mass 2×10^{-24} g, possessing a kinetic energy of 2.9×10^{-18} J interacts with a hydrogen atom in its ground state. The electron in the hydrogen atom is promoted to level 5 and the particle escapes to the surroundings. Calculate the energy of the escaping particle.

   Answer: 8.1×10^{-19} J

3. Given that kinetic energy is defined as $KE = (1/2)mv^2$, calculate the velocity of the escaping particle as described in Problem 2 (in cm/s, mi/s, mi/h).

   Answers: 2.8×10^6 cm/s; 18 mi/s; 64,000 mi/h

4. Fog lights are effective for driving under foggy conditions primarily because reflection is minimized since the wavelength of the yellow light is nearly equal to the diameter of a fog particle. A particular filament arrangement in fog lights generates a photon having an energy of 3.4×10^{-19} J. Estimate the diameter of a fog particle (in centimeters).

   Answer: 5.8×10^{-5} cm

5. Louis de Broglie, in 1924, reasoned that electromagnetic radiation possesses both wave and particle properties and suggested that moving particles may exhibit wave characteristics. This concept has been experimentally verified and has led to the development of the electron microscope and more recently the discovery and use of radiation generated by moving particles in nuclear accelerators. The equation derived by de Broglie from equating $E = h\nu$ and $E = mc^2$ relates the wavelength (λ) to the momentum ($m\nu$) of the moving particle.

$$\lambda = \frac{h}{m\nu}$$

   By using this equation, calculate the wavelength and frequency of a wave generated by a tennis ball, mass 60.0 g, moving at 90 mi/h. (60 mi/h = 88 ft/s = 2700 cm/s).

   Answer: $\lambda = 2.7 \times 10^{-32}$ cm; $\nu = 1.1 \times 10^{42}$ s^{-1}

6. The ionization potential of lithium is 5.2×10^5 J/mol. Atoms are readily ionized through interaction with high-energy radiation. Calculate the wavelength of a photon having sufficient energy just to ionize a lithium atom.

Answer: 2000 Å

UNIT 5

CHAPTER 5: CHEMICAL BONDING—GENERAL CONCEPTS

INTRODUCTION

With few exceptions, elements in the earth's crust and atmosphere are found combined with other elements. The forces holding these elements together in compounds are electrical in nature. The different kinds of bonding of atoms are responsible for the great variety of substances making up our universe, thus providing the basis for life itself.

At first sight chemical bonding seems almost to defy definition and description because of the variety and character of the bonds formed by the interactive forces involved. Scientists have developed models to represent chemical bonds that range from a simple model likening a bond to an interconnective spring between atoms to sophisticated computer simulations of electron density distributions between atoms, a model based on mathematical wave functions. Such models are used to help visualize theoretical explanations for the observed characteristics of elements and compounds. Theories and models, however, are not absolutes; they are subject to modification and refinement as the capacity to critically examine matter improves.

The discussion of bonding in this chapter specifically treats the types of bonds formed when atoms interact to form molecules or ion pairs. These bonds are interatomic or intramolecular (bonding within the same molecule), as would exist between hydrogen and oxygen in a water molecule or between sodium and chlorine in a sodium chloride ion pair. Discussed later in the text are the types of bonds formed between neighboring molecules or ion pairs (intermolecular bonds) that provide the basis for the properties and structures of liquids and solids.

The chemical bonds between atoms in molecules vary greatly in terms of their strength, bond length or distance between the atoms, and directional character. Some bonds are relatively weak. For example, the intermolecular bonds in mothballs (napthalene) holding napthalene molecules together are very weak. The mere vibrational motion of the molecules is sufficient to rupture the bonds and allow the napthalene molecules to escape the structure. Other bonds such as those holding calcium, oxygen, and silicon in cement are very strong by comparison. These atoms rarely escape the cement structure by their vibrational motion alone.

Molecular structure results from the directional nature of the bonds. Bonds are directed in ways that produce geometries that provide for the lowest or most stable energy states of the substance. On the other hand, ions in aqueous solution form bonds that are almost completely nondirectional with as many water molecules as can gather about them. In solids the distances between the nuclei of bonding atoms vary considerably according to the size and nature of the bonded atoms. Although these characteristics of bonds are introduced in this chapter, virtually all the remainder of the text deals with concepts that are directly related to chemical bonding.

In this chapter the discussion of bonding is limited to ionic or electrovalent bonding and covalent bonding. The characteristics of elements leading to these types of bonds and compounds produced through these bonds are emphasized. Methods used for naming compounds are also discussed.

FORMULAS AND DEFINITIONS

Valence electrons The electrons involved in the formation of bonds between atoms are those in the highest energy levels. For the representative elements, Group A elements, these electrons are in the outermost *s* or *s* and *p* subshells, and the number of valence electrons in a representative element equals its group number. The number of valence electrons in transition elements is not clearly defined. Bonds involving atoms of transition elements include the electrons in the outermost *s* sublevel and electrons in the *d* sublevel one quantum level below the *s* sublevel. For example, manganese, with atomic number 25 and electron configuration Ar $4s^2 3d^5$, has seven valence electrons. However, data from experiments with manganese show that manganese may use as few as one electron or as many as seven electrons in bonding with other elements.

Lewis formula The Lewis formula of an element is a means of showing the valence electrons available for bonding. The valence electrons are designated by symbols, such as •, °, or ×, as shown in the following:

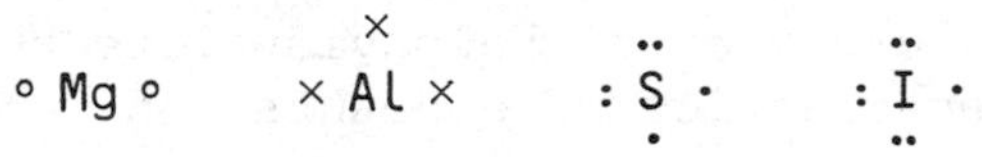

```
             ×        ••       ••
° Mg °     × Al ×    : S •    : I •
                       •        ••
```

Compound formation is shown through the use of the Lewis formula of individual atoms by bringing the symbols next to each other and arranging the valence electrons so that eight electrons (where possible) occur about each symbol. For example:

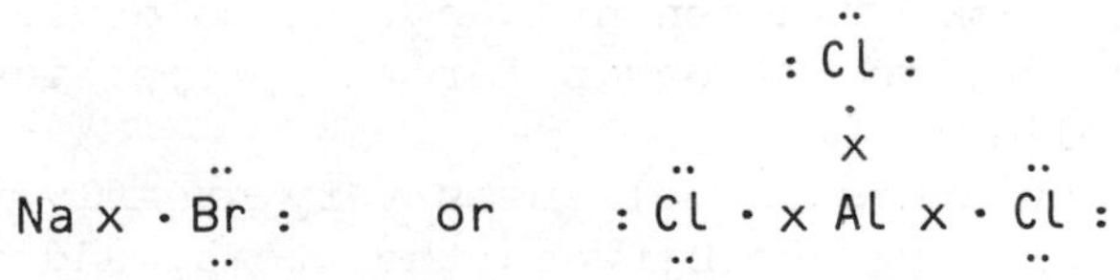

```
                               ••
                             : Cl :
                                •
                                ×
          ••             ••         ••
Na × • Br :     or    : Cl • × Al × • Cl :
          ••             ••         ••
```

Note that although different symbols may be used for convenience to designate electrons, all electrons are the same regardless of origin.

Ionic bond Ionic bonds result from electrostatic attractions between two oppositely charged particles called *ions*. These mutual attractions produce bonds between ions to form ionic substances, such as sodium chloride, magnesium fluoride, and molten salts. The formation of ionic bonds when two elements react involves the transfer of electrons from one element to the other. For example, in the formation of sodium chloride, NaCl, the single valence electron of sodium is removed and transferred to complete the valence level of chlorine. This transfer creates the ion pair as shown:

$$Na\times + :\ddot{\underset{..}{Cl}}\cdot \longrightarrow Na^{+} + :\ddot{\underset{..}{Cl}}\overset{-}{\underset{\times}{:}}$$

Anion An atom or group of atoms that possesses a greater number of electrons than protons is called an *anion*. The anion has a negative charge equal to the charge difference. Elemental anions are usually formed from nonmetals; metallic elements rarely form negative ions. The dissociation of compounds, however, often produces ion pairs — one anion and one cation.

Cation An atom or group of atoms that possesses fewer electrons than protons is called a *cation*. The cation has a positive charge that equals the charge difference. Metallic elements tend to form cations in chemical reactions.

Representative elements The representative elements are those elements found in the A groups of the Periodic Table. Reactions involving these elements utilize electrons in their outermost *s* or *s* and *p* sublevels.

Noble gas electron configuration The great majority of known chemical reactions produce compounds in which elements are bonded together by transferring or sharing electrons in such a way that each element achieves the same electron configuration as that of its closest noble gas neighbor. As an example, in the formation of the ionic bond between lithium and chlorine in lithium chloride, lithium loses an electron to chlorine. Both elements achieve electron configurations like the noble gases, Li like that of helium and Cl like that of argon.

$$\underset{1s^2 2s^1}{Li\cdot} \longrightarrow \underset{1s^2}{Li^+ + e^-} \quad (He\ 1s^2)$$

$$\underset{1s^2 2s^2 2p^6 3s^2 3p^5}{Cl} + e^- \longrightarrow \underset{1s^2 2s^2 2p^6 3s^2 3p^6}{Cl^-} \quad (Ar\ 1s^2 2s^2 2p^6 3s^2 3p^6)$$

Covalent bond Two atoms may bond together to form a stable electron structure in which the bonding electrons are shared between the two atoms. In general, these bonds form between atoms of elements that have little tendency to give up electrons. Indeed, covalent bonds may form between atoms of the same element to form diatomic molecules, including H_2, N_2, O_2, and F_2. Reactions involving different nonmetallic elements form compounds through covalent bonding; these compounds include SO_2, N_2O_4, HCl, and CH_4. Carbon is particularly noteworthy because it can readily form bonds with itself and with other elements in the same compound. This property gives rise to the field of organic chemistry, in which millions of possible combinations of elements involving carbon can be assembled. Generally speaking, covalent bonds form between nonmetallic elements close to each other in the Periodic Table. These elements will generally have electronegativity differences less than 1.7.

Electronegativity Electronegativity is based on a numerical scale representing the tendencies of neutral atoms to attract one additional electron. These values range from a low of 0.7 for cesium, representing little attraction for electrons, to a high of 4.0 for fluorine. In general, metallic elements have lower electronegativities than nonmetallic elements. Stated differently, metallic elements tend to give up electrons in chemical reactions, while nonmetallic elements tend to gain electrons. Metallic elements show almost no tendency to attract another electron; this behavior is not uncommon for nonmetallic elements.

Electronegativities can be used to predict the nature of bonding between two elements. Elements that form primarily ionic bonds have electronegativities that differ by more than 1.7, while predominantly covalent bonds are formed between elements having differences less than 1.7. In real chemical bonds, regardless of electronegativity differences, both ionic and covalent characters occur to some extent. For example, the bond between sodium and chlorine in NaCl is mostly ionic, but some sharing of electrons does exist; while in H_2, although the hydrogen atoms have little tendency to give up electrons, the bond is still about 5% ionic. These definitions of ionic and covalent bonding are inadequate because real bonds show both types of bonding.

Polar bonds Atoms do not usually share electrons equally in covalent bonds. This is quite evident for covalently bonded compounds where the electronegativity difference is rather large; the more electronegative element has the strongest attraction for the shared electrons. This unequal attraction causes a shift of the shared electrons toward the more electronegative element, resulting in the electrons being nearer the more electronegative element most of the time. Covalent bonds in which the electrons are shared unequally are called *polar bonds*. For example, in hydrogen chloride, HCl, chlorine is more electronegative than hydrogen, 3.0 vs. 2.1. The electron pair in the bond is shifted toward chlorine, making the electron cloud region about chlorine relatively negative in charge with respect to the region about hydrogen. The partial charges are indicated by the symbol δ with the appropriate charge written to the upper right. Thus,

$$\overset{\delta^+}{H} \xrightarrow{\times\cdot} \overset{\delta^-}{\underset{\times\times}{\overset{\times\times}{Cl}}}\,\overset{\times}{\times}$$

EXERCISES

1. Why does a cation have a positive charge?

Solution

Atoms or groups of atoms that have fewer electrons than protons have a net positive charge and are called *cations*. Monatomic cations are formed mostly from metallic elements; some examples indicating possible ion charges are

Atom		Ion	Protons + electrons in ion's nucleus
Na	$-1e^- \longrightarrow$	Na^+	$11p^+ + 10e^- = 1+$
Ca	$-2e^- \longrightarrow$	Ca^{2+}	$20p^+ + 18e^- = 2+$
Al	$-3e^- \longrightarrow$	Al^{3+}	$13p^+ + 10e^- = 3+$
Pb	$-2e^- \longrightarrow$	Pb^{2+}	$82p^+ + 80e^- = 2+$

Polyatomic cations, such as NH_4^+, $CH_3NH_2^+$, and $CH_3CH_2^+$ are formed by the breaking or formation of bonds in neutral species in which the more electronegative component acquires additional electrons and becomes negative, while the remainder becomes an electron-deficient species known as a *positive ion* or *cation*. As an example,

$$NH_3 + H_2O \longrightarrow NH_4^+ + OH^-$$

2. Why does an anion have a negative charge?

Solution

Atoms or groups of atoms that have more electrons than protons have a net negative charge. Monatomic anions are formed mostly from nonmetallic elements; some examples indicating the charge on each ion are

Atom	Ion	Protons + electrons in the ion's nucleus
$Cl + 1e^- \longrightarrow$	Cl^-	$17p^+ + 18e^- = -1$
$O + 2e^- \longrightarrow$	O^{2-}	$8p^+ + 10e^- = -2$
$N + 3e^- \longrightarrow$	N^{3-}	$7p^+ + 10e^- = -3$

Polyatomic anions, such as SO_4^{2-}, PO_4^{3-}, $CHOO^-$, and NO_2^-, result from ionization processes in which the portion of the species that forms the anion acquires an excess electrical charge and becomes negative.

$$CH_3COOH + H_2O \rightleftharpoons CH_3COO^- + H_3O^+$$

4. Predict the charge on the monoatomic ions formed from the following elements in ionic compounds:

(a) Ag (c) Ba (e) K (g) N (i) P

(b) Al (d) Br (f) Zn (h) O (j) Se

Solution

The charge on monatomic ions can be predicted from the oxidation number rules. In the formation of ionic compounds, metallic elements lose electrons and become positive ions, while nonmetallic elements gain electrons and become negative ions. Furthermore, elements tend to gain or lose sufficient electrons to achieve a noble gas electron configuration. The electrons gained or lost by the above elements in the formation of ionic compounds are

$$Ag(Kr\ 5s^1 4d^{10}) \longrightarrow Ag^+(Kr\ 4d^{10}) + e^-$$
$$Al(Ne\ 3s^2 3p^1) \longrightarrow Al^{3+}(Ne) + 3e^-$$
$$Ba(Xe\ 6s^2) \longrightarrow Ba^{2+}(Xe) + 2e^-$$
$$Br(Ar\ 4s^2 3d^{10} 4p^5) + e^- \longrightarrow Br^-(Kr)$$
$$K(Ar\ 4s^1) \longrightarrow K^+(Ar) + e^-$$
$$Zn(Ar\ 4s^2 3d^{10}) \longrightarrow Zn^{2+}(Ar\ 3d^{10}) + 2e^-$$
$$N(He\ 2s^2 2p^3) + 3e^- \longrightarrow N^{3-}(Ne)$$
$$O(He\ 2s^2 2p^4) + 2e^- \longrightarrow O^{2-}(Ne)$$
$$P(Ne\ 3s^2 3p^3) + 3e^- \longrightarrow P^{3-}(Ar)$$
$$Se(Ar\ 4s^2 3d^{10} 4p^4) + 2e^- \longrightarrow Se^{2-}(Kr)$$

8. Write the formulas of each of the following ionic compounds using Lewis symbols:

(a) LiF (b) Na_2S (c) $CaBr_2$ (d) AlF_3 (e) Ga_2O_3

Solution

LiF: Lithium is in family IA and by losing one electron forms a singly charged cation. Its valence shell is then empty. Fluorine is in family VIIA and forms a singly charged anion to achieve the octet arrangement. Eight electrons must be shown about it.

Metal	Nonmetal	Ionic Compound
$Li\cdot$ +	$\overset{\times\times}{\underset{\times\times}{\times F \times\!\times}}$	$Li^+ \left[\overset{\times\times}{\underset{\times\times}{\dot{\times} F \times\!\times}}\right]^-$
$2Na\cdot$ +	$\overset{\times}{\underset{\times\times}{\times S \times\!\times}}$	$Na_2{}^+ \left[\overset{\bullet\times}{\underset{\times\times}{\dot{\times} S \times\!\times}}\right]^{2-}$
$Ca:$ +	$2\overset{\times\times}{\underset{\times\times}{\times Br \times\!\times}}$	$Ca^{2+} \left[\overset{\times\times}{\underset{\times\times}{\dot{\times} Br \times\!\times}}\right]^-_2$
$\ddot{Al}\cdot$ +	$3\overset{\times\times}{\underset{\times\times}{\times F \times\!\times}}$	$Al^{3+} \left[\overset{\times\times}{\underset{\times\times}{\dot{\times} F \times\!\times}}\right]^-_3$
$2\ddot{Ga}\cdot$ +	$3\overset{\times\times}{\underset{\times\times}{\times O \times}}$	$Ga_2{}^{3+} \left[\overset{\times\times}{\underset{\times\times}{\dot{\times} O \dot{\times}}}\right]^{2-}_3$

The major advantage of writing these ionic compounds in this form is to emphasize the transfer of electrons from the metals to the nonmetals. There is no convenient way to draw the true structure of ionic compounds using the electron dot picture.

12. What are the characteristics of two atoms that will form a covalent bond?

Solution

In general, the type of bond formed between two atoms depends on their relative ability to give up electrons. Because electrons are shared in a covalent bond, the two bonding atoms must have similar electronegativities and similar high-ionization potentials. Relative to the Periodic Table, atoms having these characteristics are mostly nonmetals.

14. Predict which of the following compounds are ionic and which are covalent, using the locations of their constituent elements in the Periodic Table:

(a)	Li_2O	(d)	H_2CO	(g)	CS_2	(j)	K_2S
(b)	CCl_4	(e)	CH_4	(h)	BrCl	(k)	MnS
(c)	CuO	(f)	SCl_2	(i)	$CaBr_2$	(l)	Al_2O_3

Solution

In the following set the term *close* refers to the elements composing the bonds that are close to each other in the Periodic Table. This favors formation of covalent bonds. The term *far* refers to the elements that are comparatively far apart and therefore form predominantly ionic compounds.

(a)	Li_2O	far	ionic	(g)	CS_2	close	covalent
(b)	CCl_4	close	covalent	(h)	BrCl	close	covalent
(c)	CuO	far	ionic	(i)	$CaBr_2$	far	ionic
(d)	H_2CO	close	covalent	(j)	K_2S	far	ionic
(e)	CH_4	close	covalent	(k)	MnS	far	ionic
(f)	SCl_2	close	covalent	(l)	Al_2O_3	far	ionic

15. How do single, double, and triple bonds differ? How are they similar?

Solution

When possible, the formation of bonds between atoms results in completing the valence levels of the elements involved to achieve noble gas electronic structures. Single, double, and triple bonds refer to atoms that are covalently bonded. In ionic bonds one atom has lost one electron to another atom to form an ion pair. In contrast, both bonding atoms in a single covalent bond share a pair of electrons. Furthermore, single bonds can form between two atoms through direct atomic orbital overlap without rearrangement (hybridization) of orbitals.

Double and triple bonds are covalent bonds in which elements must share four and six electrons, respectively. In contrast to single bonds, the formation of these bonds requires that some or all of the atomic orbitals of the bonding atoms be rearranged, that is, hybridized, to create orbitals having compatible geometries for bonding.

Note that there is a gradual increase in the bond energy and decrease in the bond distance. For example, consider the carbon-carbon bonds in the following organic molecules:

Ethane	$H_3C{-}CH_3$	distance = 1.50 Å, E = 368 kJ/mol
Ethene	$H_2C{=}CH_2$	distance = 1.34 Å, E = 682 kJ/mol
Ethyne	$HC{\equiv}CH$	distance = 1.20 Å, E = 828 kJ/mol

Also note that singly bonded atoms as in ethane, C—C, are free to rotate about each other to form different conformational geometries. On the other hand, doubly or triply bonded atoms as in ethene, —C═C—, or ethyne, —C≡C—, are fixed in a planar geometry that does not permit free rotation but does allow a degree of twisting.

17. Write Lewis structures for the following:

(a) O_2 (b) CS_2 (c) Cl_2CO (d) ClNO (e) H_2CO

Solution

O_2: $\dot{\dot{O}}$:X O (with x electrons) or O═O

Note that in this case the Lewis structure is inadequate to depict the fact that experimental studies have shown an unpaired electron on each oxygen atom, a condition known as *paramagnetism*.

CS_2: :S̈ :: C :: S: or :S̈ ═ C ═ S:

$COCl_2$: :C̈l ·· C :: O: (with :C̈l: below C) or Cl_2C═O

ClNO: :C̈l: N̈ :: O: or :C̈l — N̈ ═ O:

H_2CO: H ·· C :: O: (with second H) or H_2C═O

26. Draw the resonance forms for the following:

(a) selenium dioxide, OSeO

(b) the nitrite ion, NO_2^-

(c) the nitrate ion, NO_3^-

(d) the acetate ion, $HC(H)(H)CO_2^-$ (H above and H below the first C)

Solution

Resonance forms require that the atoms remain fixed and that only electrons are moved to maintain an octet of electrons around each atom.

The nitrate ion:

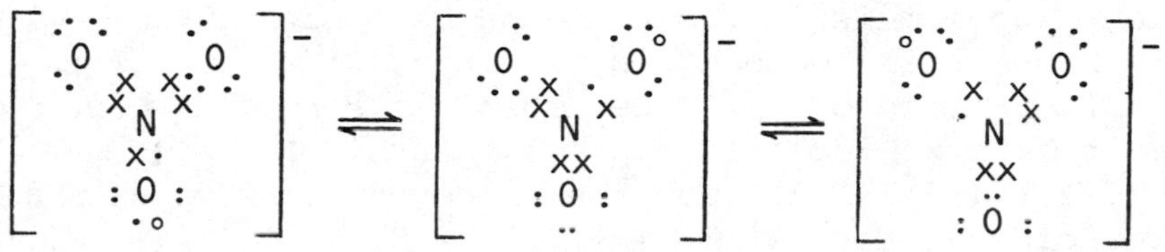

$$[\text{O}_2\text{N}(\text{O})]^- \rightleftharpoons [\text{O}_2\text{N}(\text{O})]^- \rightleftharpoons [\text{O}_2\text{N}(\text{O})]^-$$

The acetate ion:

$$[CH_3\text{—}C(O)O]^- \rightleftharpoons [CH_3\text{—}C(O)O]^-$$

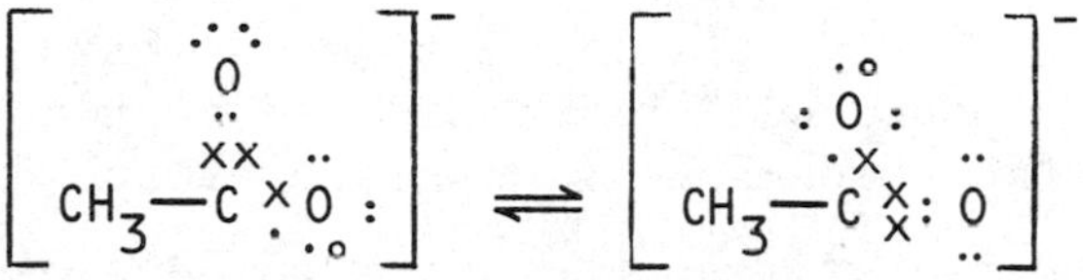

36. Many molecules that contain polar bonds are nonpolar. How can this happen?

Solution

A polar bond involves the unequal sharing of electrons by the bonding atoms. The bonding electrons shift toward the more electronegative atom, resulting in a net imbalance in electronic charge, measured by the dipole moment, in that direction. Molecules may contain several polar bonds and as a whole may be nonpolar if there are other bonds in the molecule that offset or balance the imbalance in electronic distribution. For example, in carbon dioxide both carbon-oxygen double bonds are polar and both dipole moments are in the direction of oxygen. The molecule is linear and the two dipole moments are equal but opposite in direction, and hence cancel.

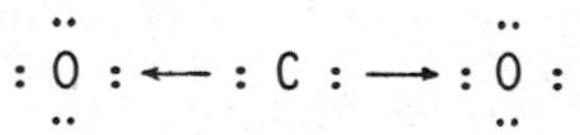

$$:\ddot{O}: \leftarrow :C: \rightarrow :\ddot{O}:$$

38. Which of the following molecules and ions contain polar bonds? F_2, P_4, SO_2, NO_3^-, N_2O, H_2S, NH_4^+, HCCH

Solution

SO_2, NO_3^-, N_2O, H_2S, NH_4^+, and HCCH all contain polar bonds because of the difference in electronegativities. SO_2, NO_3^-, and N_2O also are polar species because the charge centers of positive and negative charge do not coincide.

44. Determine the oxidation number of each of the following elements in the given compound:

(a) B in Na_3BO_3 (e) Co in $K_3[CoF_6]$

(b) C in CBr_4 (h) P in $H_2PO_3F[(HO)_2POF]$

° represents an electron that has been brought from some outside source.

Solution

(a) The oxidation number of a monoatomic ion is equal to the charge on that ion. Sodium is in Group IA and is assigned an oxidation number of +1. Since this compound is not a peroxide, the oxygen in it has an oxidation state of -2. The algebraic sum of the charges on the ions must be zero for the compound. Thus, $3Na^+ = +3$; $3O^{2-} = -6$. Then, (+3) + (-6) = -3. For the compound to be neutral, B must have a charge of +3; that is, (-3) + (+3) = 0.

(b) Bromine is in Group VIIA and is assigned an oxidation number of -1. Then 4Br = -4. For the compound to be neutral, C must have a charge of +4; that is, (-4) + (+4) = 0.

(e) Potassium has a charge of +1 (Group IA) and fluorine has a charge of -1 (Group VIIA). This compound will be considered as two units, $[K_3]^{+3}$ and $[CoF_6]^{-3}$. This assignment is possible because $3K^+ = +3$ and the neutrality requirement forces the anion to have a -3 charge. Since $6F^- = -6$ and $[CoF_6]^{-3}$, Co must have a charge of +3, that is, -3 = (+3) + (-6).

(h) Hydrogen has an oxidation number of +1, oxygen a -2, and fluorine a -1. Then, $2H^+ = +2$, $3O^{2-} = -6$, and $1F^- = -1$. The P must have a charge to balance (+2) + (-6) + (-1) = -5. Thus, P must have a charge of +5.

45. Determine the oxidation number of N in each of the following compounds: Na_3N, NH_4Cl, N_2O, N_2H_4, KNO_2, $Ca(NO_3)_2$.

Solution

The oxidation number of nitrogen in each of the compounds has been calculated from the rules described in Sections 5.11 and 5.12.

Na_3N	3(Na = +1) = +3; +3 + N = 0; N = -3
NH_4^+	4(H = +1) = +4; +4 + (N) = +1; N = -3
N_2O	O = -2; -2 + 2N = 0; N = +1
N_2H_4	4(H = +1) = +4; +4 + 2N = 0, N = -2
KNO_2	K = +1; 2(O = -2) = -4; +1 + (-4) + N = 0; N = +3
$Ca(NO_3)_2$	Ca = +2; 6(O = -2) = -12; +2 + (-12) + 2N = 0; N = +5

50. Name the following binary compounds:

(a) CaF_2	(c) Al_2O_3	(e) KH	(g) MgI_2	(i) Na_2O	(k) Mg_3N_2
(b) CuS	(d) HBr	(f) AgCl	(h) LiCl	(j) K_3P	

Solution (partial list)

CaF_2	calcium fluoride	Na_2O	sodium oxide
Al_2O_3	aluminum oxide	K_3P	potassium phosphide
KH	potassium hydride	Mg_3N_2	magnesium nitride

53. Name the following compounds (since the metal atom in each can exhibit two or more oxidation states, it will be necessary to identify the oxidation number of the metal as you name the compound):

(a) $FeBr_2$	(c) MnF_3	(e) $TlClO_4$	(g) SnI_4
(b) Cu_2O	(d) $Pb(OH)_2$	(f) $Cr(NO_3)_3$	(h) $CuSO_4$

Solution

$FeBr_2$	iron(II) bromide or ferrous bromide
Cu_2O	copper(I) oxide or cuprous oxide
MnF_3	manganese(III) fluoride
$Pb(OH)_2$	lead(II) oxide
$TlClO_4$	thallium(I) perchlorate
$Cr(NO_3)_3$	chromium(III) nitrate
SnI_4	tin(IV) iodide or stannic iodide
$CuSO_4$	copper(II) sulfate or cupric sulfate

58. Using the indicated oxidation state as a guide, write the formula of each of the following compounds:

(a) cobalt(II) chloride	(e) copper(I) sulfide
(b) lead(IV) oxide	(f) chromium(III) sulfate
(c) mercury(II) iodide	(g) tin(II) perchlorate
(d) iron(III) nitrate	

Solution

cobalt(II) chloride	$CoCl_2$	copper(I) sulfide	Cu_2S
lead(IV) oxide	PbO_2	chromium(III) sulfate	$Cr_2(SO_4)_3$
mercury(II) iodide	HgI_2	tin(II) perchlorate	$Sn(ClO_4)_2$
iron(III) nitrate	$Fe(NO_3)_3$		

63. X may indicate a different representative element in each of the following Lewis formulas. To which group does X belong in each case?

Solution

Count the number of valence electrons associated with X in each formula; the group number equals the number of valence electrons.

```
      ..
     : F :
       |   ..
(a)  : X—F :
       |   ..
     : F :
      ..
```

No. valence e^- = 3 shared with *F* plus lone pair = 5; therefore, X belongs to Group V.

```
      ..   ..
(b)  : X—Br :
      ..   ..
```

No. valence e^- = 1 shared with Br plus 3 lone pairs = 7; therefore, X belongs to Group VII.

(c) 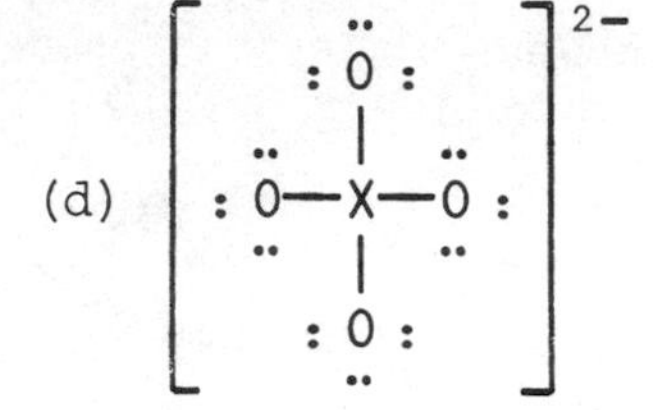

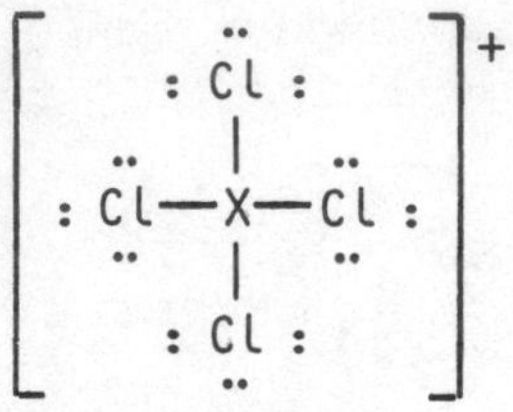

No. valence e^- = 4 shared with Cl, plus $1e^-$ given up to form the ion = 5; therefore, X belongs to Group V.

(d) Eight e^-'s are shared coordinate covalently with O, less $2e^-$ which were added from an outside source to give anion = 6 valence electrons; therefore, X belongs to Group VI.

RELATED EXERCISES

1. Write the Lewis structures for BCl_3, SF_4, IF_7, and PCl_5.

   Answers

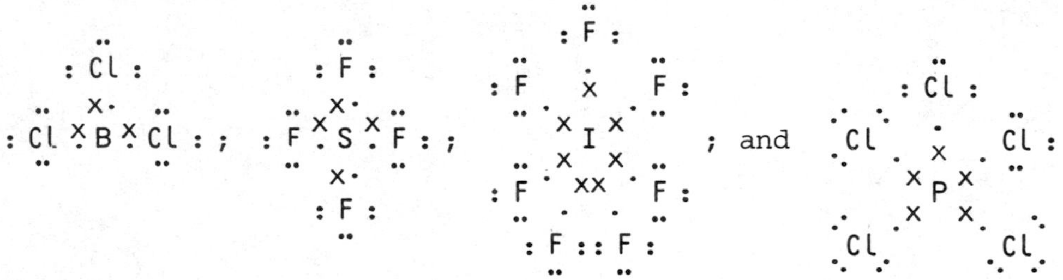

2. Indicate the direction of the dipole moment between the atoms in CH_4, $CHCl_3$, and CH_2Cl_2.

   Answers

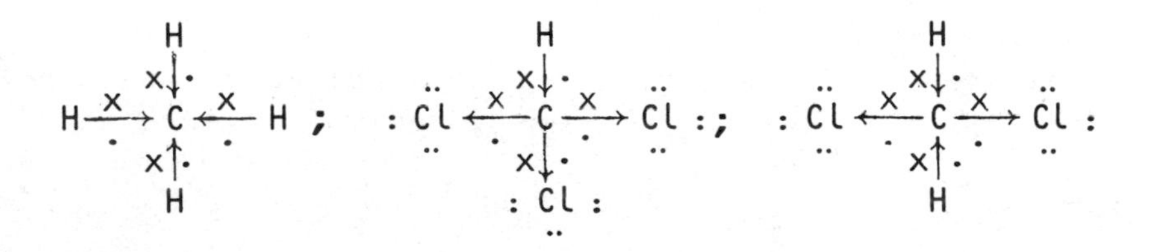

3. Calculate the oxidation state of chlorine in each of the following compounds containing chlorine: NaClO, $NaClO_2$, $NaClO_3$, $NaClO_4$

   Answer

   NaClO, +1; $NaClO_2$, +3; $NaClO_3$, +5; $NaClO_4$, +7

4. For each of the following pairs of compounds indicate the more polar compound: CuCl or CuI; PbS or FeS; and $BiCl_3$ or BiI_3.

Answer

CuCl; electronegativities of Pb and Fe are equal and polarities are about equal; $BiCl_3$

5. Draw an outline of the Periodic Table and indicate the direction of trends in atomic size, ionization potential, and electronegativity.

UNIT 6

CHAPTER 10: THE GASEOUS STATE AND THE KINETIC-MOLECULAR THEORY

INTRODUCTION

Gases have occupied a very important position in human thought and imagination from the very earliest recorded times. For example, the Greek Anaximenes (died ca. 528 B.C.) thought that air constituted the primary matter from which all other substances came into being. In this context, water could be formed by the condensation of air. Although today's perspective is different, the concept then was one of different states of matter, the gaseous state being the most important.

Although the Greeks did not develop the scientific method, experimentation with gases did occur before the seventeenth century. Hero of Alexandria (ca. A.D. 62-150) described many mechanisms operated by steam, even a steam engine; the pressure of gases was the motive power in all these systems. Hero therefore had a very clear idea of the nature of gases, and in many respects anticipated the kinetic theory of gases. His ideas on combustion were very close to those of Lavoisier, who lived at the beginning of modern chemistry. However, scientists such as Hero did not anticipate the discoveries of the eighteenth century because they tended to employ experiments only to demonstrate preconceived hypotheses.

The ability to *scientifically* study gases became possible when the Italian physicist Evangelista Torricelli invented the mercury barometer in 1643. Thus a means was finally available to measure pressure. Generally the unit of measurement of pressure is the millimeter of mercury (mmHg) or *torr* named after Torricelli.

The British chemist Robert Boyle was quick to exploit this advance and in 1662 showed that the pressure of a gas and its volume were inversely related. A French physicist, Amontons, discovered the relation between the temperature and volume of gas almost immediately after Boyle's work, but it was a century later before Charles, and still later and independently the French chemist Joseph Gay-Lussac, showed the same temperature-volume dependence.

A theoretical explanation of the behavior of gases was developed in 1738 by the Swiss mathematician Daniel Bernoulli, who laid the foundation for the kinetic-molecular theory of gases. The developments from this point are well covered in the text and no attempt to repeat them will be made here. It might be pointed out, however, that from the early 1800s, chemical science developed very rapidly.

FORMULAS AND DEFINITIONS

Pressure Gas pressure provides the most easily measured gas property. Pressure is the force exerted upon a unit area of surface. Common units for pressure are dynes per cm^2 and millimeters of mercury (mmHg) or torr (= 1 mmHg). Pressure is also expressed in terms of the standard atmosphere (atm), which equals 760 mmHg. The SI pressure unit, which will gain more recognition with time, is the pascal (Pa). The pascal is expressed in terms of kg s^{-2} m^{-1}. The relations necessary for interconversion from one unit to another are

$$1 \text{ atm} = 760 \text{ mmHg} = 760 \text{ torr} = 101{,}325 \text{ Pa}$$

Boyle's Law The volume of a given mass of gas held at constant temperature is inversely proportional to its pressure. Mathematically this can be expressed as

$$V = \text{constant} \times \frac{1}{P} = k \times \frac{1}{P} \quad \text{or} \quad PV = \text{constant} = k$$

Charles' Law The volume of a given mass of gas is directly proportional to its Kelvin temperature when holding its pressure constant. This may be stated in equation form:

$$V = \text{constant} \times T = k \times T \quad \text{or} \quad \frac{V}{T} = \text{constant} = k$$

where k is a constant different from the one in Boyle's Law.

Kelvin temperature A more fundamental temperature scale than the Celsius or Fahrenheit scale. The Kelvin scale (K) is independent of the working fluid. Its relation to the Celsius scale is

$$\text{K} = {}^\circ\text{C} + 273.15$$

The interval between degrees is the same on the Kelvin and Celsius scales. All temperatures involving the gas laws should be expressed on the Kelvin scale before attempting to work numerical problems involving multiplication and division.

Gay-Lussac's Law of Combining Volumes After studying a large number of reactions, Gay-Lussac generalized that in reactions involving gases at constant temperature and pressure, the volumes of the gases can be expressed as a ratio of small whole numbers.

Avogadro's Law Equal volumes of all gases, measured under the same conditions of temperature and pressure, contain the same number of molecules. Although this law was proposed in 1811, it was not until 1858 that this idea was generally accepted by chemists. The utility of this law is obvious once the volume of one mole of gas is determined at a particular temperature and pressure. For 0°C and 1.00 atm (STP, known as *standard temperature and pressure*), the volume of one mole of an ideal gas is 22.4 liters (L).

Ideal Gas Law This law is an equation that relates the variables P, V, T of a gas to the number of moles, n, of gas present. The equation is

$$PV = nRT$$

where R is the universal gas constant and can be expressed in several different units. Expressions showing common units are

$$P(\text{atm}) \times V(\text{L}) = n(\text{mol}) \times R\left(\frac{\text{L atm}}{\text{mol K}}\right) \times T(\text{K})$$

$$P(\text{Pa}) \times V(\text{L}) = n(\text{mol}) \times R\left(\frac{\text{L Pa}}{\text{mol K}}\right) \times T(\text{K})$$

where R has the values

0.08205 L atm/mol K or 8314 L Pa/mol K

Dalton's Law of Partial Pressure The total pressure of a mixture of gases equals the sum of the partial pressures of the component gases. If the pressures of the individual gases A, B, C, and so forth are designated P_A, P_B, P_C, and so forth, then the total pressure P_T may be written as

$$P_T = P_A + P_B + P_C + \cdots$$

This law is useful in calculating the pressure of a gas collected over water. The vapor pressure of the water is fixed for a particular temperature. Consequently, if the vapor pressure of water at the temperature of the vessel is subtracted from the total pressure, the pressure of the gas is determined.

Graham's Law, diffusion of gases This law states that the rates of diffusion of different gases are inversely proportional to the square roots of their densities or molecular weights. Mathematically,

$$\frac{\text{Rate of diffusion of gas A}}{\text{Rate of diffusion of gas B}} = \frac{\sqrt{\text{density B}}}{\sqrt{\text{density A}}} = \frac{\sqrt{\text{mol. wt B}}}{\sqrt{\text{mol. wt A}}}$$

Note that the time required for the diffusion of a gas is inversely proportional to its rate. We can write, assuming equal moles,

$$\frac{\text{Time for diffusion of A}}{\text{Time for diffusion of B}} = \frac{\sqrt{\text{density A}}}{\sqrt{\text{density B}}} = \frac{\sqrt{\text{mol. wt A}}}{\sqrt{\text{mol. wt B}}}$$

Kinetic-molecular theory As already mentioned in the introduction, ideas concerning the nature of gases arose early in the development of science. Those that stood the test of time have allowed us to develop the concept of an ideal gas (one that does not actually exist but has the generalized qualities of all gases).

The ideal gas consists of separate particles, either atoms or molecules. The volume occupied by the individual particles is small compared to the total volume of the gas. As a result, the particles in the gaseous state are relatively far apart and have no attraction for one another.

A second premise concerning the ideal gas is that the gas particles are in continuous motion and travel in straight lines with various speeds. As particles collide with each other, no net loss in average kinetic energy occurs; that is, collisions among particles occur as though the particles are perfectly elastic bodies.

Finally, the average kinetic energy ($1/2\ mu^2$) of all molecules, even of different gases, is the same at the same temperature.

These premises explain the gas laws in a qualitative way and can be shown to predict the Ideal Gas Law, $PV = nRT$, a mathematical interpretation of the behavior or characteristics of gases.

The van der Waals Equation An attempt to explain deviations from ideal gas behavior by real gases led van der Waals to formulate an equation that introduced the term a/V^2 to account for the forces of attraction in real gases and the term b to account for the finite volume real gases have. His equation is

$$\left(P + \frac{n^2 a}{V^2}\right)(V - nb) = nRT$$

EXERCISES

1. A typical barometric pressure in Kansas City is 740 torr. What is this pressure in atmospheres, millimeters of mercury, and kilopascals?

   Solution

   Conversion between different units may be accomplished by use of unit conversion factors. For conversion to atmospheres, use the relation 1 atm = 760 torr.

   $$740 \text{ torr} \times \frac{1 \text{ atm}}{760 \text{ torr}} = 0.974 \text{ atm}$$

   For conversion to millimeters of mercury, use the relation 1 torr = 1 mmHg.

   $$740 \text{ torr} \times \frac{1 \text{ mmHg}}{1 \text{ torr}} = 740 \text{ mmHg}$$

   For conversion to kilopascals, use the relation 760 torr = 101.325 kPa or 1 torr = 133.32 Pa = 0.13332 kPa.

   $$740 \text{ torr} \times \frac{0.13332 \text{ kPa}}{1 \text{ torr}} = 98.7 \text{ kPa}$$

2. European tire gauges are marked in units of kilopascals. What reading on such a gauge corresponds to 25 lb/in^2 (1 atm = 14.7 lb/in^2)?

   Solution

   This requires the use of the conversion factor: 1 atm = 101.325 kPa.

   $$25 \text{ lb/in}^2 \times \frac{1 \text{ atm}}{14.7 \text{ lb/in}^2} \times \frac{101.325 \text{ kPa}}{1 \text{ atm}} = 170 \text{ kPa}$$

5. A biochemist adds carbon dioxide, CO_2, to an evacuated bulb like the one shown on the following page [textbook page 272] and stops when the difference in the heights of the mercury columns, h, is 3.56 cm. What is the pressure of CO_2 in the bulb in atmospheres and in kilopascals?

   Solution

   Assume that the mercury levels in the two columns were at the same height before the addition of CO_2. Then the difference in heights of the columns is the pressure exerted by the CO_2. Thus convert 3.56 cm Hg into atmospheres and kilopascals using unit conversion factors. Thus

$$35.6 \text{ mm} \times 1 \text{ atm}/760 \text{ mmHg} = 0.0468 \text{ atm}$$

Then, $$0.0468 \text{ atm} \times 101.325 \text{ kPa/atm} = 4.75 \text{ kPa}$$

8. The volume of a sample of carbon monoxide, CO, is 405 mL at 10.0 atm and 467 K. What volume will it occupy at 4.29 atm and 467 K?

Solution

In a closed system there is no gain or loss in the amount of substance. Since the temperature is held constant, *P* and *V* are free to vary. The pressure on the gas is decreased from 10.0 atm to 4.29 atm. The volume is given by Boyle's Law. In a formal sense, the pressure times the volume is equal to a constant:

$$PV = k = 10.0 \text{ atm} \times 405 \text{ mL} = 4050 \text{ atm mL}$$

Using this value of the constant, the new volume can be found.

$$PV = 4.29 \text{ atm} \times V = 4050 \text{ atm mL}$$
$$V = 4050 \text{ mL}/4.29 = 944 \text{ mL}$$

In a less formal approach, from Boyle's Law volume is inversely proportional to the pressure. In other words, a decrease in pressure will cause an increase in the volume. Because the starting volume must increase, the solution of the problem is accomplished mathematically by multiplication of the original volume by an improper fraction (that is, a fraction larger than one) formed from the ratio of the two pressures. This yields the new volume.

$$V = 405 \text{ mL} \times \frac{10.0 \text{ atm}}{4.29 \text{ atm}} = 944 \text{ mL}$$

11. A typical scuba tank has a volume of 13.2 L. What volume of air in liters at 0.950 atm is required to fill such a scuba tank to a pressure of 153 atm, assuming no change in temperature?

Solution

This is a Boyle's Law problem in which both the amount of substance and its temperature remain the same. The volume change is inversely proportional to the pressure. The final conditions are known so that the constant *k* may be found.

$$P_f V_f = k = 153 \text{ atm} \times 13.2 \text{ L} = 2019.6 \text{ L atm}$$
$$P_i V_i = 0.950 \text{ atm} \times V_i = 2019.6 \text{ L atm}$$
$$V_i = 2130 \text{ L}$$

A simpler method is based on the inverse nature of the *PV* relation. The final volume is small under high pressure. The original volume must, therefore, be larger under low pressure. In order to make the volume larger, the ratio of pressures multiplying the volume must be an improper fraction, that is, one that has a larger number in the numerator than in the denominator. Thus,

$$V \text{ (air)} = 13.2 \text{ L} \times \frac{153 \text{ atm}}{0.950 \text{ atm}} = 2130 \text{ L}$$

15. What is the volume of a sample of ethane at 467 K and 1.1 atm if it occupies 405 mL at 298 K and 1.1 atm?

Solution

This is a Charles' Law problem in which the volume is directly proportional to its temperature in Kelvins. Since the temperature increases, so must the volume in proportion to the ratio of the temperatures. In order to make the volume increase, the ratio of temperatures must be an improper fraction, that is, one that has a larger number in the numerator than in the denominator. Thus,

$$V \text{ (ethane)} = 405 \text{ mL} \times \frac{467 \text{ K}}{298 \text{ K}} = 635 \text{ mL}$$

18. A 2.50-L volume of hydrogen measured at the normal boiling point of nitrogen, -210.0°C, is warmed to the normal boiling point of water, 100°C. Calculate the new volume of the gas, assuming ideal behavior and no change in pressure.

Solution

Both P and the amount of substance are held constant. According to Charles' Law, the volume is directly proportional to a change in T. The temperature increases from 63.1 K (-210.0 + 273.1) to 373.1 K (100.0 + 273.1). The volume will increase by the same proportion as the ratio of temperatures, 373.1/63.1. Mathematically,

$$V \text{ } (H_2) = 2.50 \text{ L} \times \frac{373.1 \text{ K}}{63.1 \text{ K}} = 14.8 \text{ L}$$

20. A high-altitude balloon is filled with 1.41×10^4 L of hydrogen at a temperature of 21°C and a pressure of 745 torr. What is the volume of the balloon at a height of 20 km where the temperature is -48°C and the pressure is 63.1 torr?

Solution

This problem is solved based on a combination of Boyle's and Charles' laws. The amount of hydrogen is fixed. As the temperature decreases the volume decreases. To obtain a smaller volume the original volume must be multiplied by a fraction composed of the ratio of the two temperatures. As a pressure decreases the volume increases. An increase is obtained if the original volume is multiplied by an improper fraction formed from the ratio of the pressures. Thus,

$$V = 1.41 \times 10^4 \text{ L} \times \frac{(-48°\text{C} + 273°\text{C})}{(21°\text{C} + 273°\text{C})} \times \frac{745 \text{ torr}}{63.1 \text{ torr}}$$

$$= 1.41 \times 10^4 \text{ L} \times \frac{225}{294} \times \frac{745}{63.1} = 1.27 \times 10^5 \text{ L}$$

22. A spray can is used until it is empty, except for the propellant gas, which has a pressure of 1.1 atm at 23°C. If the can is thrown into a fire ($T = 475°C$), what will be the pressure in the hot can?

Solution

Pressure varies directly with the temperature; the amount of substance in the can and its volume are constant. The new higher pressure is obtained by multiplying the original pressure by a number greater than one formed from the ratio of the two absolute temperatures.

$$P_f = 1.1 \text{ atm} \times \frac{475°\text{C} + 273.1°\text{C}}{23°\text{C} + 273.1°\text{C}} = 1.1 \text{ atm} \times \frac{748.1}{296.1} = 2.8 \text{ atm}$$

24. How many moles of hydrogen sulfide, H_2S, are contained in a 327.3-mL bulb at 48.1°C if the pressure is 149.3 kPa?

Solution

The variables, P, V, n, and T are related through the Ideal Gas Law, $PV = nRT$, where R is given in units of 8.314 L kPa/(mol K). Rearranging to solve for n followed by substitution (48.1°C = 321.2 K), yields

$$n = \frac{PV}{RT} = \frac{(149.3 \text{ kPa}) \dfrac{327.3 \text{ mL}}{1000 \text{ mL/L}}}{(8.314 \text{ L kPa/mol K})(321.2 \text{ K})} = 1.830 \times 10^{-2} \text{ mol}$$

28. What is the temperature of a 0.274-g sample of methane, CH_4, confined in a 300.0-mL bulb at a pressure of 198.7 kPa?

Solution

The Ideal Gas Law is used here because only one set of conditions is stated. The amount of substance, methane, is given by the number of grams divided by the molar mass 16.0426 g mol^{-1}. Rearrangement of $PV = nRT$ gives

$$T = \frac{PV}{nR} = \frac{198.7 \text{ kPa} \times 0.3000 \text{ L}}{\dfrac{0.274 \text{ g}}{16.0426 \text{ g/mol}} \times 8.314 \dfrac{\text{L kPa}}{\text{mol K}}} = 420 \text{ K or } 147°\text{C}$$

30. Assume that 453.6 g of dry ice (solid CO_2) is placed in an evacuated 50.0-L closed tank. What is the pressure in the tank in atmospheres at a temperature of 45°C after all the $CO_2(s)$ has been converted to gas?

Solution

The conditions $n = 10.308$ mol (453.6 g/44.0098 g mol^{-1}), 50.0 L, and 318 K (45.0°C + 273.1°C) are related to the pressure by the Ideal Gas Law.

$$P = \frac{nRT}{V} = \frac{10.308 \text{ mol} \times 8.314 \dfrac{\text{L kPa}}{\text{mol K}} \times 318.1 \text{ K}}{50.0 \text{ L}}$$

$$= 545 \text{ kPa} = 545 \text{ kPa} \times \frac{1 \text{ atm}}{101.325 \text{ kPa}} = 5.38 \text{ atm}$$

Alternatively, R may be expressed as 0.08205 L atm/mol K; in which case the value of P is given in atmospheres directly.

31. How many grams of gas are present in the following case?

(a) 0.100 L of NO at 703 torr and 62°C.

Solution

The Ideal Gas Law can be cast in the form

$$PV = \frac{\text{g}}{\text{molar mass}} RT \qquad \text{or} \qquad \text{g} = \frac{PV\ (\text{molar mass})}{RT}$$

$$\text{g} = \frac{\frac{703\ \text{torr}}{760\ \text{torr/atm}} \times 0.100\ \text{L} \times 30.0061\ \text{g/mol}}{0.08205\ \frac{\text{L atm}}{\text{mol K}} \times 335.1\ \text{K}} = 0.101\ \text{g}$$

32. Calculate the volume in liters of the following quantity of gas at STP.

(b) 0.588 g of NH_3

Solution

From the Ideal Gas Law, $PV = nRT$,

$$V = \frac{\text{g}}{\text{molar mass}} \times \frac{RT}{P}$$

$$= \frac{0.588\ \text{g}}{17.0304\ \text{g/mol}} \times \frac{0.08205\ \frac{\text{L atm}}{\text{mol K}} \times 273.1\ \text{K}}{1.00\ \text{atm}}$$

$$= 0.774\ \text{L}$$

34. Calculate the density of fluorine gas, F_2, at STP, and at 30.0°C and 725 torr.

Solution

The Ideal Gas Law may be written in the form

$$P = \frac{n}{V} RT = \frac{\frac{\text{g}}{\text{molar mass}}}{V} RT = \frac{DRT}{\text{molar mass}}$$

Then for the first set of conditions,

$$D = \frac{P(\text{molar mass})}{RT} = \frac{1.00\ \text{atm} \times 37.9968\ \text{g/mol}}{0.08205\ \frac{\text{L atm}}{\text{mol K}} \times 273.1\ \text{K}} = 1.70\ \text{g/L}$$

For the second set of conditions,

$$D = \frac{P(\text{molar mass})}{RT} = \frac{\frac{725\ \text{torr}}{760\ \text{torr}} \times 37.9968\ \text{g/mol}}{0.08205\ \frac{\text{L atm}}{\text{mol K}} \times 303.1\ \text{K}} = 1.46\ \text{g/L}$$

40. Most mixtures of hydrogen gas with oxygen gas are explosive. However, a mixture that contains less than 3.0% O_2 is not. If enough O_2 is added to a cylinder of H_2 at 33.2 atm to bring the total pressure to 34.5 atm, is the mixture explosive?

Solution

Pressure is directly proportional to the moles of gas present, so all that is required is to find the pressure of oxygen and to determine the percentage of oxygen present. From Dalton's Law of partial pressures,

$$P_T = P_{H_2} + P_{O_2} = 34.5 \text{ atm} = 33.2 \text{ atm} + P_{O_2}$$

$$P_{O_2} = 1.3 \text{ atm}$$

As a percentage of the total pressure,

$$\frac{P_{O_2}}{P_T} \times 100 = \frac{1.3 \text{ atm}}{34.5 \text{ atm}} \times 100 = 3.8\%$$

Therefore, the mixture is explosive.

41. A 5.73-L flask at 25°C contains 0.0388 mol of N_2, 0.147 mol of CO, and 0.0803 mol of H_2. What is the pressure in the flask in atmospheres, torr, and kilopascals?

Solution

This is an application of the Ideal Gas Law in which the amount of substance of each gas must be lumped together:

$$0.0388 \text{ mol} + 0.147 \text{ mol} + 0.0803 \text{ mol} = 0.2661 \text{ mol}$$

$$P = \frac{0.2661 \text{ mol} \times 0.08205 \frac{\text{L atm}}{\text{mol K}} \times 298.1 \text{ K}}{5.73 \text{ L}} = 1.136 \text{ atm rounded to } 1.14 \text{ atm}$$

Then,

$$1.136 \text{ atm} \times 760 \frac{\text{torr}}{\text{atm}} = 863 \text{ torr}$$

$$1.136 \text{ atm} \times 101.325 \frac{\text{kPa}}{\text{atm}} = 115 \text{ kPa}$$

43. A sample of carbon monoxide was collected over water at a total pressure of 764 torr and a temperature of 23°C. What is the pressure of the carbon monoxide? (See Table 10-1 for the vapor pressure of water.)

Solution

This is another application of Dalton's Law. $P_T = P_{CO} + P_{H_2O}$. At 23°C, the vapor pressure of water is 21.1 torr. Thus,

$$764 \text{ torr} = P_{CO} + 21.1 \text{ torr}$$

$$P_{CO} = 743 \text{ torr}$$

44. A sample of oxygen collected over water at a temperature of 29.0°C and a pressure of 764 torr has a volume of 0.560 L. What volume would the dry oxygen have under the same conditions of temperature and pressure?

Solution

The pressure of pure oxygen must be found from Dalton's Law of Partial Pressures. At 29°C the partial pressure of H_2O is 30.0 torr. Thus,

$$P_{O_2} = P_{Total} - P_{H_2O} = 764 \text{ torr} - 30.0 \text{ torr} = 734 \text{ torr}$$

Applying Boyle's Law, we find

$$V = 0.560 \text{ L} \times \frac{734 \text{ torr}}{764 \text{ torr}} = 0.538 \text{ L}$$

51. Calculate the volume of oxygen required to burn 7.00 L of propane gas, C_3H_8, to produce carbon dioxide and water, if the volumes of both the propane and oxygen are measured under the same conditions.

Solution

The balanced equation is

$$C_3H_8(g) + 5O_2(g) \longrightarrow 3CO_2(g) + 4H_2O(g)$$

From Gay-Lussac's Law of combining volumes, 1 vol of C_3H_8 requires 5 vol of O_2. Therefore, 7.00 L of propane requires

$$7.00 \text{ L} \times 5 = 35.0 \text{ L of } O_2$$

55. What is the molecular weight of methylamine if 0.157 g of methylamine gas occupies 125 mL with a pressure of 99.5 kPa at 22°C?

Solution

Rearrangement of $PV = nRT$ and substitution of $n = \text{g/molar mass}$, where molar mass = molecular weight, gives

$$\text{molar mass} = \frac{g\,RT}{PV} = \frac{0.157 \text{ g} \times 8.314 \frac{\text{L kPa}}{\text{mol K}} \times 295 \text{ K}}{99.5 \text{ kPa} \times 0.125 \text{ L}} = 31.0 \text{ g mol}^{-1}$$

58. The density of a certain gaseous fluoride of phosphorus is 3.93 g/L at STP. Calculate the molecular weight of this fluoride, and determine its molecular formula.

Solution

From $PV = nRT$,

$$P = \frac{\frac{g}{\text{molar mass}} RT}{V} = \frac{g\ RT}{V \times \text{molar mass}}$$

Since $D = g/V$, substitution and rearrangement gives

$$\text{molar mass} = \frac{DRT}{P} = \frac{3.93\ \text{g/L} \times 0.08205\ \frac{\text{L atm}}{\text{mol K}} \times 273.1\ \text{K}}{1.00\ \text{atm}} = 88.1\ \text{g mol}^{-1}$$

The atomic weight of phosphorus is 31; the only compound with two phosphorus atoms and one fluorine (at. wt 19) would not match the found weight of 88. The compound must, therefore, contain only one phosphorus and three fluorine atoms.

59. Cyclopropane, a gas containing only carbon and hydrogen, is often used as an anaesthetic for major surgery. Taking into account that 0.45 L of cyclopropane at 120°C and 0.72 atm reacts with O_2 to give 1.35 L of CO_2 and 1.35 L of $H_2O(g)$ at the same temperature and pressure, determine the molecular formula of cyclopropane.

Solution

This problem uses Gay-Lussac's Law of combining volumes.

$$0.45\ \text{L (cyclopropane)} + O_2 \longrightarrow 1.35\ \text{L } (CO_2) + 1.35\ \text{L } (H_2O)$$

All the carbon from the combustion must appear in the CO_2. The ratio of volumes is 1 vol (cyclopropane) $\longrightarrow$ 3 vol (CO_2), that is, 1.35 L/0.45 L = 3/1. Similarly, 1 vol (cyclopropane) $\longrightarrow$ 3 vol (H_2O). The ratio again is 1 to 3. Carbon appears only once in CO_2, so there must be 3 carbons in cyclopropane. Hydrogen occurs twice in H_2O, there must, therefore, be 2×3 or 6 hydrogens in cyclopropane. The formula is thus C_3H_6.

63. Gaseous hydrogen chloride, HCl(*g*), is prepared commercially by the reaction of NaCl with H_2SO_4 (Section 8.7, Part 1). What mass of NaCl is required to prepare enough HCl(*g*) to fill a 35.4-L cylinder to a pressure of 125 atm at 0°C?

Solution

The reactions are:

$$H_2SO_4(\ell) + NaCl(s) \longrightarrow NaHSO_4(s) + HCl(g)$$

$$NaHSO_4(s) + NaCl(s) \xrightarrow{\Delta} Na_2SO_4(s) + HCl(g)$$

From the equations, 1 mol of NaCl produces 1 mol of HCl(g). Thus, the problem reduces to one of finding the amount of HCl(g) required in the cylinder. From $PV = nRT$,

$$n = \frac{PV}{RT} = \frac{125 \text{ atm} \times 35.4 \text{ L}}{0.08205 \frac{\text{L atm}}{\text{mol K}} \times 273.1 \text{ K}} = 197.5 \text{ mol}$$

$$197.5 \text{ mol} \times 58.44277 \frac{\text{g}}{\text{mol}} \text{ (NaCl)} = 11.5 \text{ kg}$$

67. In a common freshman laboratory experiment, $KClO_3$ is decomposed by heating to give KCl and O_2. What mass of $KClO_3$ must be decomposed to give 638 mL of O_2 at a temperature of 18°C and a pressure of 752 torr?

Solution

The amount of oxygen in moles is computed first; then the mass of $KClO_3$ is determined from the balanced equation for the decomposition:

$$2KClO_3 \longrightarrow 2KCl + 3O_2$$

$$PV = nRT, \quad n = \frac{PV}{RT} = \frac{\frac{752 \text{ torr}}{760 \text{ torr}} \times 0.638 \text{ L}}{0.08205 \frac{\text{L atm}}{\text{mol K}} \times 291.1 \text{ K}} = 0.0264 \text{ mol}$$

2 mol $KClO_3 \longrightarrow$ 3 mol O_2; therefore,

$$\text{amount of } KClO_3 = 2 \times \frac{0.0264 \text{ mol}}{3} = 0.0176 \text{ mol}$$

The formula weight of $KClO_3$ is 122.5495 g/mol.

$$\text{Mass } KClO_3 = 122.5495 \text{ g/mol} \times 0.0176 \text{ mol} = 2.16 \text{ g}$$

69. As 1 g of the radioactive element radium decays over one year, it produces 1.16×10^{18} α particles (helium nuclei). Each α particle becomes an atom of helium gas. What volume of helium gas at a pressure of 56.0 kPa and a temperature of 25°C is produced?

Solution

After one year there are

$$1.16 \times 10^{18} / 6.022 \times 10^{23} \text{ mol}^{-1} = 1.926 \times 10^{-6} \text{ mol}$$

From the Ideal Gas Law, $PV = nRT$

$$V = \frac{nRT}{P} = \frac{1.926 \times 10^{-6}\ \text{mol} \times 8.314\ \frac{\text{L kPa}}{\text{mol K}} \times 298\ \text{K}}{56.0\ \text{kPa}}$$

$$= 8.52 \times 10^{-5}\ \text{L} = 8.52 \times 10^{-2}\ \text{mL}$$

76. The root-mean-square speed of H_2 molecules at 25°C is about 1.6 km/s. What is the root-mean-square speed of a CH_4 molecule at 25°C?

Solution

$$\frac{(u_1)\text{rms}}{(u_2)\text{rms}} = \sqrt{\frac{m_2}{m_1}}\ ; \qquad \frac{(1.6\ \text{km/s})_{H_2}}{(u)_{CH_4}} = \sqrt{\frac{(16.0426)_{CH_4}}{(2.0158)_{H_2}}}$$

$$(u)_{CH_4} = 1.6\ \text{km/s}\sqrt{\frac{2.0158}{16.0426}} = 0.57\ \text{km/s}$$

Keystrokes: 2.0158 ÷ 16.0426 = [√] × 1.6 = ans. 0.56716

81. A gas of unknown identity diffuses at the rate of 83.3 mL/s in a diffusion apparatus in which a second gas, whose molecular weight is 44.0, diffuses at the rate of 102 mL/s. Calculate the molecular weight of the first gas.

Solution

According to Graham's Law of effusion, the rate of diffusion is inversely proportional to the square root of its molar mass. Thus,

$$\frac{\text{Rate(1)}}{\text{Rate(2)}} = \frac{\sqrt{\text{MW(2)}}}{\sqrt{\text{MW(1)}}}\ ; \qquad \frac{83.3\ \text{mL/s}}{102\ \text{mL/s}} = \frac{\sqrt{44}}{\sqrt{\text{MW(1)}}}$$

It is perhaps easiest to square both sides in this case and then solve for the MW(1). Thus,

$$\text{MW(1)} = 44.0\ \frac{\text{g}}{\text{mol}} \times \frac{(102\ \text{mL/s})^2}{(83.3\ \text{mL/s})^2} = 44.0 \times \frac{102^2}{83.3^2}\ \frac{\text{g}}{\text{mol}} = 66.0\ \text{g/mol}$$

Keystrokes: 44 × 102 [x^2] ÷ 83.3 [x^2] = ans. 65.9725

82. (a) When two cotton plugs, one moistened with ammonia and the other with hydrochloric acid, are simultaneously inserted into opposite ends of a glass tube 87.0 cm long, a white ring of NH_4Cl forms where gaseous NH_3 and gaseous HCl first come into contact.

$$NH_3(g) + HCl(g) \longrightarrow NH_4Cl(s)$$

At what distance from the ammonia-moistened plug does this occur?

Solution

The ring forms when an equal number of molecules of the two gases meet as controlled by the rate of diffusion. Since the rate of gas diffusion is inversely proportional to $\sqrt{MW}$, we write

$$\frac{\text{Rate } NH_3}{\text{Rate } HCl} = \frac{\sqrt{\text{MW } HCl}}{\sqrt{\text{MW } NH_3}} = \frac{\sqrt{36.5}}{\sqrt{17.0}} = 1.465$$

The ammonia diffuses 1.465 times as fast as the hydrogen chloride. Therefore, equal amounts meet in the tube at the fraction of distance, 1.465/2.465, covered by the ammonia. Thus,

$$\text{Distance} = \frac{1.465}{2.465} \times 87.0 \text{ cm} = 51.7 \text{ cm}$$

Interestingly, after the initial formation of the white ring, additional rings will form on both sides of the original ring. This is one example of Liesegang phenomena; similar behavior is known in hundreds of other systems.

88. (a) What is the total volume of the $CO_2(g)$ and $H_2O(g)$ at 600°C and 735 torr produced by the combustion of 1.00 L of $C_3H_8(g)$ measured at STP?

Solution

The balanced equation for the combustion of $C_3H_8(g)$ at this temperature (water is in the vapor state) is

$$C_3H_8(g) + 5O_2(g) \longrightarrow 3CO_2(g) + 4H_2O(\ell)$$

The volume of 1.00 L of $C_3H_8(g)$ at 600°C and 735 torr can be used as the starting volume to solve this problem from Gay-Lussac's Law of combining volumes. The new volume is found from the gas laws.

$$V(C_3H_8) = 1.00 \text{ L} \times \frac{760 \text{ torr}}{735 \text{ torr}} \times \frac{(600 + 273)\text{K}}{(0 + 273)\text{K}}$$

$$= \frac{760}{735} \times \frac{873}{273} \text{ L} = 3.306 \text{ L}$$

Because 1 vol of C_3H_8 produces 7 vol of CO_2 and H_2O, 3.31 L of C_3H_8 produces 7×3.306 L = 23.1 L.

(b) What is the partial pressure of CO_2 in the product gases?

Solution

The total pressure is 735 torr and from the balanced equation CO_2 represents 3/7 of the total gases present. Thus, 3/7(735 torr) = 315 torr partial pressure of CO_2.

90. Ethanol, C_2H_5OH, is produced industrially from ethylene, C_2H_4, by the following sequence of reactions:

$$3C_2H_4 + 2H_2SO_4 \longrightarrow C_2H_5HSO_4 + (C_2H_5)_2SO_4$$

$$C_2H_5HSO_4 + (C_2H_5)_2SO_4 + 3H_2O \longrightarrow 3C_2H_5OH + 2H_2SO_4$$

What volume of ethylene at STP is required to produce 1.000 metric ton (1000 kg) of ethanol if the overall yield of ethanol is 90.1%?

Solution

At 100% production of ethanol

$$\text{kg }(C_2H_4) = 1000 \text{ kg ethanol} \times \frac{1 \text{ molar mass } C_2H_4}{1 \text{ molar mass } C_2H_5OH}$$

$$= 1000 \text{ kg} \times \frac{28.054 \text{ g ethylene}}{46.054 \text{ g}} = 609 \text{ kg}$$

At 90.1% yield, 609 kg/0.901 = 676 kg, are required. From the Ideal Gas Law at STP,

$$V = \frac{nRT}{P} = \frac{(\text{g/molar mass})RT}{P}$$

$$= \frac{\left(\frac{676{,}000}{28.054}\right) \text{mol} \times 0.08205 \frac{\text{L atm}}{\text{mol K}} \, 273 \text{ K}}{1 \text{ atm}} = 5.40 \times 10^5 \text{ L}$$

Alternatively, when the amount in moles of ethylene is found, this procedure could be used:

$$\text{mol} \times \frac{22.4 \text{ L}}{\text{mol}} = 5.40 \times 10^5 \text{ L}$$

93. Ethanol, C_2H_5OH, is often produced by the fermentation of sugars. For example, the preparation of ethanol from the sugar glucose is represented by the equation

$$C_6H_{12}O_6(aq) \xrightarrow{\text{yeast}} 2C_2H_5OH(aq) + 2CO_2(g)$$

What volume of CO_2 at STP is produced by the fermentation of 125 g of glucose if the reaction has a yield of 97.5%?

Solution

One mole of glucose produces two moles of CO_2. The amount of glucose is 125 g/(180 g/mol) = 0.694 mol.

$$\text{Amount of } CO_2 \text{ produced} = 2(0.694 \text{ mol}) = 1.388 \text{ mol}$$

$$V\,(CO_2) \text{ at STP} = 1.388 \text{ mol} \times \frac{22.4 \text{ L}}{\text{mol}} \times \frac{97.5\%}{100\%} = 30.3 \text{ L}$$

95. One method of analysis of amino acids is the van Slyke method. The characteristic amino groups ($—NH_2$) in protein material are allowed to react with nitrous acid, HNO_2, to form N_2 gas. From the volume of the gas, the amount of amino acid can be determined. A 0.0604-g sample of a biological material containing glycine, $CH_2(NH_2)COOH$, was analyzed by the van Slyke method giving 3.70 mL of N_2 collected over water at a pressure of 735 torr and 29°C. What was the percentage of glycine in the sample?

$$CH_2(NH_2)COOH + HNO_2 \longrightarrow CH_2(OH)COOH + H_2O + N_2$$

Solution

From Dalton's Law of Partial Pressures:

$$\text{Pressure of } N_2 = 735 \text{ torr} - 30.0 \text{ torr}(H_2O)$$

$$= 705 \text{ torr}$$

From the Ideal Gas Law the amount of N_2 present is

$$n = \frac{PV}{RT} = \frac{\frac{705 \text{ torr}}{760 \text{ torr}}(0.00370 \text{ L})}{\left(0.08205 \frac{\text{L atm}}{\text{mol K}}\right)(302 \text{ K})} = 0.0001385 \text{ mol}$$

From the balanced equation, 1 mol of glycine produces 1 mol of N_2. Therefore, 0.0001385 mol of glycine is present. The molecular weight of glycine is 75.0 g/mol. Thus,

$$\text{Mass glycine} = 75.0 \text{ g/mol} \times 0.0001385 \text{ mol}$$

$$= 0.01039 \text{ g}$$

$$\% \text{ glycine} = \frac{0.01039 \text{ g}}{0.0604 \text{ g}} \times 100 = 17.2\%$$

98. A sample of a compound of xenon and fluorine was confined in a bulb with a pressure of 24 torr. Hydrogen was added to the bulb until the pressure was 96 torr. Passage of an electric spark through the mixture produced Xe and HF. After the HF was removed by reaction with solid KOH, the final pressure of xenon and unreacted hydrogen in the bulb was 48 torr. What is the empirical formula of the xenon fluoride in the original sample? *Note:* Xenon fluorides contain only one xenon atom per molecule.

Solution

The reaction of hydrogen with the unknown xenon compound, assuming constant temperature, is

$$XeF_y + \frac{y}{2}H_2 \longrightarrow Xe + yHF$$

where

$$H_2 + 2F \longrightarrow 2HF$$

The pressure in the vessel due to the gaseous XeF_y is 24 torr. After the addition of hydrogen, the pressure in the vessel rises to 96 torr.

$$P_{\text{Total}} = 96 \text{ torr} = P_{XeF_y} + P_{H_2}$$

$$P_{H_2} = 96 \text{ torr} - 24 \text{ torr} = 72 \text{ torr}$$

After the reaction the vessel contained a mixture of xenon and hydrogen at a total pressure of 48 torr. But the pressure due to xenon must equal 24 torr because each mole of XeF_y contains 1 mol of Xe. The hydrogen consumed by the production of HF is 72 torr - 24 torr = 48 torr. Because the pressure exerted by a gas is proportional to the number of moles of gas, the mole ratio between reacting substances is equivalent to the ratio of pressures. In this case, the reaction of XeF_y with H_2 can be interpreted as

$$XeF_y : H_2$$

$$24 \text{ torr} : 48 \text{ torr}$$

or

$$1 \text{ mole} : 2 \text{ moles}$$

A molecule of XeF_y, therefore, must contain four atoms of fluorine, $y = 4$, to react with two molecules of hydrogen; hence, XeF_4.

RELATED EXERCISES

1. The van der Waals Equation is just one of many equations that describe the nonideal behavior of gases. Compare the pressure of 1 mol of SO_2 contained in a 5.0-L flask maintained at 500 K as given by the Ideal Gas Law and by the van der Waals Equation. For SO_2, the constant a is 6.71 L^2 atm/mol^2 and the constant b is 0.0564 L.

   Answer: From ideal gas: 8.20 atm; from van der Waals Equation: 8.02 atm

2. Water vapor behaves as a gas, and its volume can be calculated from the Ideal Gas Law. Compare the value obtained for the molar volume of water using the Ideal Gas Law with that obtained from the experimental density of 0.0005970 g mL^{-1} both under the condition of 100°C and 1 atm.

   Answer: V(observed) = 30.18 L; V(ideal) = 30.60 L

3. In the early study of air composition, Ramsey in 1894 separated water vapor, nitrogen, oxygen and carbon dioxide from air by absorption processes. Lord Rayleigh separated these gases by a different process. Both men were left with a small amount of gas with a density of 1.63 g/L at 25°C and 1 atm. What element had they discovered?

   Answer: Argon

4. A popular method for preparing oxygen is the decomposition of $KClO_3$ and collection of the gas over water. If 36.5 mL of O_2 is collected over

water at 25°C at a barometric pressure of 751 torr, what volume would the dry oxygen occupy at 0°C at the same pressure?

Answer: 32.4 mL

5. Calculate the volume of a balloon necessary to lift 100 kg at 25°C and 1 atm if the balloon is filled with helium. Assume that the composition of air is 80% nitrogen and 20% oxygen at the temperature and pressure given.

Answer: 84,750 L

UNIT 7

CHAPTER 11: THE LIQUID AND SOLID STATES

INTRODUCTION

In our daily experience we encounter three states of matter: solid, liquid, and gaseous. Solids may be in many forms, such as lumber, metal, ice, rock, and so forth. Similarly, liquids of various types occur as greases, syrups, water, alcohol, and even as a metal, mercury. Most gases, such as nitrogen in the air we breathe, are invisible. Generally, it is the effect of gases that we experience, although some gases have color, such as chlorine, and some are visible if they are present in sufficient concentration. (Smokestack and other exhaust plumes are often visible, but this is because the plumes consist of dispersions of particulate matter in the air.)

Different states of aggregation are possible for a substance; these states are called *phases* (from a Greek word meaning "appearance"). The term *phase* is used in chemistry to indicate a homogeneous region, all parts of which are the same. Water is probably the most familiar example of a substance that may exist in several different forms or phases. Under certain conditions water may exist as solid ice, liquid water, or gaseous water vapor.

Depending on the controlled condition that is imposed, another kind of variation in form or phase characterizes certain solids. For example, carbon may appear as two different solid phases, diamond and graphite.

Typically, a sufficient change in temperature will cause a change in phase or state. When solid ice is heated, for example, the temperature rises and the ice experiences a phase change as it is converted to liquid water.

Liquids have properties that are intermediate between those of the solids and gases; they are the molten or fluid form of substances. As a substance is transformed from a solid to a liquid, the rigid lattice structure of the solid is lost. Some structure does exist in liquids, but the structure becomes fluid in the sense that structures are continually forming and dispersing. In the gaseous state all structure is lost. All liquids have an ability to flow because of the continual shifting of internal structure, although some, such as glass, molten polymers, long-chain hydrocarbons, and proteins, tend to flow very slowly or not at all.

The purpose of this chapter is to examine some of the regular structural forms exhibited in solids, some of the reasons why these forms exist, and the energies associated with transformations among liquids, solids, and gases.

FORMULAS AND DEFINITIONS

Energy and change of state The transformation of any substance from one phase or state to another involves a change in the total energy of the substance. Depending on the direction of the phase change, energy is either absorbed or released by the substance during the change. The diagram represents the possible types of transformations a substance may undergo.

$$\text{Solid} \underset{\text{energy }-}{\overset{+\text{ energy}}{\rightleftarrows}} \text{Liquid} \underset{\text{energy }-}{\overset{+\text{ energy}}{\rightleftarrows}} \text{Gas}$$

The processes indicated in the diagram are reversible, that is, the energy absorbed in each forward process equals the energy released in the reverse process. The temperature of the substance remains constant during a phase change. The energies associated with these changes at the melting point and boiling point are the heat of fusion and heat of vaporization.

Heat of fusion The quantity of heat required to transform a substance, at its melting point, from a solid to a liquid. The heat of fusion of water is 333.6 J/g, or about 80 cal/g.

Heat of vaporization The quantity of heat required to transform a substance, at its boiling point, from a liquid to a gas. The heat of vaporization of water, at 100°C, is 2258 J/g, or about 540 cal/g.

Born-Haber Cycle A thermodynamic cycle that allows the calculation of the heat of a process that is otherwise difficult to measure. Only rarely can the energy associated with the lattice structure of a crystalline substance be measured directly. Instead, a basic thermochemical law must be used, which, simply stated, is that the energy associated with the final state of a substance is the same regardless of the pathway or steps used to reach that state. The Born-Haber Cycle is an application of this principle in that the formation of a crystalline substance can be analyzed into several independent steps. Because of the vast amount of thermochemical data available in the literature, the necessary steps for completion of the cycle can frequently be written even though data may not be available for the step one is interested in. Because the value of all the steps in the cycle, except one, can be found in the chemical literature, the otherwise unknown value of the thermochemical step may be resolved.

Radius ratio (r^+/r^-) The radius ratio is computed for a compound by dividing the radius of the cation (positive ion) by the radius of the anion (negative ion). Based on empirical data, the ratio is useful for predicting the structural geometry of a crystalline substance.

Bragg equation X-ray techniques are used to determine crystal patterns caused by the location of atoms in crystalline substances. The equation $\lambda = 2d \sin\theta$, developed by Bragg, relates the wavelength λ of reflected X rays to the product of the distance between atomic planes d in a crystal times the sine of the angle θ of the reflected X rays.

Pythagorean theorem This theorem relates the hypotenuse of a right triangle to the other two sides; the square of the hypotenuse equals the sum of the squares of the other two sides.

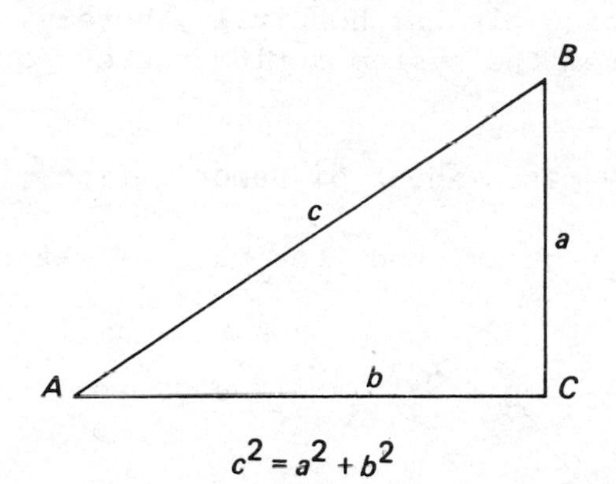

$$c^2 = a^2 + b^2$$

EXERCISES

Note: The following data for water may be useful:

Heat of vaporization of water (100°C)	2258	J/g
Heat of vaporization of water (25°C)	2443	J/g
Heat of fusion of water (0°C)	333.6	J/g
Specific heat of water (solid)	2.04	J/g K
Specific heat of water (liquid)	4.18	J/g K
Specific heat of water (gas)	2.00	J/g K

18. How much heat is required (a) to change 3.00 mol of water at 100°C to steam at 100°C?

Solution

This is an example of a phase transition. Such transitions occur at a constant temperature, in this case 100°C. The heat of vaporization of water at 100°C is 2258 J/g. The molar mass of water is 18.0152 g/mol. The heat required is

$$\text{Heat} = 3 \text{ mol} \times 18.0152 \text{ g/mol} \times 2258 \text{ J/g} = 122 \text{ kJ}$$

19. In hot, dry climates water is cooled by allowing some of it to evaporate slowly. How much water must evaporate to cool 1.00 kg from 37°C to 21°C? Assume that the heat of vaporization of water is constant in that range and is equal to the value at 25°C.

Solution

Cooling 1.00 kg of water from 37°C to 21°C requires the removal of heat as calculated from the relationship

$$\text{Heat} = \text{mass} \times \text{specific heat} \times \text{temperature change}$$
$$= M \times C \times \Delta T$$
$$= \left(1.00 \text{ kg} \times \frac{1000 \text{ g}}{\text{kg}}\right)\left(\frac{4.18 \text{ J}}{\text{g K}}\right)(37° - 21°)\text{K} = 6.7 \times 10^4 \text{ J}$$

Note that the difference in temperatures expressed in Celsius degrees is also the same expressed in Kelvins.

Cooling systems of this type involve the loss of water to the surroundings. They are commonly used for the large cooling systems needed for electrical generating plants, hospitals, hotels, and so forth. The amount of heat lost from the system during the evaporation process is given by the equation

$$\text{Heat} = \text{mass} \times \text{heat of vaporization}$$

From the amount of heat just found, the mass of water to be evaporated equals

$$\text{Mass } H_2O = \frac{6.7 \times 10^4 \text{ J}}{2443 \text{ J/g}} = 27 \text{ g}$$

20. The specific heat of copper is 0.0931 cal/g K. What weight of steam at 100°C must be converted to water at 100°C to raise the temperature of a 1.00×10^2-g copper block from 20°C to 100°C?

Solution

The heat released by the transformation of $H_2O(g)$ (100°), $H_2O(\ell)$ (100°) is transferred to the copper.

$$\text{Heat lost from steam} = \text{heat gained by copper}$$

The heat needed to raise the temperature of the copper block is

$$\text{Heat} = MC\Delta T$$

Convert C in cal/g K to J/g K:

$$C = 0.0931 \frac{\text{cal}}{\text{g K}} \times \frac{4.18 \text{ J}}{\text{cal}} = 0.389 \text{ J/(g K)}$$

$$\text{Heat} = 1.00 \times 10^2 \text{ g} \times \frac{0.389 \text{ J}}{\text{g K}} \times 80 \text{ K} = 3.11 \times 10^3 \text{ J}$$

The mass of steam required to produce this amount of heat is

$$\text{Heat} = \text{mass} \times 2258 \text{ J/g}$$

$$\text{Mass steam} = 3.11 \times 10^3 \text{ J} \times \frac{1 \text{ g}}{2258 \text{ J}} = 1.38 \text{ g}$$

22. During the fermentation step in the production of beer, 4.5×10^6 kcal of heat are evolved per 1000 gal of beer produced. How many liters of cooling water are required to maintain the optimum fermentation temperature of 58°F for 1000 gal of beer? how many gallons? The cooling water enters with a temperature of 5°C and is discharged with a temperature of 13°C.

Solution

The heat produced by the fermentation process is transferred to the cooling water. What must be determined is how much cooling water is required to absorb the heat from the 8°C (8 K) temperature rise. The heat to be removed is expressed in kcal and the constants are expressed in joules, so first convert heat units to joules:

$$\text{Heat (J)} = 4.5 \times 10^6 \text{ kcal} \times \frac{4.184 \times 10^3 \text{ J}}{\text{kcal}} = 1.88 \times 10^{10} \text{ J}$$

$$= 1.88 \times 10^7 \text{ kJ}$$

$$C = \frac{4.184 \text{ J}}{\text{g K}} = \frac{4.184 \text{ kJ}}{\text{kg K}}$$

The mass of cooling water required is determined from

$$\text{Heat} = MC\Delta T$$

$$\text{Mass } H_2O = \frac{\text{heat}}{C\ \Delta T} = \frac{1.88 \times 10^7 \text{ kJ}}{\frac{4.184 \text{ kJ}}{\text{kg K}} \times 8 \text{ K}} = 6 \times 10^5 \text{ kg}$$

Assume that the density of water is 1.0 kg/L, then the volume of water required is calculated from $V = M/D$, or

$$V = 6 \times 10^5 \text{ kg} \times \frac{1.0 \text{ L}}{1.0 \text{ kg}} = 6 \times 10^5 \text{ L}$$

and

$$V = 6 \times 10^5 \text{ L} \times \frac{1 \text{ gal}}{3.785 \text{ L}} = 2 \times 10^5 \text{ gal (1 significant figure)}$$

35. Cobalt metal crystallizes in a hexagonal closest-packed structure. What is the coordination number of a cobalt atom?

Solution

Coordination number refers to the number of nearest neighbors. A Co atom is surrounded by and in contact with six other Co atoms in a layer. It is also in contact with three Co atoms in the layer above and with three in the layer below. Consequently, it is in contact with twelve other atoms and its coordination number is 12.

39. One form of tungsten crystallizes in a body-centered cubic unit cell with a tungsten atom on each lattice point. If the edge length of the unit cell is 3.165 Å, what is the atomic radius of tungsten in this structure?

Solution

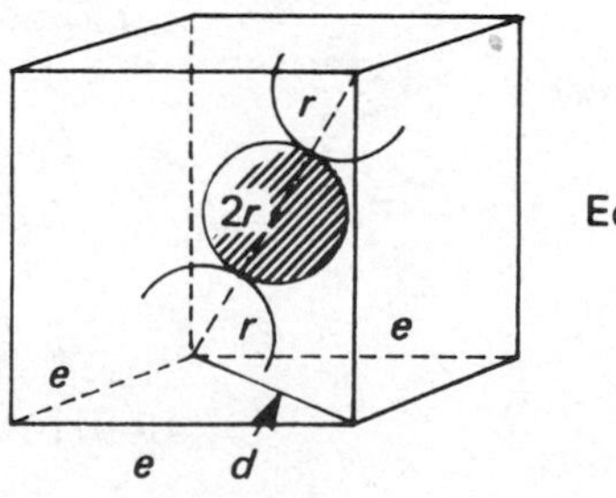

In a body-centered cubic unit cell, the metal atoms are in contact along the diagonal of the cube. The diagonal of the cube forms a right triangle with the unit cell edge and the diagonal of a face.

Use the Pythagorean theorem to determine the length of the diagonal, d, on the face of the cube in terms of e.

$$d^2 = e^2 + e^2 = 2e^2$$

$$d = \sqrt{2}e$$

The diagonal of the cube is the length of four atomic radii and can be calculated by again using the Pythagorean theorem.

$$(\text{diagonal})^2 = (4r)^2 = (\sqrt{2}e)^2 + e^2$$
$$= 16r^2 = 3e^2$$

$$\text{diagonal} = 4r = \sqrt{3}e$$
$$r = \frac{\sqrt{3}}{4}\,e = \frac{\sqrt{3}}{4}\,(3.165\ \text{Å}) = 1.370\ \text{Å}$$

Keystrokes: 3 [√] ÷ 4 x 3.165 = ans. 1.37048

Note that no calculator is needed in this problem until this final line.

41. Silver crystallizes in a face-centered cubic unit cell with a silver atom on each lattice point.

(a) If the edge length of the unit cell is 4.0862 Å, what is the atomic radius of silver?

(b) Calculate the density of silver.

Solution

(a) A face-centered cubic unit cell has the metal atoms in contact across the diagonal of the face.

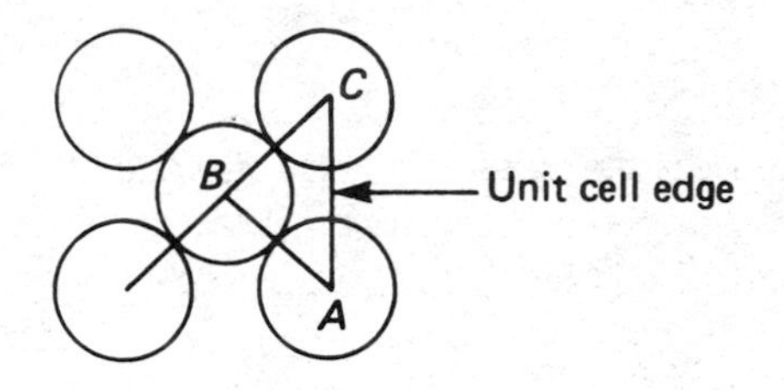

The unit cell edge length is the distance across the face along the edge, 4.0862 Å. The angle ABC is a right angle with $\overline{AB} = \overline{BC} = 2$ (silver radius). Use the Pythagorean theorem to calculate the length of $\overline{AB}$. In a right triangle the square of the hypotenuse equals the sum of squares of the other two sides. In the case of the line in question,

$$(\overline{AB})^2 + (\overline{BC})^2 = (4.0862)^2$$

or

$$2(\overline{AB})^2 = 16.697$$
$$(\overline{AB})^2 = 8.3485$$
$$\overline{AB} = 2.8894$$

$$\text{Radius Ag atom} = \frac{2.8894}{2} = 1.4447\ \text{Å}$$

(b) The density of silver is calculated from its atomic weight and the volume occupied by a mole of silver atoms. The volume of 1 mol of atoms may be found from the unit cell dimensions. A face-centered unit cell contains four atoms and has a volume of

$$V \text{ unit cell} = (4.0862\ \text{Å})^3 = (4.0862 \times 10^{-8}\ \text{cm})^3$$
$$= 6.8227 \times 10^{-23}\ \text{cm}^3$$

$$V \text{ mol (Ag atoms)} = \frac{6.022 \times 10^{23}\ \text{atoms}}{\text{mol}} \times \frac{6.8227 \times 10^{-23}\ \text{cm}^3}{4\ \text{atoms}}$$
$$= 10.272\ \frac{\text{cm}^3}{\text{mol}}$$

$$\text{Density Ag} = \frac{107.868\ \text{g}}{10.272\ \text{cm}^3} = 10.50\ \text{g/cm}^3$$

42. Rutile is a mineral that contains titanium and oxygen. The structure of rutile may be described as a closest-packed array of oxygen atoms with titanium in 1/2 of the octahedral holes. What is the formula of rutile? What is the oxidation number of titanium?

Solution

The ratio of octahedral holes to oxygen anions is 1 to 1 in a closest-packed array. Only 1/2 of the octahedral holes are occupied. This means that the titanium to oxygen ratio is 1 to 2 or that the formula is TiO_2.

The oxidation number of oxygen is -2, so the oxidation number of titanium is +4.

51. Each of the following compounds crystallizes on a structure matching that of NaCl, CsCl, ZnS, or CaF_2. From the radius ratio, predict which structure is formed by each.

(a) ZnTe (b) BaF_2 (c) KBr (d) AlP (e) CaS

Solution

The type of crystal structure formed by a compound depends on several factors, including radius ratio, type of close packing, atomic ratio, and lattice energies. Those structures listed here for comparison are common and have cations in either tetrahedral, octahedral, or cubic holes. To predict the structure of a compound, begin by calculating the radius ratio of the compound and identifying the type of close packing for the compound on the basis of data in Table 11-3. The structure of the compound will most likely be the same as that of the common structure having similar radius and atomic ratios. First calculate the radius ratios of the four basic types of structures. The radius ratio for CaF_2, 0.73, suggests either octahedral or cubic closest packing, but experimental evidence shows the Ca^{2+} ions to be in cubic holes.

	Compound	r^+/r^-	Type of Hole	Structure
	NaCl	0.95/1.81 = 0.52	octahedral	NaCl
	CsCl	1.69/1.81 = 0.93	cubic	CsCl
	ZnS	0.74/1.84 = 0.40	tetrahedral	ZnS
	CaF_2	0.99/1.36 = 0.73	cubic	CaF_2
(a)	ZnTe	0.74/2.21 = 0.33	tetrahedral	ZnS
(b)	BaF_2	1.35/1.36 = 0.99	cubic	CaF_2
(c)	KBr	1.33/1.95 = 0.68	octahedral	NaCl
(d)	AlP	0.50/2.12 = 0.24	tetrahedral	ZnS
(e)	CaS	0.99/1.84 = 0.54	octahedral	NaCl

56. LiH crystallizes with the same crystal structure as NaCl. The edge length of the unit cell of LiH is 4.08 Å. Assuming anion-anion contact, calculate the ionic radius of H^-.

Solution

The structure is face-centered cubic with the anions in contact across the diagonal d of the face as shown in the figure.

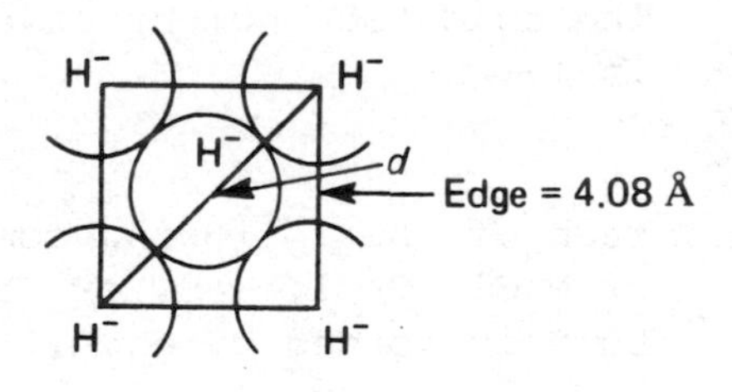

Use the Pythagorean theorem to calculate the length of the diagonal given an edge length of 4.08 Å.

$$d = (4r_{H^-})^2 = (4.08\ \text{Å})^2 + (4.08\ \text{Å})^2$$

$$r_{H^-} = 1.44\ \text{Å}$$

58. The lattice energy of LiF is 1023 kJ/mol, and the interionic distance is 2.008 Å. NaF crystallizes in the same structure as LiF but with an interionic distance of 2.31 Å. Which of the following values most closely approximates the lattice energy of NaF: 510, 890, 1023, 1175, or 4090 kJ/mol? Explain your choice.

Solution

The lattice energy is given by $U = C(Z^+Z^-/R_0)$, where R_0 is the interatomic distance. The charges are the same in both LiF and NaF. The major difference is expected to be the interatomic distance 2.008 Å vs 2.31 Å. From the data for LiF, with $Z^+Z^- = -1$,

$$C = \frac{UR_0}{Z^+Z^-} = \frac{1023 \times 2.008}{-1} = -2054$$

Then $$U_{NaF} = \frac{-2054(-1)}{2.31} = 889 \quad \text{or} \quad 890\ \text{kJ/mol}$$

60. What X-ray wavelength would give a second order reflection ($n = 2$) with a θ angle of 10.40° from planes with a spacing of 4.00 Å?

Solution

The Bragg equation is used to calculate the wavelength of the reflected X rays:

$$n\lambda = 2d \sin \theta$$

given that d is the distance between atomic planes, 4.00 Å, and the angle of reflection, θ, is 10.40°.

$$\lambda = \frac{2}{2}(4.00 \text{ Å}) \sin 10.40° = 4.00 \text{ Å } (0.1805) = 0.722 \text{ Å}$$

Keystrokes: 4.00 x 10.40 [sin] = ans. 0722076

65. How much energy is released when 75 g of steam at 135°C is converted to ice at -40°C?

Solution

The transformation of water from steam (135°C) to ice (-40°C) involves phase changes and temperature changes within the phases as shown in the following sequence:

$$H_2O(g)(135°) \xrightarrow{1} H_2O(g)(100°) \xrightarrow{2} H_2O(\ell)(100°) \xrightarrow{3} H_2O(\ell)(0°)$$

$$\xrightarrow{4} H_2O(s)(0°) \xrightarrow{5} H_2O(s)(-40°)$$

The amount of heat associated with each step is calculated through application of an expression that relates heat to the specific heat of a substance (heat = $MC\Delta T$) or to the heat required for a phase change [heat = (mass)(heat of either fusion or vaporization)].

$$H_2O(g)(135°) \longrightarrow H_2O(g)(100°)$$

$$\text{heat} = (75 \text{ g})\left(\frac{2.00 \text{ J}}{\text{g K}}\right)(35 \text{ K}) = 5250 \text{ J}$$

$$H_2O(g)(100°) \longrightarrow H_2O(\ell)(100°)$$

$$\text{heat} = (75 \text{ g})\left(2258 \frac{\text{J}}{\text{g}}\right) = 169{,}350 \text{ J}$$

$$H_2O(\ell)(100°) \longrightarrow H_2O(\ell)(0°)$$

$$\text{heat} = (75 \text{ g})\left(\frac{4.18 \text{ J}}{\text{g K}}\right)(100 \text{ K}) = 31{,}350 \text{ J}$$

$$H_2O(\ell)(0°) \longrightarrow H_2O(s)(0°)$$

$$\text{heat} = (75 \text{ g})\left(333.6 \frac{\text{J}}{\text{g}}\right) = 25{,}020 \text{ J}$$

$$H_2O(s)(0°) \longrightarrow H_2O(s)(-40°)$$

$$\text{heat} = (75\ g)\left(\frac{2.0\ J}{g\ K}\right)(40\ K) = 6000\ J$$

Total 236,970 J

The heat released to the surroundings during the transformation is the sum of the individual steps. Two significant figures are allowed, so the total heat is 240,000 J or 240 kJ.

It is important to recognize that the temperature of a substance does not change during a phase change; for example, water boiling at 100°C continues to boil at 100°C throughout the evaporation process. But temperature changes do accompany either the addition or the removal of heat from a substance remaining in a particular phase.

67. If 135 g of steam at 100°C and 475 g of ice at 0°C are combined, what is the temperature of the resultant water, assuming that no heat is lost?

Solution

This is a heat balance problem. Assume that all of the steam and all of the ice are converted to water at 100°C and 0°C, respectively. Heat released to bring steam at 100°C to water at 100°C equals

$$135\ g \times 2258\ J/g = 3.0483 \times 10^5\ J$$

Heat absorbed by ice at 0°C to become water at 0°C equals

$$475\ g \times 333.6\ J/g = 1.5846 \times 10^5\ J$$

Amount of heat remaining to heat the 0° water equals

$$3.0483 \times 10^5\ J - 1.5846 \times 10^5\ J = 1.4637 \times 10^5\ J$$

All of the heat causes a temperature increase of the water at 0°C of

$$\frac{1.4637 \times 10^5\ J}{475\ g\ H_2O \times 4.18\ J/(g°C)} = 73.72°C$$

When the two quantities are combined, the temperature is

$$\frac{(100 \times 135\ g) + (73.72 \times 475\ g)}{610\ g} = 79.5°C$$

This problem may also be done letting T_f be the final temperature

$$(135\ g)(2258\ J/g) + (135\ g)(4.18\ J/(g°C)(100 - T)$$
$$= (475\ g)(333.6\ J/g) + (475\ g)(4.18\ J/g°C)\ T_f$$

$$T_f = 79.5°C$$

69. A river is 30 ft wide and has an average depth of 5 ft and a current of 2 mi/h. A power plant dissipates 2.1×10^5 kJ of waste heat into the river every second. What is the temperature difference between the water upstream and downstream from the plant?

Solution

The amount of heat transferred to the river is known, 2.1×10^5 kJ, but the amount of water that is to absorb this heat must be calculated from the data. To calculate the volume of water flowing past the point of discharge, assume that the volume equals that of a rectangular container having dimensions of

$$\text{Width} = 30 \text{ ft} \times \frac{12 \text{ in}}{\text{ft}} \times 2.54 \frac{\text{cm}}{\text{in}} = 914.4 \text{ cm}$$

$$\text{Depth} = 5 \text{ ft} \times \frac{12 \text{ in}}{\text{ft}} \times 2.54 \frac{\text{cm}}{\text{in}} = 152.4 \text{ cm}$$

Length = distance traveled at 2 mi/h in 1 s

$$= 2 \frac{\text{mi}}{\text{h}} \times 5280 \frac{\text{ft}}{\text{mi}} \times 12 \frac{\text{in}}{\text{ft}} \times 2.54 \frac{\text{cm}}{\text{in}} \times \frac{1 \text{ h}}{3600 \text{ s}} = 89.4 \text{ cm}$$

$$V\ H_2O = (914.4 \text{ cm})(152.4 \text{ cm})(89.4 \text{ cm}) = 1.2 \times 10^7 \text{ cm}^3$$

Assume that the density of river water is 1.0 g/cm^3.

$$\text{Mass } H_2O = 1.2 \times 10^7 \text{ g} = 1.2 \times 10^4 \text{ kg}$$

Express the heat released in terms of the mass of water, the specific heat, and the temperature change. From

$$\text{Heat} = MC\Delta T$$

$$\Delta T = \frac{\text{heat}}{MC} = \frac{2.1 \times 10^5 \text{ kJ}}{1.2 \times 10^4 \text{ kg} \times 4.184 \text{ kJ/kg°C}}$$

$$= 4°\text{C} \quad \text{(1 significant figure)}$$

76. When an electron in an excited molybdenum atom falls from the L to the K shell, an X ray is emitted. These X rays are diffracted at an angle of 7.75° by planes with a separation of 2.64 Å. What is the difference in energy between the K and the L shell in molybdenum?

Solution

The energy of the X-ray photon can be calculated from its wavelength by using the equation $E = hc/\lambda$. Use the Bragg equation to calculate the wavelength.

$$\lambda = 2d \sin\theta = 2(2.64 \text{ Å}) \sin 7.75°$$

$$= 2(2.64 \text{ Å})(0.1348) = 0.712 \text{ Å}$$

$$\lambda \text{ in cm} = 0.712 \text{ Å} \times \frac{10^{-8} \text{ cm}}{\text{Å}} = 7.12 \times 10^{-9} \text{ cm}$$

$$E_{L \to K} = \frac{hc}{\lambda} = \frac{(6.627 \times 10^{-34} \text{ J s})(2.998 \times 10^{10} \text{ cm/s})}{7.12 \times 10^{-9} \text{ cm}}$$

$$= 2.79 \times 10^{-15} \text{ J} = 1.74 \times 10^4 \text{ eV}$$

RELATED EXERCISES

1. Calculate the amount of ice at $0^\circ C$ that could be melted by a 50.0-g block of iron that is at a temperature of $500.0^\circ C$, assuming that the final temperature of the iron is $0^\circ C$. The specific heat of iron is 0.448 J/g K.

   Answer: 0.672 g

2. The waste heat from electric power generators could be used for home heating. Let's assume that the heat required for a three-bedroom home, depending on outdoor temperature, is 50,000 Btu/h. The power generator discussed in Exercise 69 must dissipate 2.1×10^5 kJ/s. By using the 50,000 Btu/h as an average, calculate the number of homes that could be heated from the waste heat. (1 Btu = 252 cal)

   Answer: 1.4×10^4 homes

3. Given that the atomic radius of chromium is 1.25 Å and that it exists as a body-centered cubic lattice, calculate the length of its unit cell edge.

   Answer: 2.88 Å

4. Based on the information in Exercise 39, calculate the density of tungsten.

   Answer: 19.26 g/cm^3

5. Refer to the data in Exercise 51 and predict the type of crystal structure for KCl.

UNIT 8

CHAPTER 13: SOLUTIONS; COLLOIDS

INTRODUCTION

The nature of pure substances has been extensively discussed as well as, to some extent, their interactions. The combination of two or more nonreacting substances forms a mixture. Mixtures are very common in our surroundings and play very important roles in life processes. In this chapter a special kind of mixture, called a *solution*, is considered. A solution is a mixture that has two or more components; one component, the solvent, serves to dissolve the other component, the solute. Usually the solvent is considered to be the substance in greater amount. Many solutions are quite familiar. A few of these are listed as specific cases and then generalized:

1. The juice of sorghum cane: solid (sugar) dissolved in liquid (water)
2. Stainless steel: solid (chromium) dissolved in solid (iron)
3. Air: gas (oxygen, carbon dioxide, and other gases) dissolved in gas (nitrogen)
4. Antifreeze solution: liquid (ethylene glycol) dissolved in liquid (water)

Because matter exists in the solid, liquid, and gaseous states, it is possible for substances to form nine different types of solutions, with each state serving as a solute or as a solvent.

Solutions are homogeneous mixtures; that is, the solute is distributed evenly throughout the solvent. Some substances mix with solvents and form suspensions but do not form solutions; examples include vinegar and oil as in salad dressings, butterfat in water as in homogenized milk, and dust particles in air. Whether a solute will form a solution with a solvent depends primarily on three factors. These are (1) the size of solute particles; (2) the capacity of the solvent to form strong bonds with the solute, thereby lowering the energy of the system; and (3) the tendency of the solute particles to form a more random distribution within the solvent, further lowering the energy of the system.

Several types of interactions among solute particles and solvent particles are possible in solution. For example, solvents composed of dipolar molecules dissolve ionic substances and other dipolar substances. Thus water, a dipolar substance, dissolves ionic compounds such as NaCl and KI as well as dipolar substances such as HCl and HNO_3. Conversely, solvents composed of nonpolar molecules dissolve nonpolar substances in which the intermolecular bonding in both solvent and solute is primarily of the van der Waals type. Examples of nonpolar solvent-nonpolar solute solutions include oil in gasoline and grease in dry-cleaning fluid. In general, solvents tend to dissolve solutes that share bonding types. This fact gives rise to an often quoted phrase, "like dissolves like."

The nature of solutions and their special properties, the way they are prepared, methods for describing solutions in terms of composition, and ways for separating solutions into their component parts are all treated in this chapter. Chapter 3 treated the use of solutions in reactions and, specifically, molarity as a unit of concentration. The problems in this chapter include a review of molarity and the use of other common units of concentration to show their utility in chemical processes.

FORMULAS AND DEFINITIONS

Solute and solvent A solution consists of two parts, a solute and the solvent it is dissolved in. The solute generally is the substance in smallest concentration in a solution. The solvent is the substance in the greatest amount in the solution. The solute may be solid, liquid, or gas; the solvent may be solid, liquid, or gas. Although a solid dissolved in a liquid usually comes to mind when one thinks of a solution, there are other possibilities, as discussed in the text.

Solution A solution is a special kind of mixture in which all of the particles are of ionic or molecular dimensions not exceeding about 1.0 pm (10 Å). In addition, the solute particles are uniformly distributed throughout the solvent. This mixture is referred to as a *homogeneous* mixture and is a true solution. Solutions are distinguished from other types of mixtures primarily by properties arising from the small size of solute particles in a solution. The particles may be molecular, ionic, or atomic; they may form small clusters in solution, but usually will not exceed diameters of more than a few angstroms or up to approximately 1 pm.

Electrolytes Substances known as electrolytes dissolve in solvents to produce solutions that conduct electric current. Acids, bases, and salts that form ions in solutions are electrolytes. Other substances are considered electrolytes when, in either a molten or supercooled state, they conduct electricity. The conduction of electric current in aqueous solutions depends on the existence of ions; in molten or supercooled systems it depends on "free" electrons. Strong electrolytes are substances that form solutions that are good electrical conductors. Strong electrolytes are extensively ionized in solution. Some, like sodium chloride dissolved in water, are nearly 100% ionized:

$$\mathrm{NaCl}(s) + \mathrm{H_2O}(\ell) \xrightarrow{\sim 100\%} \mathrm{Na^+}(aq) + \mathrm{Cl^-}(aq)$$

Weak electrolytes are substances that form solutions that are poor electrical conductors. Only a small fraction of a weak electrolyte's particles are ionized in solution. Acetic acid in water is such a case:

$$\mathrm{CH_3COOH}(\ell) + \mathrm{H_2O}(\ell) \xrightarrow{\text{1 to 5\%}} \mathrm{CH_3COO^-}(aq) + \mathrm{H^+}(aq)$$

Percent composition Concentration is often expressed in terms of a percentage because percentage depends only on the easily measured quantities mass and/or volume of the components of the solution. Percent composition is commonly expressed in two ways.

1. The biological, medical, and allied health professions mostly use percentage as a ratio of mass or volume of solute to volume of solution. In the first instance, if the mass of the solute is used, we have

   Mass-volume percent

$$\% = \frac{\text{mass of solute (g)}}{\text{100 mL solution}} \times 100\%$$

   Example: A 5% solution of sodium chloride contains 5 g of NaCl in 100 mL of solution, or, 1000 mL contains 50 g of NaCl.

   Volume-volume percent If the volume of the solute is to be compared to the volume of the solution, we have

$$\% = \frac{\text{volume solute (mL)}}{\text{100 mL solution}} \times 100\%$$

   Example: A 5% ethanol solution contains 5 mL of ethanol in 100 mL of solution; or, 1000 mL of this solution contains 50 mL of ethanol.

   These percentage units are not readily adaptable to other units normally used in chemistry. They are, therefore, not commonly used in chemical calculations and are not used in the textbook.

2. In some areas of chemistry, percentage is expressed as a ratio of the mass of solute to the mass of the solvent. Sometimes referred to as weight-weight percentage, the percent by mass unit can be related to other concentration units including mole fraction and molality.

$$\% = \frac{\text{mass solute}}{\text{mass solution}} \times 100\%$$

   Example: A 5% solution of sodium chloride in water contains 5 g of NaCl in 95 g H_2O. The ratio of NaCl to solution is 5 g : 100 g.

Molarity, M A unit that expresses concentration as a ratio of moles of solute to the volume of solution expressed in liters.

$$\text{Molarity} = \frac{\text{amount of solute (mol)}}{\text{volume of solution (L)}}$$

Molality, m A unit that expresses the concentration of the solute as a ratio of moles of solute to the mass of solvent expressed in kilograms.

$$\text{Molality} = \frac{\text{amount of solute (mol)}}{\text{mass of solvent (kg)}}$$

At first glance molarity and molality appear to be quite similar. A major difference in the two units is that molarity is defined in terms of the volume of solution while molality is defined in terms of the mass of solvent. The volume of a fixed quantity of a solution is temperature dependent, but the mass of solution is not. Molarity, therefore, varies with temperature, but molality does not.

Mole fraction A unit that expresses the concentration of a component in a solution as a ratio of the number of its moles to the total number of moles of all components in that solution. The mole fraction of component A, X_A, in a solution containing components A through D, is

$$X_A = \frac{\text{amount of A (mol)}}{\text{sum of amount (mol) of each component}} = \frac{n_A}{n_A + n_B + n_C + n_D}$$

Most solutions used in laboratories are composed of two or more components, with the mole fractions of their components calculated from the masses of the components used. Mole fractions can also be calculated from molalities and mass percentage.

Raoult's Law The vapor pressure of a solution consisting of a nonvolatile solute in a volatile solvent is always lower than that of its pure solvent. Raoult's Law states that the vapor pressure of the solvent, P_{solvent}, in a dilute solution equals the mole fraction of the solvent, X_{solvent}, times the vapor pressure of the pure solvent, P°_{solvent}.

$$P_{\text{solvent}} = X_{\text{solvent}} \, P^\circ_{\text{solvent}}$$

The implications of this expression are very great. When applied to the phenomenon of boiling, the boiling point of a solution is found to be always higher than that of its solvent. Furthermore, it is found that the elevation of the boiling point is proportional to the mole fraction of solute particles in solution.

Boiling-point elevation-freezing-point depression The addition of a nonvolatile solute to a solvent to form a solution affects the boiling point and also the freezing point of the solvent. These values are changed by amounts proportional to the number of solute particles dissolved in solution. This proportionality is usually expressed in terms of the solution molality and the boiling-point elevation constant, K_b, or the freezing-point depression constant, K_f. Both K_b and K_f have specific values for each solvent but are independent of the particular solute. The boiling-point elevation of a solvent is given by

$$\Delta\text{bp} = K_b m$$

and the freezing-point depression of a solvent is given by

$$\Delta\text{fp} = K_f m$$

For water, K_b and K_f are 0.512°C per molal and 1.86°C per molal, respectively. An aqueous solution containing a nonvolatile, nonelectrolyte solute at a one-molal concentration (a solute-to-water ratio of 1 mol of solute to 1 kg of water) would have its boiling point increased by

$$\Delta\text{bp} = \left(\frac{0.512^\circ\text{C}}{m}\right)(1\ m) = 0.512^\circ\text{C}$$

The new boiling point of the solution would be

$$\text{bp} = 100.000^\circ\text{C} + 0.512^\circ\text{C} = 100.512^\circ\text{C}$$

This solution would have its freezing point lowered by

$$\Delta fp = \left(\frac{1.86°C}{m}\right)(1\ m) = 1.86°C$$

and its final freezing point would then be

$$fp = 0.00°C - 1.86°C = -1.86°C$$

Molecular weight from boiling-point or freezing-point data Because the change in boiling point or freezing point is a function of the molality of the solution, the molecular weight of the solute can be determined by observing its effect on the boiling point or the freezing point of a solvent. The calculation using the freezing-point depression is as follows:

$$\Delta fp = K_f m = K_f\left(\frac{\text{amount of solute (mol)}}{\text{mass of solvent (kg)}}\right)$$

Since

$$\text{moles solute} = \frac{\text{mass solute}}{\text{mol. wt}}$$

then

$$\Delta fp = K_f\left(\frac{\frac{\text{mass solute}}{\text{mol. wt}}}{\text{mass solvent (kg)}}\right)$$

Solving for mol. wt gives

$$\text{Mol. wt} = \frac{(K_f)(\text{mass solute})}{(\Delta fp)(\text{mass solvent in kg})}$$

A similar calculation using the change in boiling point leads to

$$\text{Mol. wt} = \frac{(K_b)(\text{mass solute})}{(\Delta bp)(\text{mass solvent in kg})}$$

Work through these derivations until you are comfortable with the equations rather than memorizing the formulas; the factors are easily misplaced in the formula.

Ion activities If the solute in a solution is an electrolyte, the changes in boiling point and freezing point depend on the number of species in the solution. For solutions that contain strong electrolytes, the boiling points and freezing points are slightly different from the values calculated using the number of ions in the formula unit. These discrepancies mainly occur because of solute particle-particle interactions in solution. The actual change in boiling point or freezing point is a measure of the effective molality of the solution. This effective molality is called the *activity* of the ions. For example, the change in freezing point is given by

$$\Delta fp(\text{actual}) = K_f m_{\text{effective}} = K_f(\text{activity})$$

Simple rearranging of this equation gives

$$\text{Activity} = \frac{\Delta fp(\text{actual})}{K_f}$$

The activities of ions in solution are related to their actual concentrations through an activity coefficient f:

$$\text{Activity} = f \times \text{molality}$$

Although in practice the situation is much more complicated than presented here, generally in solutions containing very weak electrolytes or nonelectrolytes, the activity coefficient is approximately 1.

EXERCISES

6. At 0°C, 3.36 g of CO_2 at 1.00 atm will dissolve in exactly 1 L of water. How many grams of CO_2 will dissolve in exactly 1 L of water at 0°C and 4.00 atm?

Solution

This is an application of Henry's Law. The pressure is increased by a factor of 4; and because the amount dissolved is directly proportional to the pressure, the amount dissolved is 4×3.36 g = 13.4 g.

23. There are about 10 g of calcium, as Ca^{2+}, in 1.0 L of milk. What is the molarity of Ca^{2+} in milk?

Solution

$$M = \frac{\text{amount of solute}}{\text{volume of solution}} = \frac{\text{moles of solute}}{\text{liters of solution}}$$

The amount of Ca^{2+} is

$$10 \text{ g } Ca^{2+} \times \frac{1 \text{ mol } Ca^{2+}}{40.08 \text{ g } Ca^{2+}} = 0.25 \text{ mol } Ca^{2+}$$

This is substituted into the formula for molarity or the entire calculation can be carried out in one step.

$$M = \frac{\text{mol } Ca^{2+}}{\text{liters of solution}} = \frac{\frac{10 \text{ g}}{40.08 \text{ g/mol}}}{1 \text{ L}} = 0.25\ M$$

24. Calculate the number of moles and the mass of the solute in each of the following solutions:

(a) 2.00 L of 18.5 M H_2SO_4, concentrated sulfuric acid

Solution

$$\text{Molarity} = \frac{\text{moles of solute}}{\text{liters of solution}}; \qquad 18.5\ M = \frac{\text{mol}}{2.00 \text{ L}}$$

$$\text{Moles of solute} = 18.5\ M \times 2.00 \text{ L} = 37.0 \text{ mol}$$

The mass is found from the relation

$$\text{Mass} = \text{amount of substance} \times \text{molar mass}$$

$$\text{Mass } H_2SO_4 = 37.0 \text{ mol} \times 98.07 \text{ g/mol} = 3628.6 \text{ g} = 3.63 \text{ kg}$$

(c) 500. mL of 0.30 *M* glucose, $C_6H_{12}O_6$, used for intravenous injection

Solution

$$M = \frac{\text{moles of solute}}{\text{liters of solution}}$$

$$\text{moles of solute} = 0.30\ M \times 0.500 \text{ L} = 0.15 \text{ mol}$$

The mass is found from the relation

$$\text{Mass} = \text{mol} \times \text{molar mass} = 0.15 \text{ mol} \times 180.157 \text{ g/mol} = 27 \text{ g}$$

25. Calculate the molarity of each of the following solutions:

(b) 4.25 g of NH_3 in 0.500 L of solution, the concentration of NH_3 in household ammonia

Solution

From the definition

$$M = \frac{\text{moles of solute}}{\text{liters of solution}} = \frac{\text{g/molar mass}}{\text{L}} = \frac{\frac{4.25 \text{ g}}{17.0304 \text{ g/mol}}}{0.500 \text{ L}} = 0.499\ M$$

(d) 595 g of isopropanol, C_3H_8OH, in 1.00 L of solution, the concentration of isopropanol in rubbing alcohol

Solution

As in (b)

$$M = \frac{\text{g/molar mass}}{\text{L}} = \frac{\frac{595 \text{ g}}{61.1035 \text{ g/mol}}}{1.00 \text{ L}} = 9.74\ M$$

26. Calculate the molality of each of the following solutions:

(a) 71.0 g of sodium carbonate (washing soda), Na_2CO_3, in 1.00 kg of water, a saturated solution at 0°C.

Solution

$$m = \frac{\text{moles of solute}}{\text{kilograms of solvent}} = \frac{\text{g/molar mass}}{\text{kilograms}} = \frac{\frac{71.0 \text{ g}}{105.989 \text{ g/mol}}}{1.00 \text{ kg}} = 0.670\ m$$

(b) 583 g of H_2SO_4 in 1.50 kg of water, the acid solution used in an automobile battery

Solution

$$m = \frac{\text{moles of solute}}{\text{kilograms of solvent}} = \frac{\text{g/molar mass}}{\text{kilograms}} = \frac{\frac{583\text{ g}}{98.07\text{ g/mol}}}{1.50\text{ kg}} = 3.96\ m$$

27. Calculate the mole fractions of solute and solvent in each of the solutions in Exercise 26.

Solution

(a) The mole fraction of sodium carbonate is found by calculating the amount of each substance in moles and substituting into the formula:

$$X_{Na_2CO_3} = \frac{n_{Na_2CO_3}}{n_{Na_2CO_3} + n_{H_2O}}$$

From Exercise 26

$$n_{Na_2CO_3} = g_{Na_2CO_3}/\text{molar mass} = \frac{71.0\text{ g}}{105.989\text{ g/mol}} = 0.670\text{ mol}$$

$$n_{H_2O} = \frac{1000\text{ g}}{18.0152\text{ g/mol}} = 55.51\text{ mol}$$

Substitution gives

$$X_{Na_2CO_3} = \frac{0.670\text{ mol}}{0.670\text{ mol} + 55.51\text{ mol}} = 0.0119$$

$$X_{H_2O} = 1.0000 - n_{Na_2CO_3} = 1.0000 - 0.0119 = 0.9881$$

Solution

(b)

$$n_{H_2SO_4} = g_{H_2SO_4}/\text{molar mass} = \frac{583\text{ g}}{98.07\text{ g/mol}} = 5.945$$

$$n_{H_2O} = \frac{1500\text{ g}}{18.0152\text{ g/mol}} = 83.26\text{ mol}$$

$$X_{H_2SO_4} = \frac{n_{H_2SO_4}}{n_{H_2SO_4} + n_{H_2O}} = \frac{5.945\text{ mol}}{5.945\text{ mol} + 83.26\text{ mol}} = 0.0666$$

$$X_{H_2O} = 1.0000 - X_{H_2SO_4} = 1.0000 - 0.0666 = 0.9334$$

28. What mass of nitric acid (68.0% HNO_3 by mass) is needed to prepare 400.0 g of a 10.0% solution of HNO_3 by mass?

Solution

The mass of pure nitric acid required for the solution is

$$\text{Mass } HNO_3 = \frac{10.0\%}{100.0\%} \times 400.0 \text{ g} = 40.0 \text{ g}$$

The concentrated solution contains 68.0% HNO_3 by mass; the mass of 68.0% HNO_3 required is

$$68.0\% \text{ (mass } HNO_3) = 0.680 \text{ (mass } HNO_3) = 40.0 \text{ g}$$

$$\text{Mass } HNO_3 = \frac{40.0 \text{ g}}{0.680} = 58.8 \text{ g}$$

30. What mass of a 4.00% NaOH solution by mass contains 15.0 g of NaOH?

Solution

The unknown mass of solution, when multiplied by 4.00% (0.0400), must be equal to 15.0 g. Therefore,

$$\text{Mass of solution} = \frac{15.0 \text{ g}}{0.0400} = 375 \text{ g}$$

31. What mass of HCl is contained in 45.0 mL of an HCl solution with a specific gravity of 1.19 and containing 37.21% HCl by mass?

Solution

All quantities must be converted to the same units. Mass is required, so convert mL to its mass equivalent. The specific gravity is numerically the same as the density, therefore,

$$45.0 \text{ mL} \times 1.19 \text{ g/mL} = 53.6 \text{ g}$$

$$53.6 \text{ g} \times \frac{37.21\%}{100.0\%} = 19.9 \text{ g}$$

33. The hardness of water (hardness count) is usually expressed as parts per million (by mass) of $CaCO_3$, which is equivalent to milligrams of $CaCO_3$ per liter of water. What is the molar concentration of Ca^{2+} ions in a water sample with a hardness count of 175?

Solution

The amount of $CaCO_3$ in this solution is 175 mg/L of water. The molarity is

$$\text{GFW } CaCO_3 = 100.09 \text{ g}$$

$$\text{Molarity } CaCO_3 = \frac{175 \text{ mg} \times \frac{1 \text{ g}}{1000 \text{ mg}} \times \frac{1 \text{ mol}}{100.09 \text{ g}}}{1.00 \text{ L}} = 1.75 \times 10^{-3} \ M$$

36. What volume of a 0.20 *M* K_2SO_4 solution contains 57 g of K_2SO_4?

Solution

Find the number of liters of K_2SO_4 per gram and then multiply by the number of grams desired to find the needed volume. In one equation

$$\frac{1.00 \text{ L}}{0.20 \text{ mol } K_2SO_4} \times \frac{1 \text{ mol}}{174.25 \text{ g}} \times 57 \text{ g } K_2SO_4 = 1.6 \text{ L}$$

38. Calculate what volume of a sulfuric acid solution (specific gravity = 1.070, and containing 10.00% H_2SO_4 by mass) contains 18.50 g of pure H_2SO_4 at a temperature of 25°C.

Solution

Calculate the mass of H_2SO_4 per milliliter of solution.

$$\text{Mass } H_2SO_4 = \frac{10.00\%}{100\%} \times 1.070 \text{ g/mL} = 0.1070 \text{ g/mL}$$

The volume of solution that contains 18.50 g of H_2SO_4 is

$$V \text{ } (H_2SO_4 \text{ solution}) = 18.50 \text{ g} \times \frac{1 \text{ mL}}{0.1070 \text{ g}} = 172.9 \text{ mL}$$

41. What volume of 0.600 *M* HCl is required to react completely with 2.50 g of sodium hydrogen carbonate?

$$NaHCO_3 + HCl \longrightarrow NaCl + CO_2 + H_2O$$

Solution

Because $NaHCO_3$ reacts on a 1 : 1 basis with HCl, the number of moles of $NaHCO_3$ in 2.50 g is the same as the number of moles of HCl in the 0.600 *M* HCl solution required to react with it. First find the number of moles required and then multiply by the volume, 1 L, per 0.600 mol HCl. In one expression

$$V \text{ (HCl)} = 2.50 \text{ g } NaHCO_3 \times \frac{1 \text{ mol } NaHCO_3}{84.01 \text{ g } NaHCO_3} \times \frac{1 \text{ mol HCl}}{1 \text{ mol } NaHCO_3} \times \frac{1 \text{ L}}{0.600 \text{ mol HCl}}$$

$$= 0.0496 \text{ L} = 49.6 \text{ mL}$$

42. Calculate the volume of 0.050 *M* HCl necessary to precipitate the silver contained in 12.0 mL of 0.050 *M* $AgNO_3$.

$$Ag^+ + Cl^- \longrightarrow AgCl(s)$$

Solution

Silver ion and chloride ion react on a 1 : 1 basis. Therefore, 12.0 mL of 0.050 *M* $AgNO_3$ solution requires 0.050 *M* HCl solution. More formally, find the amount of Ag^+ present, convert to moles of Cl^-, then find the volume that contains that amount of Cl^-.

$$V\ (\text{HCl}) = 12.0\ \text{mL} \times \frac{0.050\ \text{mol Ag}^+}{1\ \text{L}} \times \frac{1\ \text{mol Cl}^-}{1\ \text{mol Ag}^+} \times \frac{1\ \text{L}}{0.050\ \text{mol Cl}^-} = 12.0\ \text{mL}$$

45. A gaseous solution was found to contain 15% H_2, 10.0% CO, and 75% CO_2 by mass. What is the mole fraction of each component?

Solution

The mole fraction of each component is calculated from the definition of mole fraction as

$$X_{H_2} = \frac{n_{H_2}}{n_{\text{total}}}\ ; \qquad X_{CO} = \frac{n_{CO}}{n_{\text{total}}}\ ; \qquad X_{CO_2} = \frac{n_{CO_2}}{n_{\text{total}}}$$

For convenience, assume a sample mass of exactly 1 g, calculate the mass of each component in the sample, then the number of moles of each.

$$\text{Mass } H_2 = 0.15 \times 1.00\ \text{g} = 0.15\ \text{g}$$

$$\text{Mass CO} = (0.100)(1.00\ \text{g}) = 0.100\ \text{g}$$

$$\text{Mass } CO_2 = (0.75)(1.00\ \text{g}) = 0.75\ \text{g}$$

$$n_{H_2} = 0.15\ \text{g} \times \frac{1\ \text{mol}}{2.016} = 0.0744\ \text{mol}$$

$$n_{CO} = 0.100\ \text{g} \times \frac{1\ \text{mol}}{28.010\ \text{g}} = 0.00357\ \text{mol}$$

$$n_{CO_2} = 0.75\ \text{g} \times \frac{1\ \text{mol}}{44.010\ \text{g}} = 0.0170\ \text{mol}$$

$$n_{\text{total}} = 0.0950\ \text{mol}$$

$$X_{H_2} = \frac{0.0744\ \text{mol}}{0.0950\ \text{mol}} = 0.78$$

$$X_{CO} = \frac{0.00357\ \text{mol}}{0.0950\ \text{mol}} = 0.038$$

$$X_{CO_2} = \frac{0.017\ \text{mol}}{0.0950\ \text{mol}} = 0.18$$

48. Concentrated hydrochloric acid is 37.0% HCl by mass and has a specific gravity of 1.19. Calculate (a) the molarity, (b) the molality, and (c) the mole fraction of HCl and of H_2O.

Solution

(a) Find the mass from the specific gravity and percentage composition and convert to moles. From the amount of HCl in 1 L, the molarity may be calculated.

$$\text{Amount of HCl} = 1.19\ \frac{\text{g}}{\text{mL}} \times \frac{0.37}{1.00} \times \frac{1\ \text{mol HCl}}{36.461\ \text{g}} = 0.0121\ \frac{\text{mol}}{\text{mL}}$$

$$M = \frac{0.0121\ \text{mol/mL}}{1\ \text{L}/1000\ \text{mL}} = 12.1\ M$$

(b) Because molality depends on the number of moles per kilogram of solvent, find the mass of HCl in a convenient mass of solution, say 1.19 g, the mass of 1 mL. From the mass of HCl the number of moles present can be found. The mass of water in the 1 g of solution is converted to kilograms for substitution into the formula for molality.

$$\text{Mass (HCl)} = 1.19\ \text{g} \times \frac{0.37}{1.00} = 0.440\ \text{g}$$

$$\text{Mol (HCl)} = 0.440\ \text{g} \times \frac{1\ \text{mol HCl}}{36.461\ \text{g HCl}} = 0.0121\ \text{mol}$$

$$\text{Mass } (H_2O) = 1.19\ \text{g} - 0.440\ \text{g} = 0.750\ \text{g} = 0.000750\ \text{kg}$$

$$m\ \text{(HCl)} = \frac{0.0121\ \text{mol}}{0.000750\ \text{kg}} = 16.1\ m$$

(c) Because the moles of HCl have already been calculated in (b) and the mass of water is already known, determine the amount of H_2O (mol) present.

$$\text{Mol } (H_2O) = 0.750\ \text{g } H_2O \times \frac{1\ \text{mol}}{18.02\ \text{g } H_2O} = 0.0416\ \text{mol}$$

From the definition of mole fraction,

$$X_{HCl} = \frac{\text{mol (HCl)}}{\text{mol (HCl)} + \text{mol } (H_2O)} = \frac{0.0121}{0.0121 + 0.0416} = 0.225$$

Similarly, for H_2O

$$X_{H_2O} = \frac{0.0416}{0.0121 + 0.0416} = 0.775$$

60. A solution of 5.00 g of an organic compound in 25.00 g of carbon tetrachloride (bp 76.8°C; K_b = 5.02°C/m) boils at 81.5°C at 1 atm. What is the molecular weight of the compound?

Solution

The boiling-point elevation of a solvent is proportional to the number of solute particles dissolved in the solvent. Specifically for CCl_4, 1 mol of solute dissolved in 1.00 kg of CCl_4 changes the boiling point by 5.02°C/m, the value of K_b. The change in boiling point is expressed as

$$\Delta bp = K_b \times \text{molality} = K_b\left(\frac{\text{mol solute}}{\text{kg solvent}}\right)$$

$$\text{Mol (solute)} = \frac{\text{mass (solute)}}{\text{molar mass}}$$

and
$$\Delta bp = K_b \frac{[\text{mass (solute)/mol. wt}]}{\text{kg (solvent)}}$$

Solving this equation for mol. wt yields

$$\text{Mol. wt} = \frac{K_b \times \text{mass (solute)}}{\Delta bp \times \text{kg (solvent)}}$$

In this case,

$$\Delta bp = 81.5° - 76.8° = 4.7°C$$

$$\text{Mass (solute)} = 5.00 \text{ g}; \quad \text{kg (solvent)} = 0.02500 \text{ kg}$$

$$\text{Mol. wt (unknown)} = \frac{(5.02°C/m) \times 5.00 \text{ g}}{4.7°C \times 0.02500 \text{ kg}} = 2.1 \times 10^2 \text{ g/mol}$$

since m has units of mol/kg.

61. A solution contains 5.00 g of urea, $CO(NH_2)_2$, per 0.100 kg of water. If the vapor pressure of pure water at 25°C is 23.7 torr, what is the vapor pressure of the solution?

Solution

The vapor pressure of a solution is proportional to the mole fraction of the substance used as the solvent in the solution. This relationship is expressed as Raoult's Law, where the vapor pressure of the solution, $P_{solution}$,

$$P_{solution} = P_{solvent}X_{solvent}$$

equals the vapor pressure of the pure solvent, $P_{solvent}$, times the mole fraction of the solvent, $X_{solvent}$,

$$X_{H_2O} = \frac{n_{H_2O}}{n_{total}}$$

Molar masses: $CO(NH_2)_2 = 60.054$ g; $H_2O = 18.0152$ g

$$n_{urea} = 5.00 \text{ g} \times \frac{1 \text{ mol}}{60.054 \text{ g}} = 0.08326 \text{ mol}$$

$$n_{H_2O} = 100. \text{ g} \times \frac{1 \text{ mol}}{18.0152 \text{ g}} = 5.551 \text{ mol}$$

$$n_{total} = 5.634 \text{ mol}$$

$$X_{H_2O} = \frac{5.551}{0.08326 + 5.551} = 0.9853$$

$$\text{Vapor pressure of solution} = 23.7 \text{ torr} \times 0.9853 = 23.4 \text{ torr}$$

64. Calculate the boiling-point elevation of 0.100 kg of water containing 0.010 mol of NaCl, 0.020 mol of Na_2SO_4, and 0.030 mol of $MgCl_2$, assuming complete dissociation of these electrolytes.

Solution

The boiling point of any solution is dependent on the number of moles of solute particles in solution. In this example all solutes are ionic and are assumed to be completely dissociated in solution. The boiling-point elevation, however, will actually be less than that calculated for 100% dissociation due to ion-ion interactions in solution. Dissociation near 100% occurs only in extremely dilute solutions, < 0.001 molal. The total number of moles of particles in this solution is

$$\begin{array}{lll} 0.010 \text{ mol NaCl} & \longrightarrow 0.010 \text{ mol } Na^+ & + 0.010 \text{ mol } Cl^- \\ 0.020 \text{ mol } Na_2SO_4 & \longrightarrow 0.040 \text{ mol } Na^+ & + 0.020 \text{ mol } SO_4^{2-} \\ 0.030 \text{ mol } MgCl_2 & \longrightarrow 0.030 \text{ mol } Mg^{2+} & + 0.060 \text{ mol } Cl^- \\ \hline \text{Total} = & 0.080 \text{ mol} & + 0.090 \text{ mol} = 0.17 \text{ mol} \end{array}$$

$$\Delta bp = K_b(m) = \frac{0.512°C}{m} \times \frac{0.17 \text{ mol}}{0.100 \text{ kg}} = 0.87°C$$

66. How could you prepare a 3.08 *m* aqueous solution of glycerin, $C_3H_8O_3$? What is the freezing point of this solution?

Solution

A quantity of solution is not stated here, so it is convenient to assume that 1.0 kg of the solvent water is to be used. The amount of glycerin required is found as follows: a 3.08-*m* solution contains 3.08 mol of solute per 1.0 kg of water.

$$\text{Molar mass } (C_3H_8O_3) = 92.094 \text{ g mol}^{-1}$$

$$\text{Molality } (C_3H_8O_3) = 3.08 \text{ mol} \times 92.094 \frac{g}{mol} = 284 \text{ g}$$

and

$$\Delta fp = K_f(m) = 1.86 \frac{°C}{m} \times 3.08\ m = 5.73°$$

$$0°C - 5.73°C = -5.73°C$$

70. Lysozyme is an enzyme that cleaves cell walls. A 0.100-L sample of a solution of lysozyme that contains 0.0750 g of the enzyme exhibits an osmotic pressure of 1.32×10^{-3} atm at 25°C. What is the molecular weight of lysozyme?

Solution

The molar concentration is calculated using the expression $\pi = MRT$. The moles of lysozyme present are then calculated from which the molecular weight is obtained. Using R to match the units,

$$\pi = MRT; \quad M = \frac{\pi}{RT} = \frac{1.32 \times 10^{-3} \text{ atm}}{(0.08205 \text{ L atm/mol K})(298 \text{ K})}$$

$$M = 5.40 \times 10^{-5}$$

Thus, 0.0750 g of lysozyme in 0.100 L is equivalent to 0.750 g in 1 L or is equal to 5.40×10^{-5} mol of lysozyme.

$$\text{Molecular weight} = \frac{0.750 \text{ g}}{5.40 \times 10^{-5} \text{ mol}} = 13{,}900$$

87. A 1.80-g sample of an acid, H_2X, required 14.00 mL of KOH solution for neutralization of all the hydrogen ion. Exactly 14.2 mL of this same KOH solution was found to neutralize 10.0 mL of 0.750 *M* H_2SO_4. Calculate the molecular weight of H_2X.

Solution

The concentration of the KOH solution must be determined before the reaction with H_2X can be considered. In essence, the reaction of KOH with H_2SO_4 is a step to standardize the base before it is used in the determination with H_2X. The reaction with H_2SO_4 is

$$2KOH + H_2SO_4 \longrightarrow K_2SO_4 + 2H_2O$$
$$2 \text{ mol} + 1 \text{ mol}$$

$$\text{Mol KOH reacted} = 2(\text{mol } H_2SO_4 \text{ reacted})$$

$$\text{Mol } H_2SO_4 = (0.750\ M)(0.0100 \text{ L}) = 0.00750 \text{ mol}$$

$$\text{Mol KOH} = 2(0.00750 \text{ mol}) = (M)(L)$$

$$\text{Molarity KOH} = \frac{0.0150 \text{ mol}}{0.0142 \text{ L}} = 1.056\ M$$

The reaction of KOH with H_2X is

$$2KOH + H_2X \longrightarrow K_2X + 2H_2O$$

$$2 \text{ mol} + 1 \text{ mol}$$

$$\text{Mol } H_2X \text{ in sample} = \frac{1}{2}(\text{mol KOH reacted})$$

$$\text{Mol } H_2X = \frac{1}{2}(M)(L) = \frac{1}{2}(1.056\ M)(0.01400 \text{ L}) = 0.00739 \text{ mol}$$

Recognize that this amount of H_2X is contained in the sample, 1.80 g. Therefore,

$$0.00739 \text{ mol} = 1.80 \text{ g}$$

$$1 \text{ mol} = \frac{1.80 \text{ g}}{0.00739} = 243 \text{ g}$$

91. What is the molarity of H_3PO_4 in a solution that is prepared by dissolving 10.0 g of P_4O_{10} in sufficient water to make 0.500 L of solution?

Solution

The reaction is

$$P_4O_{10} + 6H_2O \longrightarrow 4H_3PO_4$$

The amount of H_3PO_4 is

$$10.0 \text{ g } P_4O_{10} \times \frac{1 \text{ mol } P_4O_{10}}{283.8350 \text{ g } P_4O_{10}} \times \frac{4 \text{ mol } H_3PO_4}{1 \text{ mol } P_4O_{10}} = 0.141 \text{ mol}$$

Then $$\text{Molarity} = \frac{0.141 \text{ mol}}{0.500 \text{ L}} = 0.282\ M$$

94. The sulfate in 50.0 mL of dilute sulfuric acid was precipitated using an excess of barium chloride. The mass of $BaSO_4$ formed was 0.482 g. Calculate the molarity of the sulfuric acid solution.

Solution

The equation for this reaction is

$$\begin{array}{ccc} \text{No. moles?} & & 0.482 \text{ g} \\ H_2SO_4(aq) + BaCl_2(aq) & \longrightarrow & BaSO_4(s) + HCl(aq) \\ 1 \text{ mol} & & 1 \text{ mol} \end{array}$$

Calculate the number of moles of H_2SO_4 required to produce 0.482 g of $BaSO_4$. This amount of H_2SO_4 is contained in 50.0 mL of solution, hence, it can be equated to molarity through the expression

$$\text{No. moles} = \text{molarity} \times \text{volume in L}$$

The number of moles of H_2SO_4 required is

$$\text{Molar mass } BaSO_4 = 137.3 + 32.1 + 4(16.0) = 233.4 \text{ g mol}^{-1}$$

$$\text{Mol } H_2SO_4 = \left(0.482 \text{ g } BaSO_4 \times \frac{1 \text{ mol}}{233.4 \text{ g}}\right)\left(\frac{1 \text{ mol } H_2SO_4}{1 \text{ mol } BaSO_4}\right)$$

$$= 2.065 \times 10^{-3} \text{ mol}$$

Then $$2.065 \times 10^{-3} \text{ mol} = (\text{molarity } H_2SO_4)\left(50.0 \text{ ml} \times \frac{1 \text{ L}}{1000 \text{ mL}}\right)$$

$$\text{Molarity } H_2SO_4 = \frac{2.065 \times 10^{-3} \text{ mol}}{0.050 \text{ L}} = 0.0413\ M$$

95. A sample of $HgCl_2$ weighing 9.41 g is dissolved in 32.75 g of ethanol, C_2H_5OH ($K_b = 1.20°C/m$). The boiling-point elevation of the solution is 1.27°C. Is $HgCl_2$ an electrolyte in ethanol? Show your calculations.

Solution

Calculate the number of moles of $HgCl_2$ present and from that, the molality. Then calculate the boiling-point elevation based on the molality. If the change is the same as found experimentally, it is not dissociated; otherwise, it is.

$$9.41 \text{ g } HgCl_2 \times \frac{1 \text{ mol } HgCl_2}{271.50 \text{ g } HgCl_2} = 0.03466 \text{ mol}$$

$$\frac{0.03466 \text{ mol}}{0.03275 \text{ kg}} = 1.058\ m; \quad 1.058\ m \times 1.20°C/m = 1.27°C$$

Because the change actually found is 1.27°C, $HgCl_2$ is not an electrolyte.

97. A solution of 0.045 g of an unknown organic compound in 0.550 g of camphor melts at 158.4°C. The melting point of pure camphor is 178.4°C. K_f for camphor is 37.7°C/m. The solute contains 93.46% C and 6.54% H by mass. What is the molecular formula of the solute? Show your calculations.

Solution

The molecular formula is determined from the freezing point and the elemental analysis data. The molecular weight of the unknown is calculated from the change in freezing point as

$$\text{Mol. wt} = \frac{K_f \times \text{mass solute}}{\text{No. kg camphor} \times \Delta fp} = \frac{37.7 \frac{°C}{m} \times 0.045 \text{ g}}{0.000550 \text{ kg} \times 20.0°C} = 154 \frac{\text{g}}{\text{mol}}$$

Next, determine the empirical formula of the unknown.

$$\text{C:} \quad 93.46 \text{ g} \times \frac{1 \text{ mol}}{12.011 \text{ g}} = 7.78 \text{ mol}$$

$$\text{H:} \quad 6.54 \text{ g} \times \frac{1 \text{ mol}}{1.0079 \text{ g}} = 6.49 \text{ mol}$$

$$C_{7.78}H_{6.49} = C_{1.2}H_1 \quad \text{or} \quad C_{12}H_{10}$$

The empirical formula has a formula weight of 154, which equals the molecular weight calculated from the freezing-point data and is the molecular formula of the unknown.

RELATED EXERCISES

1. Which one of the following methanol-water solutions will have the lowest freezing point? What is the freezing point of this solution?

   (a) 1.00 mol of CH_3OH in 500.0 g of water

   (b) 0.50 mol of CH_3OH in 500.0 g of water

   (c) 0.25 mol of CH_3OH in 125.0 g of water

   (d) 0.75 mol of CH_3OH in 125.0 of water

Answer: (d); fp = -11°C

2. Calculate the volume of 0.200 *M* H_2SO_4 required to completely react with 10.0 g of zinc according to the following reaction.

$$H_2SO_4(aq) + Zn(s) \longrightarrow ZnSO_4(aq) + H_2(g)$$

Answer: 764 mL

3. The assay printed on the manufacturer's label of a bottle of concentrated, reagent-grade hydrochloric acid indicates a purity of 36.7% HCl and a specific gravity of 1.186. Calculate the molarity of this solution. Calculate the volume of this solution required to prepare 5.00 mL of 0.250 *M* solution.

Answer: 11.9 M; 0.105 mL

4. Diffusion by osmosis can occur when two solutions of different concentrations are separated by a semipermeable membrane that selectively allows water molecules to pass in both directions but prevents the solute molecules or ions from doing so. A net transport of water will occur from the solution of lesser concentration (lower mole fraction) to the solution of greater concentration tending to "level" the concentrations of the two solutions. Given the following two solutions separated by such a membrane, which one will show a net gain in volume: solution A, 4.000 g KCl in 500.0 g H_2O and solution B, 3.000 g NaCl in 350.0 g H_2O?

Answer: B

5. The volume of a given mass of water varies according to the equation

$$V_t = V_0(1 - 6.427 \times 10^{-5}\ t + 8.5053 \times 10^{6}\ t^2 - 6.7900 \times 10^{-8}\ t^3)$$

between temperatures of 0°C and 30°C. Assuming that the molarity of a solution is 1.000 mol/L at 0°C, calculate the molarity of the solution at 4°C, 10°C, 20°C, 25°C, and 30°C. Show the variation in molarity with temperature by plotting *M* vs *t*.

Answer: M(4°) = 1.0001; M(10°) = 0.9999; M(20°) = 0.9984

UNIT 9

CHAPTER 14: ACIDS AND BASES

INTRODUCTION

Scientists have speculated on the nature of acids and bases for some two hundred years. Early attempts to define acids were based on easily discernible properties. Substances that tasted sour, caused vegetable dyes to change color, and reacted with metals to produce hydrogen were called *acids*. *Bases* were defined as those substances that had properties that were the opposite of acids: bases tasted bitter, felt slippery, and caused vegetable dyes to change to different colors than those produced by acids.

The first classification of acids and bases according to their chemical properties was done by Svante Arrhenius. He said that acids caused the increase of hydrogen ion in aqueous solution and bases caused the increase of hydroxide ion in solution. This was a rather limited view, because substances other than hydrogen ion and hydroxide ion that reacted like hydrogen ion or hydroxide ion were not allowed in the classification scheme.

A useful but also somewhat limited definition was proposed independently by Johannes Brönsted and Thomas Lowry. According to this theory acids are defined as proton donors and bases as proton acceptors. Acids are thought to contain ionizable protons that can be transferred to bases. This definition is broader than Arrhenius' definition, because a base can be anything that accepts a hydrogen ion rather than just a hydroxide ion.

Three industrially and physiologically important classes of compounds fit into the Brönsted-Lowry classification scheme:

1. Mineral acids-bases. These compounds include acids such as hydrochloric acid, HCl; sulfuric acid, H_2SO_4; and nitric acid, HNO_3; and bases such as sodium hydroxide, NaOH and sodium carbonate, Na_2CO_3.
2. Organic acids-bases. These compounds include fatty acids that have the group —COOH as the acid function, such as acetic acid, CH_3COOH, and stearic acid, $C_{17}H_{35}COOH$, and bases such as methylamine, $CH_3—NH_2$.
3. Amino acids. These acids contain both acid and base functional groups, giving them the potential to react with either acids or bases. Examples include glycine, $CHNH_2COOH$, and glutamic acid, $HOOCCH_2CH_2CHNH_2COOH$.

G. N. Lewis proposed in 1923 a definition of acids and bases that was much more comprehensive and encompassing. According to his definition, an acid is *any species that can accept a pair of electrons* from a donor, the base. The Lewis classification scheme includes all ionizable protonic substances and all the donors in the Brönsted-Lowry scheme. The Lewis theory has the advantage of being able to explain acid vs base characteristics of compounds not containing protons, including metal and nonmetal oxides. In contrast to the previous schemes, many reactions in organic chemistry also can be explained in terms of the Lewis concept.

A more recent definition of acids and bases involves classifying substances in terms of their capacity to form cations or anions characteristic of the solvent. A cation of the solvent is said to be acidic; an anion of the solvent is said to be basic. This definition is particularly appropriate for work in nonaqueous solvents.

The problems in the exercises mainly deal with acid-base reactions involving protonic substances. Most of the problems illustrate the principle that protons lost from an acid during a reaction are consumed by the base. The concept of equivalent weight is introduced and used in several examples.

FORMULAS AND DEFINITIONS

Gram-equivalent weight (GEW) The gram-equivalent weight of an acid or base is the mass of a substance that donates or accepts 1 mol (6.02×10^{23}) of electrons, e^-, or protons, H^+, or neutralizes 1 mol of negative or positive charges. The equivalent weight, or one equivalent (one equiv), of a substance used in a protonic acid-base reaction is the molar mass (molecular weight) of the substance divided by the number of protons donated or consumed per molecule of acid or base used, respectively:

$$\text{GEW} = \frac{\text{molar mass of acid or base}}{\text{no. of } H^+ \text{ donated or consumed}}$$

The GEW of a substance can be deduced only by examination of the chemical equation representing a specific reaction — not merely from the formula of the acid or base. The need to know the balanced chemical equation for the reaction is illustrated in the variability of the GEW's of the acids and bases in the following examples:

1. $$HCl(aq) + NaOH(aq) \longrightarrow NaCl(aq) + H_2O(\ell)$$

In this reaction one mole of protons is donated by 1 mol of HCl and is accepted by 1 mol of NaOH. The equivalent weight of both HCl and NaOH is equal to their molar mass.

$$\text{Equivalent weight (HCl)} = \frac{36.461}{1} = 36.461;$$

$$\text{Equivalent weight (NaOH)} = \frac{39.9971}{1}$$

2. $$H_2SO_4(aq) + 2NaOH(aq) \longrightarrow Na_2SO_4(aq) + 2H_2O(\ell)$$

In this reaction 1 mol of H_2SO_4 donates 2 mol of protons, which are accepted by 2 mol of NaOH. The equivalent weights of the acid and base are

$$\text{Equivalent weight } H_2SO_4 = \text{one equiv} = \frac{\text{molar mass}}{2} = \frac{98.07}{2} = 49.04$$

Equivalent weight NaOH = one equiv

$$= \frac{\text{molar mass}}{1} = \frac{39.9971}{1} = 39.9971$$

3. $$2H_2SO_4(aq) + Ca(OH)_2(aq) \longrightarrow Ca(HSO_4)_2(aq) + H_2O(\ell)$$

In this reaction only one of the protons from H_2SO_4 is reacted, while both hydroxide ions from $Ca(OH)_2$ are reacted. Their equivalent weights are

$$\text{Equivalent weight } H_2SO_4 = \frac{\text{molar mass}}{1} = 98.07$$

$$\text{Equivalent weight } Ca(OH)_2 = \frac{\text{molar mass}}{2} = \frac{74.09}{2} = 37.04$$

Normality (N) The normality of a solution is the number of equivalents of solute dissolved per liter of solution:

$$\text{Normality} = \frac{\text{no. equiv solute}}{V \text{ of soln in L}}$$

Related equations (units only):

$$\text{No. equiv} = \text{Normality} \times V \text{ in L} = \frac{\text{equiv}}{\text{L}} \times \text{L} = \text{equiv}$$

$$V \text{ (soln)} = \frac{\text{no. equiv}}{\text{no. equiv/L}} = \text{L}$$

Because the equivalent weight of a substance is specific to a reaction, the normality of a solution must be determined from a known reaction.

A useful relationship involving normality and equivalent weight follows from the definition of gram-equivalent weight. The number of equivalents of all reactants and products in a reaction is equal:

$$\text{No. GEW reactant 1} = \text{no. GEW reactant 2} = \cdots = \text{no. GEW reactant } n$$
$$= \text{no. GEW product 1} = \cdots = \text{no. GEW product } n$$

If the reactants and/or products are in solution, their quantities are related through their normalities as

$$N_{R_1}V_{R_1} = N_{R_2}V_{R_2} = \cdots = N_{R_n}V_{R_n} = N_{P_1}V_{P_1}$$
$$= N_{P_2}V_{P_2} = \cdots = N_{P_n}V_{P_n}$$

Dilution formulas A volume of concentrated solution can be diluted to form a solution of lesser concentration. A larger volume of solution is formed thereby in which the number of moles or equivalents of solute has remained constant; only the volume of solution has changed. The following two relations involving molarity and normality are extremely useful for preparing solutions.

Molarity: No. mol solute in $soln_1$ = no. mol solute in $soln_2$

$$\left(M_{soln_1}\right)\left(V_{soln_1}\right) = \left(M_{soln_2}\right)\left(V_{soln_2}\right)$$

$$\left(\frac{mol}{L}\right)_1 (no.\ L_1) = \left(\frac{mol}{L}\right)_2 (no.\ L_2)$$

Example involving molarity: Calculate the volume of 0.15 *M* H_2SO_4 solution that can be prepared by diluting 20.0 mL of 6.00 *M* H_2SO_4.

$$(M_1)(V_1) = (M_2)(V_2)$$

$$V_2 = \frac{(M_1)(V_1)}{M_2} = \frac{(6.00\ mol/L)(0.0200\ L)}{0.15\ mol/L} = 0.80\ L$$

Normality: No. equiv solute in $soln_1$ = no. equiv solute in $soln_2$

$$\left(N_{soln_1}\right)\left(V_{soln_1}\right) = \left(N_{soln_2}\right)\left(V_{soln_2}\right)$$

$$\left(\frac{equiv}{L}\right)_1 (no.\ L_1) = \left(\frac{equiv}{L}\right)_2 (no.\ L_2)$$

Example involving normality: Calculate the volume of 6.0 *N* H_2SO_4 required to prepare 4.00 L of 0.10 *N* H_2SO_4 solution.

$$N_1V_1 = N_2V_2$$

$$V_1 = \frac{N_2V_2}{N_1} = \frac{(0.10\ equiv/L)(4.00\ L)}{6.0\ equiv/L} = 0.067\ L = 67\ mL$$

Neutralization reaction Any reaction of an acid and a base in which equal equivalents of acid and base are reacted is known as a *neutralization reaction.* The resulting solution may be neutral, acidic, or basic depending on the H^+ and OH^- concentrations (or their counterparts in nonaqueous systems) that remain. These concentrations, in turn, depend on the relative strengths of the reacting substances.

End point This term, normally used with reactions taking place in aqueous solution, defines the point in a reaction at which an equal number of equivalents of the reactants have been consumed. Some type of indicator, either chemical or electronic, is used to monitor the reaction and to determine the end point.

EXERCISES

42. Calculate the mass of an equivalent of each of the reactants in the following equations to one decimal place:

(a) $Sr(OH)_2 + 2HNO_3 \longrightarrow Sr(NO_3)_2 + 2H_2O$

Solution

The mass of an equivalent (a gram-equivalent weight) of a protonic acid is defined as the mass of the acid in grams that will provide 1 mol of protons (H^+) in a reaction. This is found by dividing the molar mass by the number of hydrogen ions transferred in the reaction. The GEW of a base is the mass of the base in grams that will provide 1 mol of hydroxide ions in a reaction or react with 1 mol of protons. This is found by dividing the molar mass by the number of hydroxide ions or their equivalent transferred in the reaction.

In this problem 1 mol of $Sr(OH)_2$ provides 2 mol of OH^- to react with 2 mol of hydrogen ion from 2 mol of HNO_3.

$$\frac{121.64 \text{ g } Sr(OH)_2}{1 \text{ mol } Sr(OH)_2} \times \frac{1 \text{ mol } Sr(OH)_2}{2 \text{ equiv } Sr(OH)_2} = \frac{60.82 \text{ g } Sr(OH)_2}{1 \text{ equiv } Sr(OH)_2}$$

In simplified form:

$$\text{Equivalent weight } Sr(OH)_2 = \frac{87.62 + 2(15.9994) + 2(1.0079)}{2} = 60.8 \text{ g}$$

$$\text{Equivalent weight } HNO_3 = \frac{1.0079 + 14.0067 + 3(15.9994)}{1} = 63.0 \text{ g}$$

(c) $2Al(OH)_3 + 3H_2SO_4 \longrightarrow Al_2(SO_4)_3 + 6H_2O$

Solution

$$\text{Equivalent weight } Al(OH)_3 = \frac{\text{molar mass}}{3}$$

$$= \frac{26.98154 + 3(1.0079) + 3(15.9994)}{3} = 26.0 \text{ g}$$

$$\text{Equivalent weight } H_2SO_4 = \frac{\text{molar mass}}{2}$$

$$= \frac{2(1.0079) + 32.06 + 4(15.9994)}{2} = 49.0 \text{ g}$$

(e) $KOH + H_2SO_4 \longrightarrow KHSO_4 + H_2O$

Solution

1 mol of KOH provides one equivalent of OH^- to react with 1 equivalent of $KHSO_4$.

$$\frac{56.1056 \text{ g KOH}}{1 \text{ mol KOH}} \times \frac{1 \text{ mol KOH}}{1 \text{ equiv KOH}} = \frac{56.1056 \text{ g KOH}}{1 \text{ equiv KOH}}$$

Similarly for H_2SO_4

$$\frac{98.07 \text{ g } H_2SO_4}{1 \text{ mol } H_2SO_4} \times \frac{1 \text{ mol } H_2SO_4}{1 \text{ equiv } H_2SO_4} = \frac{98.07 \text{ g } H_2SO_4}{1 \text{ equiv } H_2SO_4}$$

In simplified terms,

$$\text{Equivalent weight KOH} = \frac{39.0983 + 1.0079 + 15.9994}{1} = 56.1\text{ g}$$

$$\text{Equivalent weight } H_2SO_4 = 98.07 = 98.1\text{ g} \quad \text{(after rounding)}$$

Note that only one of the two replaceable H^+ in H_2SO_4 actually reacted.

43. Calculate the normality of each of the following solutions:

(a) 5.0 equiv of HCl in 2.0 L of solution.

Solution

The normality of a solution equals the number of equivalents of solute per liter of solution

$$N\text{ HCl} = \frac{5.0\text{ equiv HCl}}{2.0\text{ L}} = 2.5\ N$$

(c) 0.0015 equiv of HCl in 100.0 mL of solution

Solution

$$N\text{ Ca(OH)}_2 = \frac{0.0015\text{ equiv Ca(OH)}_2}{0.1000\text{ L}} = 0.015\ N$$

44. A 0.244 N solution of KOH is titrated with H_2SO_4, producing K_2SO_4. If a 48.0-mL sample of the KOH solution is used, what volume of 0.244 M H_2SO_4 is required to reach the end point? What volume of 0.244 N H_2SO_4? (Assume both protons react.)

Solution

The end point is defined as the point in a reaction at which an equal number of equivalents of both reactants have reacted. In this case

$$\text{No. equiv KOH reacted} = \text{no. equiv } H_2SO_4 \text{ reacted}$$

The number of equivalents of KOH reacted is

$$\text{KOH} = \frac{0.244\text{ equiv}}{\text{L}} \times 0.0480\text{ L} = 0.0117\text{ equiv}$$

This number of equivalents must be equal to the number of equivalents of H_2SO_4 reacted. Because 1 M H_2SO_4 equals 2 N H_2SO_4 (both protons react),

$$0.244\ M\ H_2SO_4 = 0.488\ N\ H_2SO_4$$

then the 0.0117 equiv H_2SO_4 must have come from a 0.488 N solution. Thus,

$$0.0117\text{ equiv } H_2SO_4 = \frac{0.488\text{ equiv}}{\text{L}}\ (V)$$

$$V = \frac{0.0117}{0.488}\text{ L} = 0.0240\text{ L} = 24.0\text{ mL}$$

For the volume required of 0.244 N H_2SO_4, follow the same procedure as above:

$$0.0117 \text{ equiv } H_2SO_4 = \left(\frac{0.244 \text{ equiv}}{\text{L}}\right)(V)$$

$$V = \frac{0.0117 \text{ L}}{0.244} = 0.0480 \text{ L} = 48.0 \text{ mL}$$

45. What volume of 0.421 N HCl is required to titrate 47.0 mL of 0.204 N KOH?

Solution

Neutralization is the point in an acid-base reaction when equal equivalents of a protonic acid and a hydroxyl base have reacted. In most cases the resulting salt solutions are not neutral; the concentrations of hydronium ion and hydroxide ion in solution are unequal because of the relative strengths of the acid and base. In this case the acid and base strengths are approximately the same and the solution is neutral.

$$(N_{\text{acid}})(V_{\text{acid}}) = (N_{\text{base}})(V_{\text{base}})$$

$$(0.421\ N)(V_{\text{acid}}) = (0.204\ N)(47.0 \text{ mL})$$

$$V_{\text{acid}} = \frac{(0.204)(0.0470)}{0.421} = 0.0228 \text{ L}$$

46. Titration of 0.1500 g of an acid requires 47.00 mL of 0.0120 N NaOH to reach the end point. What is the mass of an equivalent of the acid?

Solution

At the end point an equal number of equivalents of acid (in 0.1500 g) has reacted with the same number of equivalents of base.

$$\text{No. equiv acid} = \text{no. equiv base} = \left(\frac{0.0120 \text{ equiv}}{\text{L}}\right)(0.04700 \text{ L})$$

$$= 0.000564 \text{ equiv}$$

Because 0.1500 g = 0.000564 equiv,

$$\text{One equiv} = \frac{0.1500 \text{ g}}{0.000564 \text{ equiv}} = 266 \text{ g}$$

59. The reaction of 0.871 g of sodium with an excess of liquid ammonia containing a trace of $FeCl_3$ as a catalyst produced 0.473 L of pure H_2 measured at 25°C and 745 torr. What is the equation for the reaction of sodium with liquid ammonia? Show your calculations.

Solution

In the reaction with liquid ammonia, sodium functions as a Lewis base by donating electrons to hydrogen ions formed through the ionization of ammonia:

$$NH_3 \longrightarrow NH_2^- + H^+$$

To write the equation for the overall reaction, first determine the mole ratio of sodium used to hydrogen produced.

$$\text{No. mol Na} = 0.871 \text{ g} \times \frac{1 \text{ mol}}{22.98977} = 0.0379 \text{ mol}$$

From the Ideal Gas Equation, the number of moles of hydrogen produced equals

$$PV = nRT, \qquad n = \frac{PV}{RT}$$

$$n_{H_2} = \frac{\left(\frac{745}{760} \text{ atm}\right)(0.473 \text{ L})}{\left(\frac{0.08205 \text{ L atm}}{\text{mol K}}\right)(298 \text{ K})} = 0.0190 \text{ mol}$$

Thus, 0.0379 mol of Na reacts with 0.0190 mol of H_2. The ratio of Na to H_2 is

$$0.0379 : 0.0190 \quad \text{or} \quad 2\text{Na} : 1\text{H}_2$$

Therefore, the balanced equation must be

$$2Na + 2NH_3 \longrightarrow 2NaNH_2 + H_2$$

RELATED EXERCISES

1. The molarity of a vinegar solution is found to be 0.90 *M*. Calculate the mass of acetic acid, CH_3COOH, per 100 mL of vinegar.

   Answer: 5.4 g per 100 mL vinegar

2. Calculate the volume of water that must be added to 10.1 mL of 12.0 *N* H_2SO_4 to prepare a 0.10 *N* solution

   Answer: 1190 mL

3. (a) The hydrogen ion concentration in urine from a healthy person is about 1×10^{-6} *M*. If a person eliminates 1300 mL of urine per day, calculate the number of equivalents of hydrogen ion eliminated per day.
   (b) What volume of 1.00 *M* HCl would contain the same amount of hydrogen ion as 1300 mL of urine?

   Answer: (a) 1.3×10^{-6} equiv; (b) 1.3×10^{-6} L or 1.3 μL

4. A 0.2500 *M* sulfuric acid solution is used to standardize a sodium hydroxide solution. The titration of 55.00 mL of the NaOH solution required 11.00 mL of the H_2SO_4 solution. Calculate the molarity of the NaOH solution.

   Answer: 0.100 M

5. Citric acid, found in the juices of oranges and lemons, is commonly used in the first-aid treatment of a victim having swallowed a quantity of alkaline substance. Given that a 2.00 g sample of citric acid neutralizes 313 mL of 0.100 *M* NaOH, calculate the gram-equivalent weight of citric acid.

   Answer: 63.9 g

UNIT 10

CHAPTER 15: CHEMICAL KINETICS AND CHEMICAL EQUILIBRIUM

INTRODUCTION

The concept of speed, velocity, or rate is so familiar that it may come as a surprise that the idea is relatively modern and can be traced to Sir Isaac Newton (1642-1727). So too, the study of the rate of a chemical reaction did not originate until recently. In the 1850s the Norwegian scientists Guldberg and Waage were the first to write rate equations; these equations led them to develop the equilibrium expression for a chemical reaction. This early work was quickly followed by studies of the factors on which rates depend, such as temperature and concentration of reactants.

Today, the measurement of the rates of chemical reactions plays a key role in understanding reaction mechanisms (that is, the detailed way in which the reactants come together to form the products). Indeed, the object of most of the work in kinetics today is to elucidate mechanisms.

The importance of mechanisms is easy to understand. Most reactions proceed in several steps and are said to be *complex; elementary* reactions, on the other hand, occur in only one step. Complex reactions often occur by free-radical mechanisms. (A free radical is a reactive, neutral species that contains an odd number of electrons. The hydroxyl radical, $H:\ddot{\underset{..}{O}}\cdot$, is an example.) Industry as well as science is acutely interested in the study of complex reactions, because the industrial procedures that produce the wide variety of chemical intermediates and final products involve, for the most part, such complex processes. Examples are numerous and include such diverse processes as the cracking of heavy oil fractions to usable chemicals, the process of explosion, and even the effect on chemicals by enzymes. Indeed, enzyme kinetics, an area of study in kinetics, has been instrumental in the development of the understanding of biochemical processes.

Knowledge of the equilibrium constant expression as developed by Guldberg and Waage is equally important to chemists, because the amount of any substance in a chemical reaction at equilibrium may be determined from it. Depending on how the equilibrium expression is viewed, it may be applied to solubility problems (Unit 11) or to acid-base problems (Unit 9). A knowledge of equilibrium is essential as a theoretical base for studying most areas of chemistry, especially analytical chemistry, which is discussed at a qualitative level in the Qualitative Analysis section of *College Chemistry*, by Holtzclaw et al.

The problems in this chapter primarily deal with the determination of rates, rate constants, or concentrations using rate laws or the equilibrium constant expression.

FORMULAS AND DEFINITIONS

Activated complex-activation energy Whether a product forms as a result of an interaction of reacting particles in a chemical reaction depends on several factors. The collision theory of reactions states that reacting species must collide and, if conditions are suitable, form an unstable intermediate species called an *activated complex*. The complex may proceed to form the desired product or simply dissociate and return to the initial state. The energy required to give reactants sufficient energy to form activated complexes is called the *activation energy*. Most chemical reactions require activation energy. Some examples of different forms of energy used to activate a chemical reaction include a spark or high temperature to ignite motor fuels, a sudden jar to detonate nitroglycerin, or a flame to start charcoal.

Arrhenius equation One of the most important relationships in chemical kinetics is the Arrhenius equation, which relates the rate constant, k, of a reaction to the temperature. Its form, as developed by Svante Arrhenius, is $k = A \times 10^{(-E_a/2.303\,RT)}$, where A is the frequency factor, E_a is the activation energy, R is the gas constant, 8.314 J mol^{-1} K^{-1}, and T is the Kelvin temperature. Many natural processes, even though they are complex, obey this law because they are controlled by chemical reactions. For example, the law is obeyed by the chirring of crickets and the flashing of fireflies.

Catalyst A catalyst is a substance that changes the rate of a chemical reaction but which is itself unchanged and, in theory, can be recovered at the end of the process. If a substance increases the rate of a catalyzed reaction, it is called an *accelerator*; and if it slows the rate of the reaction, it is called an *inhibitor*.

Chain reaction This term refers to a reaction that, once started, occurs in a series of steps in which one or more reactions are repeated, thereby providing additional reagent to keep the overall reaction continuing. Such a reaction is characterized by an *initiation step*, a *propagation step*, and a *termination step*. The propagation step generates the species (reactive intermediate) necessary for the initiation step to continue. Thus the reaction continues until the termination step finally consumes all of the reactive intermediate.

Elementary reaction Reactions that occur in one step are termed *elementary*. By contrast, the *molecularity of the reaction* is referred to as the number of molecules that participate in the single step. Thus, *unimolecular* refers to a reaction in which only one molecule is needed. Similarly, *bimolecular* means two molecules participate, and *termolecular* refers to a simultaneous reaction of three molecules. This last type of reaction is comparatively rare.

Equilibrium This is a condition in a reacting system when the two opposing (forward and reverse) reactions occur simultaneously at the same rate.

Chemical systems at equilibrium are dynamic, but no net change in the amounts of reactants or products occurs.

Law of Mass Action For a general equation of a chemical reaction written as $aA + bB + \cdots \rightleftharpoons cC + dD + \cdots$, the Law of Mass Action is, at equilibrium,

$$\frac{[C]^c[D]^d \cdots}{[A]^a[B]^b \cdots} = K_e$$

This Law of Chemical Equilibrium or Mass Action applies at a particular temperature, and the brackets indicate the molar concentrations of the different chemical species raised to a power equal to their stoichiometric coefficients. Pressure is proportional to concentration, so an equilibrium constant expressed in terms of pressure, K_p, is defined as

$$K_p = \frac{(P_C)^c(P_D)^d}{(P_A)^a(P_B)^b}$$

Le Châtelier's principle If some applied action causes a stress to occur in a chemical system at equilibrium, such as a change in concentration, pressure, or temperature, then the equilibrium shifts in such a way that tends to undo the effect of the stress.

Rate of chemical reaction The amount of a substance (moles) that disappears or is formed by a reaction in unit time is the rate of reaction. In a reaction in which the stoichiometric coefficients are all 1 for the products and reactants, the rate R is directly proportional to the concentrations of the reacting substances. For the reaction $A + B \rightleftharpoons C + D$, the rate equation is written as $R_1 = k_1[A][B]$, where the brackets indicate molar concentrations and k_1 is the rate constant for the formation of product. The opposing reaction is the formation of reactants; its rate is $R_2 = k_2[C][D]$. At equilibrium, the opposing rates are equal. Setting the rates equal gives

$$k_1[A][B] = k_2[C][D] \quad \text{or} \quad \frac{k_1}{k_2} = \frac{[C][D]}{[A][B]}$$

Because the ratio of the two constants is a constant, $K_e = k_1/k_2$, where K_e is the equilibrium constant.

Order of a reaction In the general reaction $aA + bB \rightleftharpoons cC + dD$, if the rate of the reaction is $R = k[C]^c[D]^d$, then the order of the reaction is the sum of the exponents, $c + d$. Several common cases occur: If the value of the sum is 1, the reaction is first order and the rate depends on the concentration of only one species. If the value is 2, the reaction is second order and depends on the concentration of two species. If the sum is 0 (that is, if the rate is independent of the concentration), the reaction is zeroth order.

Half-life of a reaction The half-life, $t_{1/2}$ (time required for one-half of the original amount of limiting reactant to be converted to product), can be expressed in simple terms only for first-order reactions

$$t_{1/2} = \frac{0.693}{k}$$

where k is a rate constant. Elements undergoing radioactive decay are one class of substances that decompose via first-order kinetics.

van't Hoff's Law This law is a special case of Le Châtelier's principle. It states that when the temperature of a system in equilibrium is raised, the equilibrium is displaced in the direction that heat can be absorbed.

EXERCISES

5. The rate constant for the decomposition at 45°C of dinitrogen pentoxide, N_2O_5, dissolved in chloroform, $CHCl_3$, is 6.2×10^{-4} min^{-1}.

$$2N_2O_5 \longrightarrow 4NO_2 + O_2$$

The decomposition is first order in N_2O_5.

(a) What is the rate of the reaction when $[N_2O_5] = 0.40\ M$?

(b) What is the concentration of N_2O_5 remaining at the end of 1 h if the initial concentration of N_2O_5 was 0.40 M?

Solution

(a) The rate of reaction for a first-order reaction in N_2O_5 may be written as

$$\text{Rate} = k[N_2O_5]$$

where k, the rate constant at 45°C, is 6.2×10^{-4} min^{-1}. When $[N_2O_5] = 0.40\ M$,

$$\text{Rate} = 6.2 \times 10^{-4}\ \text{min}^{-1}\left(0.40\ \frac{\text{mol}}{\text{L}}\right)$$
$$= 2.48 \times 10^{-4} = 2.5 \times 10^{-4}\ \text{mol L}^{-1}\ \text{min}^{-1}$$

(b) In 60 min (1 h), the amount of N_2O_5 that decomposes is

$$2.5 \times 10^{-4}\ \text{mol L}^{-1}\ \text{min}^{-1}(60\ \text{min}) = 1.5 \times 10^{-2}\ \text{mol L}^{-1}$$

The amount remaining is

$$0.40\ M - 0.015\ M = 0.38\ M$$

7. Most of the 17.1 billion pounds of HNO_3 produced in the United States during 1980 was prepared by the following sequence of reactions, each run in a separate reaction vessel:

$$4NH_3(g) + 5O_2(g) \longrightarrow 4NO(g) + 6H_2O(g) \quad (1)$$

$$2NO(g) + O_2(g) \longrightarrow 2NO_2(g) \quad (2)$$

$$3NO_2(g) + H_2O(\ell) \longrightarrow 2HNO_3(aq) + NO(g) \quad (3)$$

The first reaction is run by burning ammonia in air over a platinum catalyst. This reaction is fast. The reaction in Equation (3) is also fast. The second reaction limits the rate at which nitric acid can be prepared from ammonia. If Equation (2) is second order in NO and first order in O_2, what is the rate of formation of NO_2 when the oxygen concentration is 0.50 *M* and the nitric oxide concentration is 0.75 *M*? The rate constant for the reaction is $5.8 \times 10^{-6}\ L^2\ mol^{-2}\ s^{-1}$.

Solution

The rate law governing the formation of HNO_3 is

$$\text{Rate} = k[NO]^2[O_2]$$

From the data given,

$$\text{Rate} = 5.8 \times 10^{-6}\ L^2\ mol^{-2}\ s^{-1}\left[0.75\ \frac{mol}{L}\right]^2\left[0.50\ \frac{mol}{L}\right]$$

$$= 1.6 \times 10^{-6}\ mol\ L^{-1}\ s^{-1}$$

Keystrokes: 5.8 [EE] [+/-] 6 × .75 [x^2] × .50 = ans. 1.63125 - 06

9. Hydrogen reacts with nitric oxide to form nitrous oxide, laughing gas, according to the equation

$$H_2(g) + NO(g) \longrightarrow N_2O(g) + H_2O(g)$$

Determine the rate equation and the rate constant for the reaction from the following data:

Experiment	1	2	3
[NO] *M*	0.30	0.60	0.60
$[H_2]$ *M*	0.35	0.35	0.70
Rate, mol $L^{-1}\ s^{-1}$	2.835×10^{-3}	1.134×10^{-2}	2.268×10^{-2}

Solution

It is intended that the data for each separate experiment given in this problem be substituted into the general rate equation (Rate = $k[NO]^m[H_2]^n$) to determine how n and m, the order of the reaction with respect to each component, varies. The three reactions, after substituting the data, are

Exp. 1: 2.835×10^{-3} mol L^{-1} s^{-1}

$$= k[0.30 \text{ mol L}^{-1}]^m[0.35 \text{ mol L}^{-1}]^n$$

Exp. 2: 1.134×10^{-2} mol L^{-1} s^{-1}

$$= k[0.60 \text{ mol L}^{-1}]^m[0.35 \text{ mol L}^{-1}]^n$$

Exp. 3: 2.268×10^{-2} mol L^{-1} s^{-1}

$$= k[0.60 \text{ mol L}^{-1}]^m[0.70 \text{ mol L}^{-1}]^n$$

Because k, m, and n refer to the same reaction, they must retain the same values throughout the three experiments. Going from Experiment 1 to Experiment 2, holding $[H_2]$ constant, shows that doubling the concentration of NO causes a fourfold increase in the rate. This can only occur if the value of m is 2 since $[2]^2 = 4$. In a similar manner, holding the concentration of NO constant in Experiment 2 and Experiment 3 with subsequent doubling of the concentrations of H_2 doubles the rate. The rate equation holds for this latter group of two experiments only if n equals 1. Therefore, the overall rate is second order in NO and first order in H_2.

To determine the value of the rate constants, the data of Experiment 1 is used in the rate law.

$$\text{Rate} = k[\text{NO}]^2[\text{H}_2]^1$$

$$2.835 \times 10^{-3} \text{ mol L}^{-1} \text{ s}^{-1} = k[0.30 \text{ mol L}^{-1}]^2[0.35 \text{ mol L}^{-1}]$$

$$k = \frac{2.835 \times 10^{-3} \text{ mol L}^{-1} \text{ s}^{-1}}{(0.09 \text{ mol}^2 \text{ L}^{-2})(0.35 \text{ mol L}^{-1})} = 9.0 \times 10^{-2} \text{ L}^2 \text{ mol}^{-2} \text{ s}^{-1}$$

14. The decomposition of SO_2Cl_2 to SO_2 and Cl_2 is a first-order reaction with $k = 2.2 \times 10^{-5}$ s^{-1} at 320°C. Determine the half-life of this reaction. At 320°C, how much $SO_2Cl_2(g)$ would remain in a 1.00-L flask 90.0 min after the introduction of 0.0238 mol of SO_2Cl_2? Assume the rate of the reverse reaction is so slow that it can be ignored.

Solution

The half-life of a first-order reaction is determined from the expression

$$t_{1/2} = \frac{0.693}{k} = \frac{0.693}{2.2 \times 10^{-5} \text{ s}^{-1}}$$

$$= 3.2 \times 10^4 \text{ s}$$

For a first-order reaction,

$$\log \frac{[A_0]}{[A]} = \frac{kt}{2.303}$$

The concentration of A_0 is 0.0238 M, and substitution gives

$$\log \frac{[0.0238]}{[A]} = \frac{2.2 \times 10^{-5} \text{ s}^{-1}\left(90.0 \text{ min} \times \frac{60 \text{ s}}{\text{min}}\right)}{2.303}$$

$$\log 0.0238 - \log A = 0.0516$$

$$-\log A = 0.0516 + 1.623$$

$$\log A = -1.675$$

$$A = 0.0211\ M = 2.11 \times 10^{-2}\ \text{mol in the 1.00-L flask}$$

20. If the rate of a reaction doubles for every 10°C rise in temperature, how much faster would the reaction proceed at 45°C than at 25°C? at 95°C than at 25°C?

Solution

The rate doubles for each 10°C rise in temperature. Forty-five degrees is a 20° increase over 25°. Thus, the rate doubles two times, or 2^2 (rate at 25°) = 4 times faster. Ninety-five degrees is a 70° increase over 25°. Thus, the rate doubles 7 times, or 2^7 (rate at 25°) = 128 times faster.

22. The rate constant at 325°C for the reaction $C_4H_8 \longrightarrow 2C_2H_4$ (Section 15.11) is $6.1 \times 10^{-8}\ s^{-1}$, and the activation energy is 261 kJ per mole of C_4H_8. Determine the frequency factor for the reaction.

Solution

The rate constant, k, is related to the activation energy, E_a, by a relationship known as Arrhenius' Law. Its form is

$$k = A \times 10^{-(E_a/2.303\ RT)}$$

where A is the frequency factor. Using the data above, and converting kJ to joules,

$$6.1 \times 10^{-8}\ s^{-1} = A \times 10^{-[+261{,}000\ J/2.303(8.314\ J\ K^{-1})(325 + 273)K]}$$

$$= A \times 10^{-22.8}$$

Convert $10^{-22.8}$ to its equivalent expressed as a whole power of 10:

Keystrokes: 10 [y^x] 22.8 [+/-] = ans. 1.5848932 - 23

$$A = \frac{6.1 \times 10^{-8}\ s^{-1}}{1.58 \times 10^{-23}} = 3.8 \times 10^{15}\ s^{-1}$$

35. The rate of the reaction $H_2(g) + I_2(g) \longrightarrow 2HI(g)$ at 25°C is given by

$$\text{Rate} = 1.7 \times 10^{-18}\,[H_2][I_2]$$

The rate of decomposition of gaseous HI to $H_2(g)$ and $I_2(g)$ at 25°C is given by

$$\text{Rate} = 2.4 \times 10^{-21}\,[HI]^2$$

What is the equilibrium constant for the formation of gaseous HI from the gaseous elements at 25°C?

Solution

The equilibrium constant may be written as

$$K = \frac{k_{\text{forward}}}{k_{\text{reverse}}}$$

since at equilibrium, $\text{Rate}_1 = \text{Rate}_2$, or

$$k_{\text{forward}}[H_2][I_2] = k_{\text{reverse}}[HI]^2$$

This may be expressed as

$$\frac{k_{\text{forward}}}{k_{\text{reverse}}} = \frac{[HI]^2}{[H_2][I_2]}$$

which is recognized as an equilibrium constant. Substitution gives

$$K = \frac{k_{\text{forward}}}{k_{\text{reverse}}} = \frac{1.7 \times 10^{-18}}{2.4 \times 10^{-21}} = 7.1 \times 10^2$$

36. A sample of $NH_3(g)$ was formed from $H_2(g)$ and $N_2(g)$ at 500°C. If the equilibrium mixture was found to contain 1.35 mol H_2 per liter, 1.15 mol N_2 per liter and 4.12×10^{-1} mol NH_3 per liter, what is the value of the equilibrium constant for the formation of NH_3?

Solution

The reaction may be written as

$$N_2 + 3H_2 \rightleftharpoons 2NH_3$$

The equilibrium constant for the reaction is

$$K_e = \frac{[NH_3]^2}{[N_2][H_2]^3} = \frac{[4.12 \times 10^{-1}\ \text{mol L}^{-1}]^2}{[1.15\ \text{mol L}^{-1}][1.35\ \text{mol L}^{-1}]^3}$$

$$= \frac{0.170\ \text{mol}^2\ \text{L}^{-2}}{(1.15\ \text{mol L}^{-1})(2.46\ \text{mol}^3\ \text{L}^{-3})}$$

$$= 0.0600\ \text{mol}^{-2}\ \text{L}^2 = 6.00 \times 10^{-2}\ \text{mol}^{-2}\ \text{L}^2$$

38. A 0.72-mol sample of PCl_5 is put into a 1.00-L vessel and heated. At equilibrium the vessel contains 0.40 mol of $PCl_3(g)$ as well as $Cl_2(g)$ and undissociated $PCl_5(g)$. What is the equilibrium constant for the decomposition of PCl_5 to PCl_3 and Cl_2 at this temperature?

Solution

The reaction is

$$PCl_5(g) \rightleftharpoons PCl_3(g) + Cl_2(g)$$

The equilibrium constant is

$$K_e = \frac{[PCl_3][Cl_2]}{[PCl_5]}$$

The amount of PCl_5 that had to decompose is 0.40 mol. This must also be the amount of Cl_2 present. The amount of PCl_5 remaining is 0.72 - 0.40 mol = 0.32 mol. Substitution gives

$$K_e = \frac{[0.40 \text{ mol L}^{-1}][0.40 \text{ mol L}^{-1}]}{[0.32 \text{ mol L}^{-1}]}$$

$$= 0.50 \text{ mol L}^{-1} = 0.50\ M$$

39. The vapor pressure of water is 0.196 atm at 60°C. What is the equilibrium constant for the transformation $H_2O(\ell) \rightleftharpoons H_2O(g)$?

Solution

The equilibrium constant may be expressed in terms of concentrations or in terms of pressures. For the transformation $H_2O(\ell) \longrightarrow H_2O(g)$ the equilibrium may be expressed most easily in terms of pressures. The equilibrium constant depends only on the substance in the gaseous phase at equilibrium. Thus,

$$K_p = P(H_2O(g)) = 0.196 \text{ atm}$$

42. Sodium sulfate 10-hydrate, $Na_2SO_4 \cdot 10H_2O$, dehydrates according to the equation

$$Na_2SO_4 \cdot 10H_2O(s) \rightleftharpoons Na_2SO_4(s) + 10H_2O(g)$$

with $K = 4.08 \times 10^{-25}$ at 25°C. What is the pressure of water vapor in equilibrium with a sample of $Na_2SO_4 \cdot 10H_2O$?

Solution

Because two of the substances involved in the equilibrium are solids, their concentrations are constant and do not appear in the equilibrium expression. Thus,

$$K = 4.08 \times 10^{-25} = \left[P_{H_2O}\right]^{10}$$

$$P_{H_2O} = \sqrt[10]{4.08 \times 10^{-25}} = 3.64 \times 10^{-3} \text{ atm}$$

Keystrokes: 4.08 [EE] 25 [+/-] [$\sqrt[x]{y}$] 10 = ans. 3.639703 - 03

If your calculator has only an y^x key, one approach is to recognize that $\sqrt[10]{y} = y^{0.10}$.

44. The equilibrium constant K for the reaction

$$PCl_5(g) \rightleftharpoons PCl_3(g) + Cl_2(g)$$

is 0.0211 at a certain temperature. What are the equilibrium concentrations of PCl_5, PCl_3, and Cl_2 starting with a concentration of PCl_5 of 1.00 M?

Solution

The concentrations of these substances at equilibrium depend on the initial amount of PCl_5. Let x equal the amount of PCl_5 that dissociates. The amount of PCl_5 at equilibrium is 1.00 mol $L^{-1} - x$. The decomposition of 1 mol of PCl_5 produces 1 mol of PCl_3 and 1 mol of Cl_2, so the amount of PCl_3 formed is x and the amount of Cl_2 formed is x. Substitution into the equilibrium constant and solving for x follows the sequence

$$K_e = \frac{[PCl_3][Cl_2]}{[PCl_5]} = \frac{(x)(x)}{1-x} = 0.0211$$

$$x^2 = 0.0211(1-x)$$

$$= 0.0211x - 0.0211 = 0$$

See page 161 and Exercise 8, page 165, for information on how to solve a quadratic equation. Use the quadratic equation to solve for x.

$$x = \frac{-b \pm \sqrt{b^2 - 4ac}}{2a} = \frac{-0.0211 \pm \sqrt{(0.0211)^2 + (0.0844)}}{2}$$

$$x = \frac{-0.0211 \pm \sqrt{0.0848}}{2} = \frac{-0.0211 + 0.2912}{2} = 0.135 \text{ mol } L^{-1}$$

The final concentrations are

$$PCl_5 = 1.00 - 0.135 = 0.865\ M$$

$$Cl_2 = PCl_3 = 0.135\ M$$

46. The equilibrium constant K_p for the decomposition of nitrosyl bromide

$$2NOBr(g) \rightleftharpoons 2NO(g) + Br_2(g)$$

is 1.0×10^{-2} atm at 25°C. What percentage of NOBr is decomposed at 25°C and a total pressure of 0.25 atm?

Solution

The equilibrium constant in terms of pressures for the above reaction is

$$K_p = \frac{(P_{NO})^2 \left(P_{Br_2}\right)}{(P_{NOBr})^2} = 1.0 \times 10^{-2} \text{ atm}$$

Since $K_e < 1$, the reaction favors the reactants; that is, much of the initial amount of NOBr will remain undissociated at equilibrium. The chemical equation indicates that for each mole of NOBr decomposed, 1 mol of NO and 1/2 mol of Br_2 are formed. All of the components are gases, so the pressure of each component will vary as its number of moles. Let x = the pressure in atm of the NOBr decomposed. Then, in terms of pressure, the partial pressures of the species at equilibrium are

$$P_{NOBr} = 0.25 \text{ atm} - x - 0.5\,x = 0.25 \text{ atm} - 1.5\,x \text{ atm}$$

$$P_{NO} = x \text{ atm} \qquad P_{Br_2} = 0.5\,x \text{ atm}$$

Substitution into the equilibrium expression gives

$$K_p = \frac{[x]^2[0.5\,x]}{[0.25 - 1.5\,x]^2} = 1.0 \times 10^{-2} \text{ atm}$$

Solving for x,

$$x^3 - 4.5 \times 10^{-2}x^2 + 1.5 \times 10^{-2}x - 1.25 \times 10^{-3} = 0$$

This is a cubic equation, which may be solved in several ways. One technique involves using a cubic formula, but this formula is not readily available nor always applicable, as is the familiar quadratic equation. The method chosen for use here is one of successive approximations. This technique for quadratic equations is discussed in Chapter 16 of the textbook. The method is based on the idea that the value of x that solves this equation will give a value of the equation exactly equal to zero. Values of x that are too large will be positive and those too small will be negative. In order to shorten the process as much as possible, the best choice of x should be made using the information available. In this case, no more than 0.25 nor less than 0.0 can be dissociated, and, therefore, the range of x must be 0 to 0.25. A value of 0.1 for x is selected and substituted into the equation. Thus,

$$x^3 - 4.5 \times 10^{-2}x^2 + 1.5 \times 10^{-2}x - 1.25 \times 10^{-3} = 0 \qquad \text{for } x = 0.10$$

$$f(0.10) = 10^{-3} - 4.5 \times 10^{-4} + 1.5 \times 10^{-3} - 1.25 \times 10^{-3} = 8.00 \times 10^{-4}$$

The value obtained from this guess is close to zero but is positive. This means that the selection is on the high side of the true value of x. A negative value will mean that the value chosen is too low. Additional choices of x give the following values:

x	$f(x)$, Value of Cubic Equation Expression Above
0.10	8.00×10^{-4}
0.05	-4.875×10^{-4}
0.08	1.74×10^{-4}
0.07	-7.75×10^{-5}
0.073	$+1.118 \times 10^{-3}$
0.071	-5.39×10^{-5}
0.072	-3.03×10^{-5}
0.0725	-1.8×10^{-5}

From these obtained values, it is clear that the values for x of 0.072 and 0.073 bracket the transition from negative values to positive values. Consequently, the value 0.0725 is chosen to narrow the value of x. By comparison of the values $f(x)$ for 0.0725 and 0.073, the value of x that solves the cubic equation (that is, gives a value of zero) is close to 0.0725. This value could be refined further still, but this last value is sufficiently accurate to substitute and to determine the value of the partial pressures. Thus,

$$P_{NO} = 0.0725 \text{ atm}$$

$$P_{Br_2} = 0.0362 \text{ atm}$$

$$P_{NOBr} = 0.250 - 0.109 = 0.141 \text{ atm}$$

$$\text{Total pressure at equilibrium} = 0.0725 \text{ atm} + 0.0362 \text{ atm} + 0.141 \text{ atm} = 0.250 \text{ atm}$$

Substitution of these values into the original equation proves that $x = 0.0725$ atm is a solution and that the amount that decomposes is responsible for 0.109 atm of the total pressure.

The problem does not end here. To determine the amount of NOBr originally present, needed in order to calculate the percent decomposition, it must be considered that had the NOBr not decomposed, the total pressure due to it alone would have been larger than the 0.141 atm present at equilibrium. That is, the original pressure is 0.141 atm plus the pressure that would have been present had NOBr not decomposed. Using the values above, the original pressure is 0.141 atm + 0.0725 atm = 0.213 atm. Thus, the percent decomposition is

$$\frac{\text{Amount decomposed}}{\text{original amount}} \times 100\% = \frac{0.0725 \text{ atm}}{0.213 \text{ atm}} \times 100\% = 34\%$$

55. The equilibrium constant for the reaction

$$CO + H_2O \rightleftharpoons CO_2 + H_2$$

is 5.0 at a given temperature.

(a) On analysis, an equilibrium mixture of the substances present at the given temperature was found to contain 0.20 mol of CO, 0.30 mol of water vapor, and 0.90 mol of H_2 in a liter. How many moles of CO_2 were there in the equilibrium mixture?

(b) Maintaining the same temperature, additional H_2 was added to the system, and some water vapor was removed by drying. A new equilibrium mixture was thereby established containing 0.40 mol of CO, 0.30 mol of water vapor, and 1.2 mol of H_2 in a liter. How many moles of CO_2 were in the new equilibrium mixture? Compare this with the quantity in part (a) and discuss whether the second value is reasonable. Explain how it is possible for the water vapor concentration to be the same in the two equilibrium solutions even though some vapor was removed before the second equilibrium was established.

Solution

(a) For the above reaction,

$$K_e = \frac{[CO_2][H_2]}{[CO][H_2O]} = 5.0$$

The concentrations at equilibrium are 0.20 *M* CO, 0.30 *M* H_2O 0.90 *M* H_2. Substitution gives

$$K_e = 5.0 = \frac{[CO_2][0.90]}{[0.20][0.30]}$$

$$[CO_2] = \frac{5.0(0.20)(0.30)}{0.90} = 0.33\ M$$

$$\text{No. moles } CO_2 = 0.33 \text{ mol L}^{-1} \times 1 \text{ L} = 0.33 \text{ mol}$$

(b) At the particular temperature of reaction, K_e remains constant at 5.0. The new concentrations are 0.40 *M* CO, 0.30 *M* H_2O, 1.2 *M* H_2.

$$\frac{[CO_2][1.2]}{[0.40][0.30]} = 5.0$$

$$[CO_2] = [0.50\ M]$$

$$\text{No. moles } CO_2 = 0.50 \text{ mol L}^{-1} \times 1 \text{ L} = 0.50 \text{ mol}$$

56. A 1.00-L vessel at 400°C contains the following equilibrium concentrations: N_2, 1.00 *M*; H_2, 0.50 *M*; and NH_3, 0.50 *M*. How many moles of hydrogen must be removed from the vessel in order to increase the concentration of nitrogen to 1.2 *M*?

Solution

The reaction is $N_2 + 3H_2 \rightleftharpoons 2NH_3$. The equilibrium constant for this equilibrium is calculated from the original data.

$$K_e = \frac{[NH_3]^2}{[N_2][H_2]^3} = \frac{(0.50)^2}{(1.00)(0.50)^3} = 2.0 \text{ mol}^{-2} \text{ L}^2$$

The concentration of N_2 increases by $1.2\ M - 1.0\ M = 0.2\ M$. From the balanced equation, 1 mol $N_2 \longrightarrow$ 2 mol NH_3. Thus, $0.2\ M\ N_2$ must have come from the decomposition of $2(0.2\ M\ NH_3) = 0.4\ M\ NH_3$. The amount of NH_3 remaining is $0.5\ M - 0.4\ M = 0.1\ M\ NH_3$. In addition to the nitrogen formed, hydrogen was also formed. Because $NH_3 \longrightarrow (3/2)H_2$, the decomposition of $0.4\ M\ NH_3$ results in the formation of $0.60\ M\ H_2$. Thus, a total of $0.50\ M + 0.60\ M = 1.10\ M\ H_2$ is the total amount of H_2.

In order to maintain equilibrium, hydrogen must be removed from the reaction mixture to allow the concentration of N_2 to increase to 1.2 M. Let x = the amount of H_2 in the final mixture. Then

$$K_e = \frac{[0.1 \text{ mol L}^{-1}]^2}{[1.20 \text{ mol L}^{-1}][x]^3} = 2.0 \text{ mol}^{-2} \text{ L}^2$$

$$2.4\ x^3 \text{ mol}^{-1} \text{ L}^1 = 0.01 \text{ mol}^2 \text{ L}^{-2}$$

$$x^3 = 4.17 \times 10^{-3} \text{ mol}^3 \text{ L}^{-3}$$

$$x = 0.16 \text{ mol L}^{-1}$$

Only 0.16 M remains, so

$$1.10\ M - 0.16\ M = 0.94\ M\ H_2$$

were removed. This is not unreasonable even though more hydrogen was removed than originally present, because the decomposition of NH_3 formed additional H_2.

59. The hydrolysis of the sugar sucrose to the sugars glucose and fructose

$$C_{12}H_{22}O_{11} + H_2O \longrightarrow C_6H_{12}O_6 + C_6H_{12}O_6$$

follows a first-order rate equation for the disappearance of sucrose.

$$\text{Rate} = k[C_{12}H_{22}O_{11}]$$

(The products of the reaction have the same molecular formulas but differ in the arrangement of the atoms in their molecules.)

(a) In neutral solution, $k = 2.1 \times 10^{-11}\ s^{-1}$ at 27°C and $8.5 \times 10^{-11}\ s^{-1}$ at 37°C. Determine the activation energy, the frequency factor, and the rate constant for this equation at 47°C.

Solution

The text demonstrates that the value of E_a may be determined from a plot of log k against $1/T$ that gives a straight line whose slope is $-E_a/2.303\ R$. This is based on the equation

$$\log k = \log A - \frac{E_a}{2.303\ RT}$$

Only two data points are given and these must determine a straight line when log k is plotted against $1/T$. The values needed are

$$k_1 = 2.1 \times 10^{-11} \qquad \log k_1 = -10.6778$$
$$k_2 = 8.5 \times 10^{-11} \qquad \log k_2 = -10.0706$$
$$T_1 = 27^\circ\mathrm{C} = 300\ \mathrm{K} \qquad 1/T_1 = 3.3333 \times 10^{-3}$$
$$T_2 = 37^\circ\mathrm{C} = 310\ \mathrm{K} \qquad 1/T_2 = 3.2258 \times 10^{-3}$$

The slope of the line determined by these points is given by

$$\mathrm{Slope} = \frac{\Delta(\log k)}{\Delta(1/T)} = \frac{(-10.0706) - (-10.6778)}{(3.2258 \times 10^{-3}) - (3.3333 \times 10^{-3})}$$
$$= \frac{0.6072}{-0.1075 \times 10^{-3}} = -5648$$
$$E_a = 2.303(8.314\ \mathrm{J\ mol^{-1}})(-5648)$$
$$= 108{,}100\ \mathrm{J} = 108\ \mathrm{kJ}$$

Whenever differences of very small numbers are taken, such as the reciprocals of T above, an inherent problem occurs. In order to have accurate differences, a larger number of significant figures than justified by the data must be used. Thus five figures were used to obtain the value $E_a = 108$ kJ. This difficulty may be alleviated by the approach shown below.

For only two data points Arrhenius' Law

$$k = A \times 10^{-E_a/2.303\,RT}$$

may be used in an equally accurate, analytical solution for E_a. This is possible because the value of A will be the same throughout the course of the reaction. Once the value of E_a is determined, the value of A may be determined from either Equations (1) or (2) shown below. Then k at 47°C may be determined using the value of E_A and A so determined. The procedure is as follows:

$$k = A \times 10^{-E_a/2.303\,RT}$$

$$2.1 \times 10^{-11}\ \mathrm{s^{-1}} = A \times 10^{-E_a/2.303(8.314\ \mathrm{J\ K^{-1}})(300\ \mathrm{K})} \qquad (1)$$

$$8.5 \times 10^{-11}\ \mathrm{s^{-1}} = A \times 10^{-E_a/2.303(8.314\ \mathrm{J\ K^{-1}})(310\ \mathrm{K})} \qquad (2)$$

Equating the values of A as solved from (1) and (2),

$$2.1 \times 10^{-11}\ \mathrm{s^{-1}} \times 10^{+E_a/2.303(8.314\ \mathrm{J\ K^{-1}})(300\ \mathrm{K})}$$
$$= 8.5 \times 10^{-11}\ \mathrm{s^{-1}} \times 10^{+E_a/2.303(8.314\ \mathrm{J\ K^{-1}})(310\ \mathrm{K})}$$

or $$2.1 \times 10^{-11} \times 10^{+E_a/5744} = 8.5 \times 10^{-11} \times 10^{+E_a/5936}$$

Taking common logs of both sides gives

$$(\log 2.1 \times 10^{-11}) + \frac{E_a}{5744} = (\log 8.5 \times 10^{-11}) + \frac{E_a}{5936}$$

$$-10.68 + \frac{E_a}{5744} = -10.07 + \frac{E_a}{5936}$$

$$E_a\left(\frac{1}{5744} - \frac{1}{5936}\right) = -10.07 + 10.68$$

$$E_a(1.741 \times 10^{-4} - 1.685 \times 10^{-4}) = 0.61$$

$$E_a = \frac{0.61}{0.056 \times 10^{-4}} = 109 \text{ kJ}$$

The value of A may be found from either Equation (1) or (2). Using (1)

$$2.1 \times 10^{-11}\ s^{-1} = A \times 10^{-109,000/2.303(8.314)(300)} = A \times 10^{-18.98}$$

$$A = 2.1 \times 10^{-11}\ s^{-1} \times 10^{+18.98} = 2.1 \times 10^{-11}(9.55 \times 10^{18}\ s^{-1}) = 2.0 \times 10^{8}\ s^{-1}$$

Keystrokes: 2.1 [EE] [+/-] 11 × 10 [y^x] 18.98 = ans. 2.0054844 08

The value of k at 47°C may be determined from the Arrhenius equation now that the value of E_a and A have been calculated.

$$k = A \times 10^{-E_a/2.303\,RT}$$

$$= 2.0 \times 10^{8}\ s^{-1} \times 10^{-109,000\ J/2.303(8.314\ J\ K^{-1})(320\ K)}$$

$$= 2.0 \times 10^{8}\ s^{-1} \times 10^{-17.79} = 2.0 \times 10^{8}\ s^{-1}(1.62 \times 10^{-18})$$

$$= 3.2 \times 10^{-10}\ s^{-1}$$

61. Assume that the rate equations given below apply to each reaction at equilibrium. Determine the rate equation for each reverse reaction.

(a) $NO(g) + O_3(g) \longrightarrow NO_2(g) + O_2(g)$

Rate$_1$ = $k_1[NO][O_3]$

(d) $2NO(g) + 2H_2(g) \longrightarrow N_2(g) + 2H_2O(g)$

Rate$_1$ = $k_1[NO][H_2]$

Solution

(a) The overall equilibrium constant is

$$K_e = \frac{[NO_2][O_2]}{[NO][O_3]} = \frac{k_1}{k_2}$$

From the rate expression,

$$k_1 = \frac{\text{Rate}_1}{[\text{NO}][\text{O}_3]}$$

At equilibrium, $\text{Rate}_1 = \text{Rate}_2$, and therefore, substituting for k_1,

$$\frac{k_1}{k_2} = \frac{\frac{\text{Rate}_1}{[\text{NO}][\text{O}_3]}}{k_2} = \frac{\frac{\text{Rate}_2}{[\text{NO}][\text{O}_3]}}{k_2} = \frac{[\text{NO}_2][\text{O}_2]}{[\text{NO}][\text{O}_3]}$$

Using the last two terms,

$$\frac{\text{Rate}_2}{k_2} = [\text{NO}_2][\text{O}_2] \quad \text{or} \quad \text{Rate}_2 = k_2[\text{NO}_2][\text{O}_2]$$

(d)

$$K_e = \frac{[\text{N}_2][\text{H}_2\text{O}]^2}{[\text{NO}]^2[\text{H}_2]^2} = \frac{k_1}{k_2}$$

Using the same procedure as in part (a),

$$\text{Rate}_1 = \text{Rate}_2$$

$$\text{Rate}_1 = k_1[\text{NO}][\text{H}_2]$$

$$k_1 = \frac{\text{Rate}_1}{[\text{NO}][\text{H}_2]} = \frac{\text{Rate}_2}{[\text{NO}][\text{H}_2]}$$

Substitution into the value for K_e gives

$$\frac{k_1}{k_2} = \frac{\frac{\text{Rate}_2}{[\text{NO}][\text{H}_2]}}{k_2} = \frac{[\text{N}_2][\text{H}_2\text{O}]^2}{[\text{NO}]^2[\text{H}_2]^2}$$

$$\text{Rate}_2 = k_2\frac{[\text{N}_2][\text{H}_2\text{O}]^2}{[\text{NO}][\text{H}_2]}$$

RELATED EXERCISES

1. Hemoglobin thermally denatures in a first-order reaction that has been found to take 3460 s at 60.0°C and 530 s at 65.0°C. Assume that Arrhenius' Law governs the reaction. Calculate the energy of activation at 60.0°C.

   Answer: $E_a = 35.1$ *kJ*

2. A certain reaction

$$2\text{X}^- + \text{A} + 2\text{H}^+ \longrightarrow \text{X}_2 + 2\text{HA}$$

occurs in several stages. The rate equation is

$$\text{Rate} = k[\text{A}][\text{H}^+][\text{X}^-]$$

(a) If the concentration of A is increased by a factor of 4, by what factor is the rate of disappearance of X^- ions increased?
(b) How is k changed by increasing the concentration of X^- ions?
(c) If the rate of disappearance of X^- ions is 3.8×10^{-3} mol L^{-1} s^{-1}, what is the rate of appearance of X_2?

Answer: (a) 3; (b) none; (c) 1.9×10^{-3} mol L^{-1} s^{-1}

3. The following rate data for the formation of C were taken for the reaction

$$\text{A} + 2\text{B} \longrightarrow \text{C}$$

[A]/mol L^{-1}	[B]/mol L^{-1}	Rate/mol L^{-1} s^{-1}
3.50×10^{-2}	2.3×10^{-2}	5.0×10^{-7}
7.0×10^{-2}	4.6×10^{-2}	2.0×10^{-6}
7.0×10^{-2}	9.2×10^{-2}	4.0×10^{-6}

What are m and n in the rate equation

$$\text{Rate} = k[\text{A}]^m[\text{B}]^n$$

and what is the rate constant k?

Answer: $m = 1$, $n = 1$; $k = 6.21 \times 10^{-4}$ L $mol^{-1}s^{-1}$

4. A 1.00-L vessel contains at equilibrium 0.300 mol of N_2, 0.400 mol of H_2, and 0.100 mol of NH_3. If the temperature is maintained constant, how many moles of H_2 must be introduced into the vessel in order to double the equilibrium concentration of NH_3?

Answer: 0.425 mol

5. At a certain temperature, the gas-phase reaction $N_2O_4(g) \rightleftharpoons 2NO_2(g)$ has a K_e of 1.1×10^{-5}. If 0.20 mol of N_2O_4 is dissolved in 400 mL of chloroform and the reaction allowed to come to equilibrium, (a) what is the new NO_2 concentration, and (b) what will be the percent dissociation of the original N_2O_4?

Answer: (a) 2.3×10^{-3} mol L^{-1} NO_2; (b) 0.23% dissociation

UNIT 11

CHAPTER 16: IONIC EQUILIBRIA OF WEAK ELECTROLYTES

INTRODUCTION

Weak electrolytes are substances, such as acids, bases, and salts, that only slightly conduct an electric current in aqueous solution. Examples include the various substances that carbonates form in the oceans, in ground water, and even in blood; organic acids that make up the fats of plants and animals; and the vast amount of sulfides and sulfites formed by the oxidation of sulfur. One important aspect of weak electrolytes is that they have a major role in the control of acidity in our natural environment.

Some weak electrolytes in water function as either weak acids or weak bases that undergo dissociation, thereby reaching an equilibrium that involves both ions of the electrolyte and the undissociated electrolyte molecules. Aqueous solutions of weak acids, HA, and weak bases, WB, are represented by the following equilibria:

$$HA(aq) + H_2O(\ell) \rightleftharpoons H_3O^+(aq) + A^-(aq) \qquad K_a << 1$$

$$WB(aq) + H_2O(\ell) \rightleftharpoons WBH^+(aq) + OH^-(aq) \qquad K_b << 1$$

In general, the fraction of electrolyte molecules so dissociated at equilibrium ranges from a high of about 20% to a low of less than 1%.

The salts formed by the reaction of weak acids with weak bases are themselves weak bases or weak acids; they are called the *conjugate base* or *conjugate acid* of the respective acid or base. In combination with the parent acid or base, these salts form solutions, called *buffer solutions*, that are resistant to changes in acidity. If these salts are dissolved by themselves in water, they hydrolyze: that is, they react with water to form solutions that are acidic or basic depending on their origin.

The discussion in this chapter treats the equilibria of weak electrolytes in aqueous solution. Specifically, the equilibria include weak electrolytes, buffer solutions, and hydrolytic solutions. Exercises have been included to represent each type of equilibrium.

FORMULAS AND DEFINITIONS

Acid and base ionization constants (K_a for weak acids and K_b for weak bases)

An ionization constant is a measure of the dissociation of an acid or base in aqueous solution into $H_3O^+(aq)$ or $OH^-(aq)$. In general, aqueous solutions of weak acids and weak bases are ionized according to the reactions

$$HA(aq) + H_2O(\ell) \rightleftharpoons H_3O^+(aq) + A^-(aq)$$

$$WB(aq) + H_2O(\ell) \rightleftharpoons WBH^+(aq) + OH^-(aq)$$

where HA stands for a weak acid and WB stands for a weak base. The corresponding equilibrium expressions are

$$K_a = K[H_2O] = \frac{[H_3O^+][A^-]}{[HA]} \quad ; \quad K_b = K'[H_2O] = \frac{[WBH^+][OH^-]}{[WB]}$$

in which the concentration of water has been multiplied times the equilibrium constant and has been redefined as K_a or K_b. The brackets indicate molar concentrations of the ions or molecular species in solution. Typical values for K_a and K_b are considerably less than 1 (see Appendixes G and H). The larger the value of K_a or K_b, the greater the extent of ionization or dissociation; hence the acid or base is correspondingly stronger.

Buffer solution A buffer solution contains either a weak acid or a weak base that is in solution with a salt having an ion common to the weak acid or weak base. Such solutions resist changes in acidity produced by the addition of more acid or base. This is accomplished by reaction of the added acid or base with either the ions A^-, WBH^+, or with the molecular acid or base already in solution. The reactions that can occur upon the addition of acid or base to a buffer solution are as follows:

Weak acid-salt:

$$A^-(aq) + H^+(aq) \longrightarrow HA(aq)$$

$$HA(aq) + OH^-(aq) \longrightarrow A^-(aq) + H_2O$$

Weak base-salt:

$$WBH^+(aq) + OH^-(aq) \longrightarrow WB(aq) + H_2O$$

$$WB(aq) + H^+(aq) \longrightarrow WBH^+(aq)$$

Buffer solutions are described algebraically in terms of the K_a or K_b defining the weak acid or weak base. The general expressions defining buffers are

Weak acid-salt: $$K_a = [H^+] \frac{[\text{salt}]}{[\text{acid}]}$$

Weak base-salt: $$K_b = [OH^-] \frac{[\text{salt}]}{[\text{base}]}$$

Typical buffer solutions are composed of weak acids and their sodium salts or weak bases and their chlorides. Examples are acetic acid-sodium acetate and ammonia-ammonium chloride.

Ion-product constant for water (K_w) The equilibrium constant that describes the relation between the hydrogen ion and hydroxyl ion concentrations in aqueous solution is

$$K_w = [H^+][OH^-] = 1 \times 10^{-14} \quad \text{at } 25°C$$

In pure water, the concentration of H^+ and of OH^- equals 1×10^{-7} *M* at 25°C.

pH The pH of a solution is a measure of the hydrogen ion concentration in solution and is defined as

$$pH = -\log [H^+]$$

The pH values of most aqueous solutions range from a low of about 0 to a high of about 14; at a pH of 0, $[H^+] = 1.0$ *M*, and at 14, $[H^+] = 1 \times 10^{-14}$ *M*.

Example 1: Calculate the pH of a solution that has a hydrogen ion concentration of 0.0025 *M*.

Solution

If you use a calculator with logarithm capability, merely enter the concentration value, take the log, and change the sign to get pH. However, if you use log tables, the $[H^+]$ should be rewritten in scientific notation and the log taken as follows:

$$\begin{aligned} pH &= -\log 0.0025 = -\log 2.5 \times 10^{-3} \\ &= -[\log 2.5 + \log 10^{-3}] \\ &= -[0.3979 + (-3)] \\ &= -(-2.60) = 2.60 \end{aligned}$$

The numbers to the right of the decimal point are the significant figures in the pH value; the 2, in this case, represents the power of 10 in the log function. Further help with logs is available in Part III, p. 257.

Hydrogen ion concentrations are frequently needed in calculations involving equilibrium constants. Given the pH of a solution, the $[H^+]$ can be calculated by using the definition of pH.

$$\begin{aligned} pH &= -\log [H^+] \\ \log [H^+] &= -pH \\ [H^+] &= 10^{-pH} \end{aligned}$$

By using a calculator with log capability, you should be able to enter the exponent, -pH, and calculate directly. Log tables, however, contain only positive values, and negative log values must be rewritten as a product of two exponential base-10 terms for conversion to a rational term.

Example 2: Calculate the hydrogen ion concentration in a solution with a pH of 9.26.

Solution

$[H^+] = 10^{-pH} = 10^{-9.26}$. To evaluate this exponential, rewrite $10^{-9.26}$ as a product of two log base-10 exponentials — one with a positive decimal exponent, the other with a negative integer exponent. In order to maintain the same value as the original term, calculate the difference between -9.26 and the next smaller integer exponent. The result gives

$$10^{-9.26} = 10^{0.74} \times 10^{-10}$$

The $[H^+]$ is the antilog of 0.74 times 10^{-10}:

$$10^{0.74} = 5.5$$

and, therefore, $[H^+] = 5.5 \times 10^{-10}$

pOH The pOH of a solution is a measure of the hydroxide ion (OH^-) concentration in solution and is defined as

$$pOH = -\log [OH^-]$$

The expression for K_w relates pOH and pH through the concentration of OH^- and H^+ in the following way:

$$K_w = [H^+][OH^-] = 1 \times 10^{-14}$$

Take logs of both sides and multiply by (-1).

$$-\log [H^+] + (-\log [OH^-]) = -(-14)$$

Substitution from the definitions above yields

$$pH + pOH = 14$$

This relation is especially useful when working with basic solutions in which the $[OH^-]$ is known or is to be calculated.

Hydrolysis Salts formed from weak acids or weak bases interact with water to form solutions that are acidic or basic depending on the nature of the parent substance. The conjugate bases of weak acids and the conjugate acids of weak bases are relatively strong bases and acids, respectively. For the general case these substances undergo hydrolysis with water in the following ways:

$$A^-(aq) + H_2O(\ell) \rightleftharpoons HA(aq) + OH^-(aq) \quad (1)$$

$$WBH^+(aq) + H_2O(\ell) \rightleftharpoons WB(aq) + H_3O^+(aq) \quad (2)$$

Water functions as an acid in the reaction with A^- and as a base with WBH^+.

Hydrolysis ionization constant (K_h) This constant describes the equilibrium established in a reaction involving hydrolysis. For the reactions shown above in the definition of hydrolysis, the expressions for K_h are

For (1), $A^-(aq) + H_2O(\ell) \rightleftharpoons HA(aq) + OH^-(aq)$

$$K_h = \frac{[HA][OH^-]}{[A^-]}$$

Substitution of $[OH^-] = \frac{K_w}{H^+}$

gives $$K_h = \frac{[HA]K_w}{[A^-][H^+]} = \frac{K_w}{K_a\text{(weak acid)}}$$

Example 1: Calculate the hydrolysis constant for the cyanide ion, CN^-.

Solution

The ion, CN^-, is the conjugate base of the weak acid, HCN, $K_a = 4 \times 10^{-10}$. Hydrolysis will occur as follows:

$$CN^-(aq) + H_2O(\ell) \rightleftharpoons HCN(aq) + OH^-(aq)$$

$$K_h = \frac{[HCN][OH^-]}{[CN^-]} = \frac{[HCN]K_w}{[CN^-][H^+]} = \frac{1 \times 10^{-14}}{4 \times 10^{-10}} = 2 \times 10^{-5}$$

For (2), $$WBH^+(aq) + H_2O(\ell) \rightleftharpoons WB(aq) + H_3O^+(aq)$$

$$K_h = \frac{[WB][H^+]}{[WBH^+]}, \quad \text{where} \quad [H^+] = \frac{K_w}{[OH^-]}$$

Then, $$K_h = \frac{[WB]K_w}{[WBH^+][OH^-]} = \frac{K_w}{K_b\text{(weak base)}}$$

Example 2: Calculate the hydrolysis constant for the methyl ammonium ion, $CH_3NH_3^+$.

Solution

The ion, $CH_3NH_3^+$, is the conjugate acid of the weak base, CH_3NH_2, $K_i = 4.4 \times 10^{-4}$. Hydrolysis will occur as follows:

$$CH_3NH_3^+(aq) + H_2O(\ell) \rightleftharpoons CH_3NH_2(aq) + H_3O^+(aq)$$

$$K_h = \frac{[CH_3NH_2][H^+]}{[CH_3NH_3^+]} = \frac{[CH_3NH_2]K_w}{[CH_3NH_3^+][OH^-]}$$

$$= \frac{1.0 \times 10^{-14}}{4.4 \times 10^{-4}} = 2.3 \times 10^{-11}$$

To summarize, $$K_h = K_w/K_a\text{(weak acid)}$$

$$K_h = K_w/K_b\text{(weak base)}$$

Quadratic equations A quadratic equation is defined as any equation that can be written in the form

$$ax^2 + bx + c = 0$$

in which a, b, and c are constants and $a \neq 0$. Because these equations are second order, x^2, two solutions are possible, that is, a negative and a positive solution. In chemistry, reactions involve real amounts of substances; hence equations rarely are written in such a way that negative values of x are plausible.

Several methods can be used to solve quadratic equations, but the most convenient method for solving equations dealing with chemical processes is to use the quadratic formula (shown below), which is based on the general form of the quadratic equation

$$x = \frac{-b \pm \sqrt{b^2 - 4ac}}{2a}$$

EXERCISES

Note: Values of K_a and K_b not given in the exercises may be found in Appendixes G and H. In many cases the abbreviation HOAc is used for acetic acid.

3. What ionic and molecular species are present in an aqueous solution of hydrogen fluoride, HF? a solution of sulfuric acid? a solution of SO_2 in water (sulfurous acid)?

Solution

HF: Hydrogen fluoride is a weak acid in aqueous solution. As such, it ionizes to a slight extent to form ions as shown by the equation:

$$HF(aq) \rightleftharpoons H^+(aq) + F^-(aq)$$

H_2SO_4: Sulfuric acid is normally considered as a strong acid. It does, however, ionize in two steps in which the first step is complete in relatively dilute solutions. The second ionization is inhibited by the existence of hydrogen ion from the first reaction and it occurs only to about 10% while the HSO_4^- ion in water without H^+ as the counter ion would ionize to about 30%. The successive ionizations occur according to the following reactions.

$$H_2SO_4(aq) \xrightarrow{100\%} H^+(aq) + HSO_4^-(aq)$$

$$HSO_4^-(aq) \xrightleftharpoons{10\%} H^+(aq) + SO_4^{2-}(aq)$$

SO_2: Sulfur dioxide is the acid anhydride of sulfurous acid, H_2SO_3. Mixed with water, SO_2 reacts to form H_2SO_3 as shown by the equation

$$SO_2(g) + H_2O(\ell) \rightleftharpoons H_2SO_3(aq)$$

Sulfurous acid is a weak acid which ionizes to about 35% in a 0.1 *M* solution. It is fairly strong among the weak acids and, like H_2SO_4, ionizes in two steps.

$$H_2SO_3(aq) \xrightleftharpoons{\sim 35\%} H^+(aq) + HSO_3^-(aq) \qquad K_a = 1.2 \times 10^{-2}$$

The ionization of HSO_3^- occurs only slightly.

$$HSO_3^-(aq) \xrightleftharpoons{\leq 1\%} H^+(aq) + SO_3^{2-}(aq)$$

In solutions of polyprotic acids, all possible ions are present, but those generated by the second and third ionizations in sequence are present in much smaller amounts than those from the first step.

4. Calculate the concentration of each of the ions in the following solutions of strong electrolytes:

(a) 0.085 *M* HNO_3 (c) 0.107 *M* $[Fe(H_2O)_6]_2(SO_4)_3$

Solution

(a) Nitric acid, HNO_3, is a strong electrolyte and is fully ionized in aqueous solution. In the ionization one molecule of HNO_3 ionizes to produce one ion each of hydrogen ion, H^+, and nitrate ion, NO_3^-. Each ion concentration, therefore, equals the molarity of HNO_3.

$$HNO_3(aq) \xrightarrow{100\%} H^+(aq) + NO_3^-(aq)$$

$$0.085\ M \longrightarrow 0.085\ M + 0.085\ M$$

(c) Some metal ions in aqueous solution bind readily to one or more water molecules to form complex salts. The solid form of the complex, when placed in water, dissociates or ionizes fully to form the hydrated metal ions and the associated anions. The complex salt, hexaquoiron(III) sulfate, dissociates according to the following equation:

$$[Fe(H_2O)_6]_2(SO_4)_3(aq) \longrightarrow 2[Fe(H_2O)_6{}^{3+}](aq) + 3SO_4{}^{2-}(aq)$$

$[Fe(H_2O)_6]_2(SO_4)_3$	$[Fe(H_2O)_6{}^{3+}]$	$SO_4{}^{2-}$
1 mol	2 mol	3 mol
0.107 *M*	2(0.107 *M*)	3(0.107 *M*)
	0.214 *M*	0.321 *M*

5. From the equilibrium concentrations given, calculate K_a for each of the following weak acids or weak bases:

(a) CH_3CO_2H: $[H^+] = 1.34 \times 10^{-3}\ M$; $[CH_3CO_2^-] = 1.34 \times 10^{-3}\ M$;

$[CH_3CO_2H] = 9.866 \times 10^{-2}\ M$

Solution

Acetic acid, CH_3CO_2H, a weak acid, ionizes to the acetate ion, $CH_3CO_2^-$, and the hydrogen ion, H^+, according to the equation

$$CH_3CO_2H(aq) \rightleftharpoons H^+(aq) + CH_3CO_2^-(aq)$$

The equilibrium constant for the ionization, K_a, called the *acid ionization constant*, is defined as

$$K_a = \frac{[H^+][CH_3CO_2^-]}{[CH_3CO_2H]}$$

where the molar concentrations of the three species are the concentrations at equilibrium. Substitution of the component concentrations gives

$$K_a = \frac{[1.34 \times 10^{-3}][1.34 \times 10^{-3}]}{[9.866 \times 10^{-2}]} = 1.82 \times 10^{-5}$$

Keystrokes: 1.34 [EE] [+/−] 3 [x^2] ÷ 9.866 [EE] [+/−] 2 = ans. 1.8199878 − 05

Although most values of K_a have units related to molarity, the units are considered to be understood and are omitted for brevity. In this case, for example, the value of K_a with units is: 1.82×10^{-5} mol/L.

6. Using the ionization constants in Appendix G, calculate the hydrogen ion concentration and the percent ionization in each of the following solutions:

(a) 0.0092 *M* HClO

Solution

(a) The hydrogen ion concentration must be calculated from the ionization constant of hypochlorous acid, HClO. The acid ionizes as

$$HClO(aq) \rightleftharpoons H^+(aq) + ClO^-(aq)$$

$$K_a = \frac{[H^+][ClO^-]}{[HClO]} = 3.5 \times 10^{-8}$$

The small value of K_a indicates that only a small fraction of the initial HClO ionizes in solution. It is helpful to set up a tabular arrangement as shown below to better visualize the concentrations of the solution components present initially and at equilibrium.

	[HClO]	$[H^+]$	$[ClO^-]$
initial	0.0092	0	0
equilibrium	$0.0092 - x$	x	x

In most cases involving weak acids or bases, and certainly strong acids or bases, the amount of hydrogen ion or hydroxide ion in solution that is contributed by water is negligibly small and can be ignored. How much HClO remains at equilibrium? Let x equal the concentration of HClO that ionizes; therefore,

$$x = [H^+] = [ClO^-]$$

$$[HClO] = 0.0092 - x$$

Substitute these values into the expression for K_a and solve for x.

$$K_a = 3.5 \times 10^{-8} = \frac{(x)(x)}{(0.0092 - x)}$$

In the factor $(0.0092 - x)$, x is very small relative to 0.0092 and can be neglected, which greatly simplifies the calculation. In general, if the initial concentration of the ionizing substance and K_a for the substance differ by an exponential factor of 3 or more, x can be neglected in the denominator. In this case the [HClO] is 9.2×10^{-3} and K_a is 3.5×10^{-8}; the exponents differ by 5. Hence x can be neglected. This generalization is premised on the fact that K values usually have two significant figures, and the factor of x in the rigorous solution of the quadratic equation does not significantly add or subtract from the equation. Solving the equation for x and neglecting the x in the term $(0.0092 - x)$, gives

$$\frac{x^2}{0.0092} = 3.5 \times 10^{-8} \quad \text{or} \quad x^2 = 0.0092(3.5 \times 10^{-8}) = 3.22 \times 10^{-10}$$

Taking the square root of both sides gives

$$x = \pm 1.8 \times 10^{-5}$$

Notice that the negative value of x is not a plausible solution to the equation; only real amounts of substance can be ionized.

$$x = 1.8 \times 10^{-5}\ M = [H^+] = [ClO^-]$$

The percent of the initial HClO ionized at equilibrium is

$$\%\ \text{ionization} = \frac{x}{[HClO]\text{initial}} \times 100\%$$

$$= \frac{1.8 \times 10^{-5}}{0.0092} \times 100\% = 0.20\%$$

8. Calculate the hydrogen ion concentration and percent ionization of the weak acid in each of the following solutions. Note that the ionization constants may be such that the change in electrolyte concentration cannot be neglected and the quadratic formula or successive approximations may be required.

   (a) 0.0184 M HCNO ($K_a = 3.46 \times 10^{-4}$)

   (d) 0.02173 M CH_2ClCO_2H ($K_a = 1.4 \times 10^{-3}$)

Solution

The procedure for calculating the hydrogen ion concentration and percent ionization for the weak acids in this exercise is the same as in Exercise 6; the same rules apply.

(a) 0.0184 M HCNO ($K_a = 3.46 \times 10^{-4}$)

$$HCNO(aq) \rightleftharpoons H^+(aq) + CNO^-(aq)$$

$$K_a = 3.46 \times 10^{-4} = \frac{[H^+][CNO^-]}{[HCNO]}$$

Let x equal the concentration of HCNO that ionizes. At equilibrium the concentrations of the solution components are

$$x = [H^+] = [CNO^-]$$

$$[HCNO] = 0.0184 - x$$

Substitution gives

$$K_a = 3.46 \times 10^{-4} = \frac{(x)(x)}{0.0184 - x}$$

Can the x in $(0.0184 - x)$ be neglected?

$$[HCNO] = 1.84 \times 10^{-2} \quad \text{and} \quad K_a = 3.46 \times 10^{-4}$$

Exponential difference equals 2.

Answer: x cannot be neglected. Therefore, solve the equation for x by using the quadratic formula. The process is as follows:

$$x^2 = 3.46 \times 10^{-4}(0.0184 - x)$$
$$= 6.37 \times 10^{-6} - 3.46 \times 10^{-4}x$$

Rearrangement to the standard quadratic form gives

$$x^2 + 3.46 \times 10^{-4}x - 6.37 \times 10^{-6} = 0$$

The equation is now in the form $ax^2 + bx + c = 0$, and the quadratic formula applies:

$$x = \frac{-b \pm \sqrt{b^2 - 4ac}}{2a}$$

Substitution into the formula gives

$$x = \frac{-3.46 \times 10^{-4} \pm \sqrt{(3.46 \times 10^{-4})^2 - 4(1)(-6.37 \times 10^{-6})}}{2(1)}$$

and

$$x = \frac{-3.46 \times 10^{-4} \pm \sqrt{1.197 \times 10^{-7} + 2.548 \times 10^{-5}}}{2}$$

$$x = \frac{-3.46 \times 10^{-4} \pm \sqrt{2.56 \times 10^{-5}}}{2} = \frac{-3.46 \times 10^{-4} \pm 5.06 \times 10^{-3}}{2}$$

The positive root of x is

$$x = \frac{4.71 \times 10^{-3}}{2} = 2.36 \times 10^{-3}\ M = [H^+] = [CNO^-]$$

If the simplification had been made, the value of $[H^+]$ would have been 2.52×10^{-3}.

$$\%\ \text{ionization} = \frac{2.36 \times 10^{-3}}{0.0184} \times 100\% = 12.8\%$$

(d) $0.02173\ M\ CH_2ClCO_2H$ $(K_a = 1.4 \times 10^{-3})$

$$CH_2ClCO_2H(aq) \rightleftharpoons H^+(aq) + CH_2ClCO_2^-(aq)$$

$$K_a = 1.4 \times 10^{-3} = \frac{[H^+][CH_2ClCO_2^-]}{[CH_2ClCO_2H]}$$

Let x = moles of CH_2ClCO_2H that ionize at equilibrium; on a 1 L basis this is also the concentration.

$$x = [H^+] = [CH_2ClO_2^-]$$

$$[CH_2ClCO_2H] = 0.02173 - x$$

and $$K_a = 1.4 \times 10^{-3} = \frac{(x)(x)}{0.02173 - x}$$

Can x in $(0.02173 - x)$ be neglected? No! Solving the equation for x via the quadratic formula gives

$$x^2 = 1.4 \times 10^{-3}(0.02173 - x) = 3.042 \times 10^{-5} - 1.4 \times 10^{-3}x$$

$$x^2 = 1.4 \times 10^{-3}x - 3.042 \times 10^{-5} = 0$$

$$x = \frac{-1.4 \times 10^{-3} \pm \sqrt{(1.4 \times 10^{-3})^2 - 4(1)(-3.042 \times 10^{-5})}}{2(1)}$$

$$x = \frac{-1.4 \times 10^{-3} \pm \sqrt{1.236 \times 10^{-4}}}{2} = \frac{-1.4 \times 10^{-3} \pm 1.1 \times 10^{-2}}{2}$$

The positive root of x is

$$x = 4.9 \times 10^{-3}\ M = [H^+]$$

$$\%\ \text{ionization} = \frac{4.9 \times 10^{-3}}{0.02173} \times 100\% = 22\%$$

11. The ionization constant of lactic acid, $CH_3CHOHCO_2H$, is 1.36×10^{-4}. If 20.0 g of lactic acid is used to make a solution with a volume of 1.00 L, what is the concentration of hydrogen ion in the solution?

Solution

The hydrogen ion concentration is calculated from the expression for K_a after the concentration of lactic acid is calculated in units of molarity.

$$\text{Molar mass } CH_3CHOHCO_2H = 3(12.011) + 6(1.0079) + 3(15.9994) = 90.079 \text{ g}$$

$$\text{Molarity lactic acid} = 20.0 \text{ g} \times \frac{1 \text{ mol}}{90.079 \text{ g}} \times \frac{1}{1.00 \text{ L}} = 0.222\ M$$

The ionization of lactic acid is

$$CH_3CHOHCO_2H(aq) \rightleftharpoons CH_3CHOHCO_2^-(aq) + H^+(aq)$$

$$LA \rightleftharpoons LA^- + H^+$$

$$K_a = 1.36 \times 10^{-4} = \frac{[H^+][LA^-]}{[LA]}$$

Let x = moles of LA that ionize; this is also the concentration of LA that ionizes.

At equilibrium: $$x = [H^+] = [LA^-]$$

$$[LA] = 0.222 - x$$

and $$K_a = 1.36 \times 10^{-4} = \frac{(x)(x)}{0.222 - x}$$

Can x in $(0.222 - x)$ be neglected? This is a borderline situation — K_a is fairly large and has 3 significant figures. The value of x^2 as determined from the equation after dropping x in the denominator is

$$x^2 = (1.36 \times 10^{-4})(0.222) = 3.02 \times 10^{-5}$$

The positive root of x^2 gives 5.49×10^{-3} *M*. Without simplification, x is 5.43×10^{-3}, a difference of only 1%. In most laboratory work involving measurements with a pH meter, this small difference would not be detected.

14. Calculate the hydroxide ion concentration and the percent ionization of the weak base in each of the following solutions:

(a) 0.0784 *M* $C_6H_5NH_2$ ($K_b = 4.6 \times 10^{-10}$)

Solution

The hydroxide ion concentration is calculated from the equilibrium expression for K_b by using the same approach as in Exercise 11. Use of the quadratic formula may be necessary.

(a) 0.0784 *M* $C_6H_5NH_2$ ($K_b = 4.6 \times 10^{-10}$)

$$C_6H_5NH_2(aq) + H_2O(\ell) \rightleftharpoons C_6H_5NH_3^+(aq) + OH^-(aq)$$

$$K_b = K[H_2O] = 4.6 \times 10^{-10} = \frac{[C_6H_5NH_3^+][OH^-]}{[C_6H_5NH_2]}$$

Let x = moles of $C_6H_5NH_2$ that ionize at equilibrium:

$$x = [OH^-] = [C_6H_5NH_3^+]$$

$$[C_6H_5NH_2] = 0.0784 - x$$

Substitution into the expression for K_b gives

$$K_b = 4.6 \times 10^{-10} = \frac{(x)(x)}{0.0784 - x}$$

Can x in $(0.0784 - x)$ be neglected? Yes, 10^{-10} is more than 3 orders of magnitude smaller than 7.8×10^{-2}. Solving for the positive root of x gives

$$x^2 = (0.0784)(4.6 \times 10^{-10}) = 3.61 \times 10^{-11}$$

$$x = [OH^-] = 6.0 \times 10^{-6}\ M$$

$$\%\ \text{ionization} = \frac{6.0 \times 10^{-6}}{0.0784} \times 100\% = 7.6 \times 10^{-3}\%$$

(c) 0.222 *M* CN^- ($K_b = 2.5 \times 10^{-5}$)

$$CN^-(aq) + H_2O(\ell) \rightleftharpoons HCN(aq) + OH^-(aq)$$

$$K_b = \frac{[HCN][OH^-]}{[CN^-]} = 2.5 \times 10^{-5}$$

Let x = moles of CN^- that react with water to form HCN and OH^-

$$x = [OH^-] = [HCN]$$

$$[CN^-] = 0.222 - x$$

Substitution into the expression for K_b gives

$$K_b = \frac{(x)(x)}{(0.222 - x)} = 2.5 \times 10^{-5}$$

Can x in $(0.222 - x)$ be neglected? Yes, because the concentration is more than 3 orders of magnitude greater than K_b. Solving for the positive root of x from the expression gives

$$x^2 = (2.5 \times 10^{-5})(0.222) = 5.55 \times 10^{-6}$$

$$x = 2.4 \times 10^{-3}\ M$$

$$\%\ \text{ionization} = \frac{2.4 \times 10^{-3}}{0.222} \times 100\% = 1.1\%$$

17. Calculate the hydroxide ion concentration and the percent ionization of the weak base in each of the following solutions. Note that the ionization constants are sufficiently large that it may be necessary to use the quadratic formula or successive approximations.

(a) 4.113×10^{-3} *M* CH_3NH_2 ($K_b = 4.4 \times 10^{-4}$)

Solution

$$CH_3NH_2(aq) + H_2O(\ell) \rightleftharpoons CH_3NH_3^+(aq) + OH^-(aq)$$

$$K_b = 4.4 \times 10^{-4} = \frac{[CH_3NH_3^+][OH^-]}{[CH_3NH_2]}$$

Let x = moles of CH_3NH_2 that ionize.

At equilibrium:

$$x = [OH^-] = [CH_3NH_3^+]$$

$$[CH_3NH_2] = 4.113 \times 10^{-3} - x$$

Substituting these values into K_b gives

$$4.4 \times 10^{-4} = \frac{(x)(x)}{(4.113 \times 10^{-3} - x)}$$

Can x in $(4.113 \times 10^{-3} - x)$ be neglected? No, the positive root of x, when solved by use of the quadratic formula, is

$$x = [OH^-] = 1.14 \times 10^{-3}\ M$$

$$\%\ \text{ionization} = \frac{1.14 \times 10^{-3}}{4.11 \times 10^{-3}} \times 100\% = 28\%$$

18. Calculate the ionization constants for each of the following solutes from the percent ionization and the concentration of the solute:

(a) 0.050 M HClO, 8.4×10^{-2}% ionized

(b) 0.010 M HNO_2, 19% ionized

Solution

The ionization constant for a weak acid or base is calculated from the equilibrium concentration of the solution components. The initial concentration of the ionizing substance is given in each of the following examples along with the percent ionized. Multiplication of the initial amount by the percentage gives the equilibrium amount of ionized species. Subtraction of the equilibrium amount from the initial amount of material gives its equilibrium amount. Then the equilibrium constant can be calculated.

(a) 0.050 M HClO, 8.4×10^{-2}% ionized

$$HClO(aq) \rightleftharpoons H^+(aq) + ClO^-(aq)$$

At equilibrium:

$$[H^+] = 8.4 \times 10^{-2}\%\ \text{of}\ 0.050\ M = [ClO^-]$$
$$= (0.00084)(0.050) = 4.2 \times 10^{-5}\ M$$

Since 1 mol of HClO ionizes to produce 1 mol each of H^+ and ClO^-, the equilibrium concentration of HClO is

$$[HClO] = [0.050 - 4.2 \times 10^{-5}]\ M$$

$$K_a(HClO) = \frac{[H^+][ClO^-]}{[HClO]}$$

Substitution of the concentrations into the expression gives

$$K_a = \frac{(4.2 \times 10^{-5})(4.2 \times 10^{-5})}{(0.050 - 4.2 \times 10^{-5})} = 3.5 \times 10^{-8}$$

(b) 0.010 M HNO_2, 19% ionized

$$HNO_2(aq) \rightleftharpoons H^+(aq) + NO_2^-(aq)$$

$$K_a = \frac{[H^+][NO_2^-]}{[HNO_2]}$$

At equilibrium:

$$[H^+] = [NO_2^-] = 19\% \text{ of } 0.010\ M$$
$$= 0.19(0.010)M = 0.0019\ M$$
$$[HNO_2] = (0.010 - 0.0019)M$$
$$K_a = \frac{(0.0019)(0.0019)}{(0.010 - 0.0019)} = 4.4 \times 10^{-4}$$

23. Calculate the pH and pOH of pure water at 25°C.

Solution

Water self-ionizes

$$H_2O \rightleftharpoons H^+ + OH^-$$

The equilibrium constant, K, for this reaction is normally combined with the concentration of water and the product is defined as K_w (= 1.0×10^{-14} at 25°C).

$$K_w = [H^+][OH^-] = 1.0 \times 10^{-14}$$

According to the equilibrium, $[H^+] = [OH^-]$ and the concentrations are obtained from the expression for K_w.

$$[H^+] = [OH^-] = \sqrt{1.0 \times 10^{-14}} = 1.0 \times 10^{-7}$$

The pH of any solution or even pure water is defined as the negative logarithm of the hydrogen ion concentration.

$$pH = -\log [H^+]$$

For pure water: $pH = -\log [1.0 \times 10^{-7}]$

On a calculator equipped with a logarithm function, merely enter the $[H^+]$ as 1.0×10^{-7}, take the log by depressing the log key, and change the sign; this value is the pH of pure water.

Keystrokes: 1.0 [EE] [+/-] 7 [2nd] [log] [+/-] = ans. 7.00

However, if you are finding logs from a log table, the $[H^+]$ should be written in exponential notation and logs taken in the following way:

$$pH = -\log [H^+] = -\log 1.0 \times 10^{-7} = -[\log 1.0 + \log 10^{-7.0}]$$
$$= -[0.0000 + (-7.00)] = -(-7.00) = 7.0$$

Here the log of the rational term (1.0) has been found in the log table and added to the log of the exponential term, which is simply the value of the exponent. For further help see the introductory material to this chapter. See also "Exponential Notation and Logarithms" in Part Two of this manual for a programmed unit for self-help in this area.

The pOH of a solution or of pure water is defined as the negative logarithm of the hydroxide ion concentration.

$$pOH = -\log [OH^-] = -\log [1.0 \times 10^{-7}]$$

Compute the pOH in the same manner as was done for the pH.

$$pOH = -\log 1.0 \times 10^{-7} = -(-7.00) = 7.0$$

Alternatively, the pH plus the pOH for an aqueous solution at 25°C equals 14.0. By knowing either pH or the pOH, the other factor can be calculated simply by subtracting the known value from 14.0 to obtain the unknown. Given that the pH is 7.0, for example, the pOH is:

$$pH + pOH = 14.0$$

$$pOH = 14.0 - pH = 14.0 - 7.0 = 7.0$$

25. Calculate the pH and pOH of each of the following solutions:

(a) 0.200 *M* HCl (c) 0.0071 *M* $Ca(OH)_2$

Solution

Strong acids, such as HCl and HNO_3, and strong bases, such as NaOH and $Ca(OH)_2$, are completely ionized. The concentrations of $[H^+]$ and of $[OH^-]$ in these solutions equal their strong acid and strong base molarities, respectively.

(a) 0.200 *M* HCl

Hydrochloric acid is completely ionized. The H^+ concentration, therefore, is equal to the HCl molarity.

$$[H^+] = [HCl] = 0.200\ M$$

$$pH = -\log [H^+] = -\log 0.200 = -(-0.699) = 0.699$$

Keystrokes: 0.200 [2nd] [log] [+/-] = ans. .6989700043

$$pOH = 14.000 - 0.699 = 13.301$$

(c) 0.0071 *M* $Ca(OH)_2$

Assume that $Ca(OH)_2$ is fully ionized:

$$Ca(OH)_2(aq) \longrightarrow Ca^{2+}(aq) + 2OH^-(aq)$$

After ionization is complete, the OH^- concentration is equal to two times the initial molarity of $Ca(OH)_2$.

$$[OH^-] = 2(0.0071) = 0.0142\ M$$

The hydroxide ion concentration is known, so the pOH of the solution will be calculated first and then the pH.

$$pOH = -\log [OH^-] = -\log(0.0142) = -(-1.85) = 1.85$$

$$pH = 14.00 - pOH = 14.00 - 1.85 = 12.15$$

30. Calculate the pH and pOH of the following solution:

(c) 0.407 *M* $HC_2O_4^-$

Solution

This ion is a weak acid derived from the first ionization of oxalic acid, $H_2C_2O_4$. To calculate the pH, first determine the H^+ concentration from the acid ionization constant and the equilibrium expression.

$$HC_2O_4^-(aq) \rightleftharpoons H^+(aq) + C_2O_4^{2-}(aq) \qquad K_a = 6.4 \times 10^{-5}$$

$$K_a = \frac{[H^+][C_2O_4^{2-}]}{[HC_2O_4^-]} = 6.4 \times 10^{-5}$$

Let x = moles of $HC_2O_4^-$ that ionize.

At equilibrium:

$$x = [H^+] = [C_2O_4^{2-}]; \qquad [HC_2O_4^-] = 0.407 - x$$

Substitution of these values into the expression for K_a gives

$$K_a = \frac{(x)(x)}{0.407 - x} = 6.4 \times 10^{-5}$$

Simplifying the expression gives

$$x^2 = (6.4 \times 10^{-5})(0.407) = 2.60 \times 10^{-5}$$

$$x = [H^+] = 5.1 \times 10^{-3}$$

$$pH = -\log 5.1 \times 10^{-3} = -(-2.29) = 2.29$$

$$pOH = 14.00 - pH = 14.00 - 2.29 = 11.71$$

Keystrokes: 5.1 [EE] [+/-] 3 [2nd] [log] [+/-] = ans. 2.2924298

34. Calculate the pH of each of the following solutions containing two solutes in the concentrations indicated.

(a) 0.50 *M* HOAc, 0.50 *M* NaOAc
(e) 0.125 *M* NH_3, 1.00 *M* NaOH
(f) 0.400 *M* $NaHSO_4$, 0.400 *M* Na_2SO_4

Solution

(a) 0.50 *M* HOAc, 0.50 *M* NaOAc

This solution is a buffer solution containing the weak acid, HOAc, and a salt, NaOAc. The salt contains the anion of the acid. The acetic acid equilibrium describes the solution; the added NaOAc merely causes a shift in the equilibrium.

$$HOAc(aq) \rightleftharpoons H^+ + OAc^-(aq)$$

Let x = moles of HOAc that ionize. At equilibrium:

$$[H^+] = x$$

The total amount of OAc^- in solution equals the amount of OAc^- added by way of NaOAc plus the amount x produced by the ionization of HOAc:

$$[OAc^-] = (x + 0.50)M$$

$$[HOAc] = (0.50 - x)M$$

$$K_a = \frac{[H^+][OAc^-]}{[HOAc]} = 1.8 \times 10^{-5}$$

Substitution of the concentration into the equation gives

$$K_a = 1.8 \times 10^{-5} = \frac{(x)(x + 0.50)}{(0.50 - x)}$$

The rules for neglecting x in the denominator also apply to buffer or hydrolysis problems. Can x be neglected? Yes. Because x is small relative to 0.50 in $(0.50 - x)$, it also is small relative to 0.50 in $(x + 0.50)$. The expression is simplified to

$$1.8 \times 10^{-5} = \frac{(x)(0.50)}{(0.50)}$$

Solving for x gives

$$x = 1.8 \times 10^{-5} = [H_3O^+]$$

Also,

$$pH = -\log [H^+] = -\log(1.8 \times 10^{-5}) = -[\log 1.8 + \log 10^{-5}]$$
$$= -[0.2553 + (-5)] = 4.74$$

(e) 0.125 M NH_3, 1.00 M NaOH

This solution is a combination of the weak base, NH_3, and the strong base, NaOH. At equilibrium in a solution containing 0.125 M NH_3 only, the OH^- concentration is approximately 1.5×10^{-3} M. Addition of 1 mol of NaOH (1.0 mol of OH^-) to the solution simply overwhelms the equilibrium, and the $[OH^-]$ approximately equals the $[OH^-]$ from NaOH.

At equilibrium:

$$NH_3(aq) \rightleftharpoons NH_4^+(aq) + OH^-(aq)$$

$$[OH^-] \approx [NaOH] = 1.00\ M$$

$$pOH = -\log 1.00 = 0.00$$

$$pH = 14.00$$

(f) 0.400 *M* $NaHSO_4$, 0.400 *M* Na_2SO_4

In this solution $NaHSO_4$ is a moderately strong acid that ionizes as

$$HSO_4^-(aq) \rightleftharpoons H^+(aq) + SO_4^{2-}(aq)$$

$$K_a = \frac{[H^+][SO_4^{2-}]}{[HSO_4^-]} = 1.2 \times 10^{-2}$$

Let x = moles of HSO_4^- that ionize.

At equilibrium:

$$[H^+] = x$$

$$[SO_4^{2-}] = (x + 0.400)M$$

$$[HSO_4^-] = (0.400 - x)M$$

Substituting the concentrations into the expression gives

$$\frac{(x)(x + 0.400)}{(0.400 - x)} = 1.2 \times 10^{-2}$$

Can x in $(0.400 - x)$ be neglected? No. Clear the fraction and transpose terms to get

$$x^2 + 0.412x - 0.0048 = 0$$

Solving for x by way of the quadratic formula gives

$$x = \frac{-0.412 \pm \sqrt{(0.412)^2 - 4(1)(-0.0048)}}{2}$$

$$= \frac{-0.412 \pm 0.435}{2} = 0.0113 = [H_3O^+]$$

$$pH = -\log(0.0113) = -\log 1.13 \times 10^{-2} = 1.95$$

37. Calculate the concentration of each species present in a 0.050 *M* solution of H_2S.

Solution

Hydrogen sulfide is a weak acid that ionizes in two steps:

1. $$H_2S(aq) \rightleftharpoons H^+(aq) + HS^- \quad K_1 = 1.0 \times 10^{-7}$$

2. $$HS^-(aq) \rightleftharpoons H^+(aq) + S^{2-} \quad K_2 = 1.3 \times 10^{-13}$$

The total hydrogen ion concentration approximately equals the H^+ generated from the first ionization. K_2 is smaller than K_1 by almost 10^6 times, so very little of the HS^- formed in reaction 1 ionizes to give additional hydrogen ions and S^{2-}. The second step is further depressed by the existence of H^+ from reaction 1. It can be assumed that the concentrations of H^+ and of HS^- are practically equal.

From reaction 1:

$$K_1 = 1.0 \times 10^{-7} = \frac{[H^+][HS^-]}{[H_2S]}$$

Let x = the concentration or amount (mol) of H_2S that ionizes.

$$x = [H^+] = [HS^-]; \qquad 0.05 - x = [H_2S]$$

$$\frac{(x)(x)}{(0.5 - x)} = 1.0 \times 10^{-7}$$

Simplify to get

$$x^2 = (1.0 \times 10^{-7})(0.05) = 5.0 \times 10^{-9}$$

$$x = [H^+] = [HS^-] = 7.1 \times 10^{-5}\ M$$

Next, substitute this data into the expression for K_2.

$$K_2 = 1.3 \times 10^{-13} = \frac{[H^+][S^{2-}]}{[HS^-]}$$

The concentration of S^{2-} equals K_2 because $[H^+]$ equals $[HS^-]$ and $[H_2S] = 0.050\ M$ because the amount that dissociates is so small. The $[OH^-]$ is calculated from K_w as

$$[OH^-] = \frac{K_w}{[H^+]} = \frac{1.0 \times 10^{-14}}{7.1 \times 10^{-5}} = 1.4 \times 10^{-10}\ M$$

42. Calculate the pH of the following buffer solution, containing two solutes in the concentrations indicated:

(d) 0.25 M H_3PO_4; 0.15 M NaH_2PO_4

Solution

Assume that NaH_2PO_4 ionizes completely to give

$$NaH_2PO_4(aq) \longrightarrow Na^+(aq) + H_2PO_4^-(aq)$$

The $H_2PO_4^-$ is the ion in common with the first ionization of H_3PO_4; it serves as the first ionization of H_3PO_4 and as the salt-acid pair required for the solution to have buffering capacity. The amount of hydrogen ion in solution is controlled by the equilibrium of H_3PO_4, and it is calculated from the expression for K_a of H_3PO_4.

$$H_3PO_4(aq) \rightleftharpoons H^+(aq) + H_2PO_4^-(aq) \qquad K_a = 7.5 \times 10^{-3}$$

Let x = moles of H_3PO_4 that ionize.

At equilibrium:

$$x = [H^+]$$

The total amount of $H_2PO_4^-$ in solution equals the amount of $H_2PO_4^-$ added plus that produced by the ionization of H_3PO_4.

$$[H_2PO_4^-] = (x + 0.15)M$$

$$[H_3PO_4] = (0.25 - x)M$$

Substitution into the expression for K_a gives

$$K_a = \frac{[H^+][H_2PO_4^-]}{[H_3PO_4]} = \frac{(x)(x+0.15)}{(0.25-x)} = 7.5 \times 10^{-3}$$

We can assume that x is very small relative to 0.25 and to 0.15 and can be ignored. The expression simplifies to

$$K_a = 7.5 \times 10^{-3} = \frac{(x)(0.15)}{(0.25)}$$

Solving for x gives

$$x = [H^+] = \frac{(7.5 \times 10^{-3})(0.25)}{0.15} = 1.25 \times 10^{-2}$$

$$pH = -\log [H^+] = -\log 1.25 \times 10^{-2} = -(-1.90) = 1.90$$

53. How many moles of NH_4Cl must be added to 1.0 L of 1.0 M solution of NH_3 to prepare a buffer solution with a pH of 9.00? of 9.50?

Solution

pH = 9.00; the equilibrium expression for ammonia defines this buffer system: NH_4^+ from NH_4Cl is the common ion.

$$NH_3(aq) + H_2O(\ell) \rightleftharpoons NH_4^+(aq) + OH^-(aq) \qquad K_a = 1.8 \times 10^{-5}$$

$$K_a = \frac{[NH_4^+][OH^-]}{[NH_3]} = 1.8 \times 10^{-5}$$

Given a pH of 9.00, the pOH equals 5.00 and the $[OH^-]$ equals 1.0×10^{-5}. Substituting this OH^- concentration into the expression for K_a gives

$$\frac{[NH_4^+](1.0 \times 10^{-5})}{(1.0)} = 1.8 \times 10^{-5}$$

Solving for $[NH_4^+]$ gives

$$[NH_4^+] = 1.8\ M$$

for 1.0 L, 1.8 moles of NH_4Cl must be added.

55. A buffer solution is made up of equal volumes of 0.100 *M* acetic acid and 0.500 *M* sodium acetate.

    (a) What is the pH of this solution?

    (b) What is the pH that results from adding 1.00 mL of 0.100 *M* HCl to 0.200 L of the buffer solution? (Use 1.80×10^{-5} for the ionization constant of acetic acid.)

Solution

0.100/2 *M* HOAc and 0.500/2 *M* NaOAc

(a) The equilibrium expression for HOAc defines this buffer system.

$$HOAc(aq) \rightleftharpoons H^+(aq) + OAc^-(aq)$$

$$K_a = 1.80 \times 10^{-5} = \frac{[H^+][OAc^-]}{[HOAc]}$$

Let x = moles of HOAc that ionize.
At equilibrium:

$$[H_3O^+] = x$$

$$[HOAc] = (0.0500 - x)M$$

$$[OAc^-] = (0.250 + x)M$$

Substitution into the K_a expression gives

$$K_a = 1.80 \times 10^{-5} = \frac{(x)(0.250 + x)}{(0.0500 - x)}$$

The value of x in $(0.0500 - x)$ may be dropped on the basis of our guidelines. However, the value of x is even smaller in a buffer solution than in a solution containing only the acid because there is a shift in the equilibrium toward HOAc. By the same reasoning, x can be neglected in $(0.250 + x)$. Rearranging and clearing the fraction gives

$$x = [H_3O^+] = \frac{1.80 \times 10^{-5}}{5} = 3.60 \times 10^{-6}$$

$$pH = -\log 3.6 \times 10^{-6} = 5.444$$

(b) Because a buffer solution contains both an acidic substance and a basic substance, the solution has the capacity to react with either added acid or base. In this case HCl is being added to the buffer system. The effect on the equilibrium and, subsequently, the pH may be found as follows. First, the HCl exists as ions in aqueous solution and the NaOAc dissociates.

$$HCl(aq) \longrightarrow H_3O^+(aq) + Cl^-(aq)$$

$$NaOAc(aq) \longrightarrow Na^+(aq) + OAc^-(aq)$$

$Cl^-(aq)$ and $Na^+(aq)$ remain unchanged in the next step.

The hydronium ion then reacts quantitatively with acetate ions forming an equilibrium leaving only small amounts of H_3O^+ and OAc^- in solution,

$$\underset{1\ \text{mol}}{H_3O^+(aq)} + \underset{1\ \text{mol}}{OAc^-(aq)} \rightleftharpoons \underset{1\ \text{mol}}{HOAc(aq)} + H_2O$$

Again, letting x equal the moles of HOAc that ionize according to the sequence above, the initial amounts of solution components are

$$\text{Moles HCl} = (0.00100\ \text{L})\left(0.100\ \frac{\text{mol}}{\text{L}}\right) = 0.00010\ \text{mol}$$

$$\text{Moles OAc}^- = (0.200\ \text{L})\left(0.250\ \frac{\text{mol}}{\text{L}}\right) + x = 0.0500\ \text{mol} + x$$

$$\text{Moles HOAc} = (0.200\ \text{L})\left(0.0500\ \frac{\text{mol}}{\text{L}}\right) - x = 0.0100\ \text{mol} - x$$

After the reaction with HCl, the equilibrium is reestablished, and the concentrations are

$$[HCl] = 0; \qquad [H_3O^+] = x$$

$$[OAc^-] = \frac{(0.0500 + x - 0.00010)\ \text{mol}}{0.201\ \text{L}} = 0.248\ M + x \approx 0.248\ M$$

$$[HOAc] = \frac{(0.0100 - x + 0.00010)\ \text{mol}}{0.201\ \text{L}} = 0.0502\ M - x \approx 0.0502\ M$$

Quite clearly, the concentrations of OAc^- and HOAc are only slightly different after the addition of a small amount of HCl. The value of $[H_3O^+]$ is found from the expression

$$K_a = 1.80 \times 10^{-5} = \frac{[H_3O^+][OAc^-]}{[HOAc]} = \frac{(x)(0.248)}{(0.0502)} = x(4.94)$$

$$x = \frac{1.80 \times 10^{-5}}{4.94} = 3.64 \times 10^{-6} = [H_3O^+]$$

Therefore,

$$\text{pH} = -\log 3.64 \times 10^{-6} = 5.438$$

Thus the pH of the solution is only very slightly changed, as shown.

74. Calculate the ionization constant (hydrolysis constant) for each of the following acids or bases:

(a) F^- (d) NH_4^+

Solution

F^-: The fluoride ion hydrolyzes in the following way:

$$F^-(aq) + H_2O(\ell) \rightleftharpoons HF(aq) + OH^-(aq)$$

An expression for the equilibrium is written in the usual way, but the ionization constant shall now be defined as the hydrolysis constant, K_h.

$$K_h = \frac{[HF][OH^-]}{[F^-]}$$

This expression is related to the ionization of HF:

$$HF(aq) \rightleftharpoons H^+(aq) + F^-(aq)$$

$$K_a = \frac{[H^+][F^-]}{[HF]} = 7.2 \times 10^{-4}$$

Inspection of the expressions for K_h and for K_a indicates an inverse relationship between K_a and K_h with the substitution of $[H^+] = K_w/[OH^-]$ in the expression for K_a.

$$K_a = \frac{[H^+][F^-]}{[HF]} = \frac{\frac{K_w}{[OH^-]} \times [F^-]}{[HF]} = \frac{K_w[F^-]}{[HF][OH^-]}$$

$$K_h = K_w \times \frac{1}{K_a} = \frac{K_w}{K_a}$$

This final relationship turns out to be the general case for the hydrolysis of an ion from either a weak acid or a weak base.

For the F^- ion:

$$K_h = \frac{1.0 \times 10^{-14}}{7.2 \times 10^{-4}} = 1.4 \times 10^{-11}$$

(d) NH_4^+: This ion is the conjugate acid of the weak base ammonia. Hydrolysis of NH_4^+ gives

$$NH_4^+(aq) + H_2O(\ell) \rightleftharpoons NH_3(aq) + H_3O^+(aq)$$

$$K_h = \frac{[NH_3][H_3O^+]}{[NH_4^+]} = \frac{K_w}{K_b(NH_3)} = \frac{1.0 \times 10^{-14}}{1.8 \times 10^{-5}} = 5.6 \times 10^{-10}$$

75. Calculate the pH of each of the following solutions:

(a) 0.4735 *M* NaCN (d) 0.333 *M* $[(CH_3)_2NH_2]_2SO_4$

[Note that $(CH_3)_2NH_2^+$ is the conjugate acid of the weak base $(CH_3)_2NH$, just as NH_4^+ is the conjugate acid of the weak base NH_3.]

Solution

The solute in each of these solutions is salt composed of ions in which one of the ions originated from either a weak acid or a weak base. These ions undergo hydrolysis to produce acidic or basic solutions depending on their origin.

(a) 0.4735 *M* NaCN. This salt dissociates in water, yielding

$$NaCN(aq) \rightleftharpoons Na^+(aq) + CN^-(aq)$$

Hydrolysis of the CN^- occurs because it is the anion of a *weak* acid. Similar hydrolysis of Na^+ does not occur because Na^+ is a cation of a *strong* base and is fully dissociated in aqueous solution. Thus

$$CN^-(aq) + H_2O(\ell) \rightleftharpoons HCN(aq) + OH^-(aq)$$

The hydroxide ion is a strong base and HCN is a weak acid; the reaction increases the concentration of hydroxide ion, thereby causing the solution to be basic. The equilibrium constant for this process is called *K*-hydrolysis and is written K_h. Its value is

$$K_h = \frac{[HCN][OH^-]}{[CN^-]}$$

where the concentration of water is absorbed into the value of K_h. Since $[OH^-] = K_w/[H^+]$, substitution gives

$$K_h = \frac{[HCN]K_w}{[CN^-][H^+]} = \frac{K_w}{K_a(HCN)}$$

where

$$K_a(HCN) = \frac{[H^+][CN^-]}{[HCN]} = 4 \times 10^{-10}$$

Therefore,

$$K_h = \frac{1 \times 10^{-14}}{4 \times 10^{-10}} = 2.5 \times 10^{-5}$$

If in our reaction x equals the moles of CN^- that hydrolyze, then, at equilibrium,

$$x = [HCN] = [OH^-]$$

and

$$[CN^-] = (0.4735 - x)M$$

$$K_h = 2.5 \times 10^{-5} = \frac{(x)(x)}{0.4735 - x}$$

Neglect x in $(0.4735 - x)$.

$$x^2 = 1.18 \times 10^{-5} \quad \text{or} \quad x = 3.44 \times 10^{-3} = [OH^-]$$

$$pOH = -\log 3.44 \times 10^{-3} = 2.46$$

$$pH = 14.00 - pOH = 14.00 - 2.46 = 11.5$$

(d) 0.333 *M* $[(CH_3)_2NH_2]_2SO_4$ (dimethylamine sulfate). The dimethylammonium ion will hydrolyze as

$$(CH_3)_2NH_2^+(aq) + H_2O(\ell) \rightleftharpoons (CH_3)_2NH(aq) + H_3O^+(aq)$$

$$K_h = \frac{K_w}{K_i((CH_3)_2NH)} = \frac{1.0 \times 10^{-14}}{7.4 \times 10^{-4}} = \frac{[(CH_3)_2NH][H_3O^+]}{[(CH_3)_2NH_2^+]} = 1.35 \times 10^{-11}$$

Let x = moles of $(CH_3)_2NH_2^+$ that hydrolyze.

At equilibrium:

$$x = [H_3O^+] = [(CH_3)_2NH]$$

Because the dissociation of the sulfate produces 2 ions of $(CH_3)_2NH_2^+$,

$$[(CH_3)_2NH_2^+] = 2(0.333) - x$$

$$K_h = 1.35 \times 10^{-11} = \frac{(x)(x)}{(0.666 - x)}$$

x is sufficiently small compared to 0.666 that it can be neglected.

$$1.35 \times 10^{-11} = \frac{x^2}{0.666}$$

$$x^2 = 8.99 \times 10^{-12}$$

$$x = 3.00 \times 10^{-6} = [H_3O^+]$$

$$pH = -\log(3.00 \times 10^{-6}) = 5.52$$

81. A 0.0010 *M* solution of KCN is hydrolyzed to the extent of 14.0%. Calculate the value of the ionization constant of HCN.

Solution

The cyanide ion, CN^-, is the conjugate base of the weak acid, HCN, and it hydrolyzes as

$$CN^-(aq) + H_2O(\ell) \rightleftharpoons HCN(aq) + OH^-(aq)$$

$$K_h = \frac{[HCN][OH^-]}{[CN^-]} = \frac{K_w}{K_a(HCN)}$$

Each concentration factor in the expression for K_h can be determined from the degree of hydrolysis and the initial concentration of KCN.

	$[CN^-]$	$[HCN]$	$[OH^-]$
Initial:	0.0010	0	0
	0.0010 − (0.14)(0.0010)	(0.14)(0.0010)	(0.14)(0.0010)
Final:	0.00086	0.00014	0.00014

$$K_h = \frac{(0.00014)(0.00014)}{(0.00086)} = 2.28 \times 10^{-5}$$

$$K_h = \frac{K_w}{K_a(\text{HCN})} \quad \text{or} \quad K_a(\text{HCN}) = \frac{K_w}{K_h} = \frac{1.0 \times 10^{-14}}{2.28 \times 10^{-5}} = 4.4 \times 10^{-10}$$

90. In many detergents phosphates have been replaced with silicates as water conditioners. If 125 g of a detergent that contains 8.0% Na_2SiO_3 by weight is used in 4.0 L of water, what are the pH and the hydroxide ion concentration in the wash water?

$$SiO_3^{2-} + H_2O \rightleftharpoons SiO_3H^- + OH^- \quad K_b = 1.6 \times 10^{-3} \quad (1)$$

$$SiO_3H^- + H_2O \rightleftharpoons SiO_3H_2 + OH^- \quad K_b = 3.1 \times 10^{-5} \quad (2)$$

Solution

The hydrolysis of SiO_3^{2-} occurs in two steps. However, owing to the small amount of SiO_3H^- produced in reaction (1) and the low K_b in (2), the amount of OH^- in the final solution is due almost entirely to the first hydrolysis. Calculate the molarity of $NaSiO_3$ and let x equal the moles of SiO_3^{2-} that hydrolyze.

$$\text{Molarity } Na_2SiO_3 = 0.080(125\text{ g}) \times \frac{1\text{ mol}/122.1\text{ g}}{4.0\text{ L}} = 0.0205\ M$$

At equilibrium:

$$x = [OH^-] \approx [SiO_3H^-]$$

$$[SiO_3^{2-}] \approx (0.0205 - x)$$

$$K_b = 1.6 \times 10^{-3} = \frac{(x)(x)}{(0.0205 - x)}$$

or

$$x^2 = 1.6 \times 10^{-3}(0.0205 - x)$$

$$x^2 = 1.6 \times 10^{-3}x - 3.28 \times 10^{-5} = 0$$

$$x = \frac{-1.6 \times 10^{-3} \pm \sqrt{2.56 \times 10^{-6} + 4(3.28 \times 10^{-5})}}{2}$$

$$= \frac{-1.6 \times 10^{-3} \pm 1.16 \times 10^{-2}}{2} = \frac{10.0 \times 10^{-3}}{2}$$

$$= [OH^-] = 5.0 \times 10^{-3}\ M$$

$$pOH = -\log 5.0 \times 10^{-3} = 2.30$$

$$pH = 11.70$$

96. Lime juice is among the most acidic of fruit juices, with a pH of 1.92. If the acidity is due to citric acid, which we can abbreviate as H_3Cit, what is the ratio of each of the following to $[Cit^{3-}]$: $[H_3Cit]$; $[H_2Cit^-]$; $[HCit^{2-}]$?

$$H_3Cit \rightleftharpoons H^+ + H_2Cit^- \qquad K_a = 8.4 \times 10^{-4} \qquad (1)$$

$$H_2Cit^- \rightleftharpoons H^+ + HCit^{2-} \qquad K_a = 1.8 \times 10^{-5} \qquad (2)$$

$$HCit^{2-} \rightleftharpoons H^+ + Cit^{3-} \qquad K_a = 4.0 \times 10^{-6} \qquad (3)$$

Solution

The pH of the solution and the K_a values for the three ionization steps are known, so a series of ratios written in terms of $[Cit^{3-}]$ can be written. Starting with reaction (3),

$$pH = 1.92; \qquad [H_3O^+] = 10^{-1.92} = 1.20 \times 10^{-2}\ M$$

$$\frac{[H^+][Cit^{3-}]}{[HCit^{2-}]} = 4.0 \times 10^{-6}$$

and $$[HCit^{2-}] = \frac{(1.20 \times 10^{-2})[Cit^{3-}]}{4.0 \times 10^{-6}} = 3.0 \times 10^3[Cit^{3-}]$$

For reaction (2),

$$\frac{[H^+][HCit^{2-}]}{[H_2Cit^-]} = 1.8 \times 10^{-5}$$

$$[H_2Cit^-] = \frac{(1.20 \times 10^{-2})[HCit^{2-}]}{1.8 \times 10^{-5}}$$

$$= 6.67 \times 10^2[HCit^{2-}] = (6.67 \times 10^2)(3.0 \times 10^3)[Cit^{3-}]$$

$$= 2.0 \times 10^6[Cit^{3-}]$$

For reaction (1),

$$\frac{[H^+][H_2Cit^-]}{[H_3Cit]} = 8.4 \times 10^{-4}$$

$$[H_3Cit] = \frac{1.20 \times 10^{-2}[H_2Cit^-]}{8.4 \times 10^{-4}}$$

$$= 14.3[H_2Cit^-] = (14.3)(2.0 \times 10^6)[Cit^{3-}]$$

$$= 2.9 \times 10^7[Cit^{3-}]$$

$[H_3Cit]$	: $[H_2Cit^-]$	: $[HCit^{2-}]$	: $[Cit^{3-}]$
2.9×10^7	2.0×10^6	3.0×10^3	1

RELATED EXERCISES

1. Calculate the pH of each of the following solutions:

   (a) 35.00 mL of 1.000 *M* KOH mixed with 65.00 mL of 0.7500 *M* HCl.
   (b) 20.0 g of NaCN diluted with distilled water to 500.0 mL.
   (c) 15.0 g of benzoic acid, C_6H_5COOH, in 1.00 L of solution.

$$K_i(C_6H_5COOH) = 6.46 \times 10^{-5}$$

(d) 3.00 g of CH_3NH_2 in 250. mL of solution.

Answer: (a) 0.86; (b) 8.7; (c) 2.5; (d) 11.5

2. The procedure for the preparation of a buffer solution directs that 83.33 mL of 6.00 *M* acetic acid be dissolved with 205.0 g of sodium acetate and the mixture diluted with distilled water to 1.00 L. Calculate the pH of this buffer solution.

Answer: pH = 5.44

3. Although several buffer combinations exist in human blood, the blood is mainly buffered by the H_2CO_3—HCO_3^- system. Calculate the ratio of HCO_3^- to H_2CO_3 in blood given that the pH of blood is 7.4.

Answer: HCO_3—H_2CO_3 = *10.8 : 1*

4. Urine produced by a human whose diet has been only fruits and vegetables will have a pH that is basic, about 7.6. Given that normal elimination of urine is about 1300 mL per day, calculate the number of equivalents of hydrogen ion eliminated per day.

Answer: Equiv $H^+ = 3.3 \times 10^{-8}$

5. At one point in the titration of a 20.0 mL sample of vinegar, the pH of the solution is 5.10. Given that the initial pH of the vinegar (acetic acid in water) was 2.80, calculate the number of moles of NaOH required to produce this change in pH.

Answer: Moles NaOH $= 1.93 \times 10^{-3}$ *mol*

UNIT 12

CHAPTER 17: THE SOLUBILITY PRODUCT PRINCIPLE

INTRODUCTION

Many ionic substances function as strong electrolytes in water although they are only slightly soluble in it. Because of their slight solubility in water, such substances often form mineral deposits that are valuable to miners. These substances include carbonates — such as limestone, $CaCO_3$ — and sulfides, such as pyrite, FeS_2, and cinnabar, HgS. Mercury(II) sulfide, HgS, is one of the least soluble ionic substances: only about 1.3×10^{-24} g will dissolve in 1 L of water.

The occurrence of mineral deposits can be understood by considering the mode of formation of slightly soluble salts. In reactions involving the formation of a slightly soluble substance, the solubility is usually exceeded and a precipitate forms. For example, the salt $BaSO_4$ exists in water as $Ba^{2+}(aq)$ and $SO_4^{2-}(aq)$, but the maximum concentration at 25°C is only 1×10^{-5} *M*. When a solution containing barium ions, such as $BaCl_2$, is added to one containing sulfate ions, such as H_2SO_4, the solubility limit is exceeded and a white precipitate of $BaSO_4$ forms and settles to the bottom. The solid $BaSO_4$ is then in equilibrium with the ions in solution. This equilibrium is defined in terms of the product of ion concentrations, that is, the solubility product, K_{sp}.

$$BaSO_4(s) \rightleftharpoons Ba^{2+}(aq) + SO_4^{2-}(aq)$$

$$K_{sp} = [Ba^{2+}][SO_4^{2-}] = 1.08 \times 10^{-10}$$

Thus a precipitate will form when the ion concentrations in the solubility product expression exceed the K_{sp}. Precipitation will continue until the product of the concentrations equals the K_{sp}.

Slightly soluble substances that contain ions derived from weak acids or weak bases dissolve and undergo hydrolysis or association depending on the pH of the solution. Two or more simultaneous equilibria that exist in solution may compete for the ions. In such cases all equilibria must be considered in the determination of the solubility.

Several concepts involving solubility are treated in this chapter of the textbook. Solubility product constants with several examples are presented and should be studied carefully before attempting to solve problems at the end of the chapter. The K_{sp} concept is extended to the dissolution of precipitates through the formation of complex ions, which ultimately involve simultaneous equilibria. Many examples that involve K_{sp} also involve the concepts of weak acids, buffers, and hydrolysis as treated in Chapter 16.

FORMULAS AND DEFINITIONS

Saturated solution A saturated solution contains a slightly soluble solute in a two-phase system in which dissolved solute is in contact with undissolved solute. The system is at equilibrium, meaning that undissolved solute continues to dissolve while simultaneously, and at the same rate, ions in solution associate and crystallize.

Solubility product constant (K_{sp}) An equilibrium constant based on the molar concentrations of ions in a saturated solution containing a slightly soluble electrolyte. Each ion concentration is raised to the power corresponding to its stoichiometric coefficient as shown in the equilibrium expression. In general, the expression for K_{sp} is written as

$$A_xB_y(s) \rightleftharpoons xA(aq)^{(\text{positive})} + yB(aq)^{(\text{negative})}$$

$$K_{sp} = \left[A^{\text{positive}}\right]^x \left[B^{\text{negative}}\right]^y$$

Solubility product constants are published without units for the sake of brevity. But always keep in mind that units for constants are in terms of concentrations raised to some power and sometimes must be used in conjunction with equilibrium calculations.

Complex ion-formation constant Certain metal ions associate with ligands to form ions containing two or more species. Examples include $Ag(NH_3)_2^+$, $Cu(CN)_2^-$, $Al(OH)_4^-$, and $Sn(OH)_3^-$. Many of the complex ions are quite soluble in water, and they provide a mechanism for dissolving precipitates of slightly soluble electrolytes.

In solution, ions associate reversibly to form complex ions and to reach an equilibrium defined by the formation constant, K_{form}. The value of K_{form} is a measure of the tendency to form the complex ion and the chemical stability of the complex. Typical values for formation constants are very large, as indicated for the formation of $Ag(NH_3)_2^+$.

$$Ag^+(aq) + 2NH_3(aq) \rightleftharpoons Ag(NH_3)_2^+$$

$$K_{form} = \frac{[Ag(NH_3)_2^+]}{[Ag^+][NH_3]^2} = 1.6 \times 10^7$$

Calculations involving formation constants are similar to those for ionic equilibria.

EXERCISES

2. How do the concentrations of Ag^+ and CrO_4^{2-} in a liter of water above 1.0 g of solid Ag_2CrO_4 change when 100 g of solid Ag_2CrO_4 is added to the system? Explain.

   Solution

   This is an example of a saturated solution in which solid solute is in contact with its water solution. The solution contains the maximum possible amount of solute. Addition of Ag_2CrO_4 to this system will not change the amount of Ag_2CrO_4 that dissolves and dissociates. Any additional Ag_2CrO_4 merely precipitates and becomes part of the dynamic equilibrium.

8. Which of the following compounds will precipitate from a solution initially containing the indicated concentrations of ions (see Appendix E for K_{sp} values)?

   (a) $CaCO_3$: $[Ca^{2+}] = 0.003\ M$, $[CO_3^{2-}] = 0.003\ M$

   (b) $Co(OH)_2$: $[Co^{2+}] = 0.01\ M$, $[OH^-] = 1 \times 10^{-7}\ M$

   Solution

   (a) $CaCO_3$: Precipitation will occur if the ion concentrations exceed the amounts allowed by the solubility product constant. To determine whether this occurs, write the expression for K_{sp}, substitute the individual concentrations into the expression, perform the indicated mathematics, and then compare the product to K_{sp}.

$$CaCO_3(s) \rightleftharpoons Ca^{2+}(aq) + CO_3^{2-}(aq)$$

$$K_{sp} = [Ca^{2+}][CO_3^{2-}] = 4.8 \times 10^{-9}$$

Test K_{sp} against $[Ca^{2+}][CO_3^{2-}]$:

$$[Ca^{2+}][CO_3^{2-}] = (0.003)(0.003) = 9 \times 10^{-6}$$

$$K_{sp} = 4.8 \times 10^{-9} < 9 \times 10^{-6}$$

Precipitates.

(b) Follow the same procedure as in part (a).

$$Co(OH)_2(s) \rightleftharpoons Co^{2+}(aq) + 2OH^-(aq)$$

$$K_{sp} = [Co^{2+}][OH^-]^2 = 2 \times 10^{-16}$$

Test K_{sp} against $[Co^{2+}][OH^-]^2$

$$[Co^{2+}][OH^-]^2 = (0.01)(1 \times 10^{-7})^2 = 1 \times 10^{-16}$$

$$K_{sp} = 2 \times 10^{-16} > 1 \times 10^{-16}$$

Does not precipitate.

9. Calculate the solubility product of each of the following from the solubility given:

(a) AgBr, 5.7×10^{-7} mol/L
(c) PbF_2, 2.1×10^{-3} mol/L
(d) Ag_2CrO_4, 4.3×10^{-2} g/L

Solution

(a) AgBr, 5.7×10^{-7} mol/L:

Silver bromide, AgBr, is a very slightly soluble salt that dissolves in water in the following way:

$$AgBr(s) \rightleftharpoons Ag^+(aq) + Br^-(aq)$$

The K_{sp} expression includes only the ions in the solution phase and is written as

$$K_{sp} = [Ag^+][Br^-]$$

In a solution of a 1 : 1 electrolyte such as AgBr, the concentration of Ag^+ and of Br^- equals the solubility of AgBr. The concentration of AgBr that dissolves is 5.7×10^{-7} mol/L and, therefore, the concentration of Ag^+ and Br^- is 5.7×10^{-7} mol/L. Substitution into the K_{sp} expression gives

$$K_{sp}(AgBr) = (5.7 \times 10^{-7})(5.7 \times 10^{-7}) = 3.2 \times 10^{-13}$$

(c) PbF_2, 2.1×10^{-3} mol/L:

This salt dissociates in water to produce

$$PbF_2(s) \rightleftharpoons Pb^{2+}(aq) + 2F^-(aq)$$

From the solubility of PbF_2, the concentrations of Pb^{2+} and of F^- are determined.

$$[Pb^{2+}] = 2.1 \times 10^{-3} \text{ mol/L}$$

$$[F^-] = 2(2.1 \times 10^{-3}\ mol/L) = 4.2 \times 10^{-3}\ mol/L$$

$$K_{sp} = [Pb^{2+}][F^-]^2 = (2.1 \times 10^{-3})(4.2 \times 10^{-3})^2 = 3.7 \times 10^{-8}$$

(d) Ag_2CrO_4, 4.3×10^{-2} g/L:

Since equilibrium values are calculated in terms of molar quantities, the solubility must be in terms of molarity.

$$\text{Molar mass } Ag_2CrO_4 = 2(107.868) + 51.996 + 4(15.9994)$$

$$= 331.730\ g\ mol^{-1}$$

$$\text{Molar solubility} = 4.3 \times 10^{-2}\ g/L \times \frac{1\ mol}{331.7\ g}$$

$$= 1.296 \times 10^{-4}\ M$$

Dissolving one formula unit of Ag_2CrO_4 produces 2 Ag^+ and 1 CrO_4^{2-}. Therefore,

$$K_{sp} = [Ag^+]^2[CrO_4^{2-}]$$

$$[Ag^+] = 2(1.296 \times 10^{-4}) = 2.59 \times 10^{-4}$$

$$[CrO_4^{2-}] = 1.296 \times 10^{-4}$$

$$K_{sp} = (2.59 \times 10^{-4})^2(1.296 \times 10^{-4}) = 8.7 \times 10^{-12}$$

11. Calculate the concentrations of ions in a saturated solution of each of the following (see Appendix E) for solubility product constants):

(a) AgI (b) Ag_2SO_4 (d) $Sr(OH)_2 \cdot 8H_2O$

Solution

(a) The dissolution of AgI produces one each of the ions Ag^+ and I^-.

$$AgI(s) \rightleftharpoons Ag^+(aq) + I^-(aq)$$

$$K_{sp} = [Ag^+][I^-] = 1.5 \times 10^{-16}$$

Let x equal the solubility of AgI in moles per liter.

$$x = [Ag^+] = [I^-]$$

$$K_{sp} = (x)(x) = 1.5 \times 10^{-16}$$

$$x^2 = 1.5 \times 10^{-16}$$

$$[Ag^+] = [I^-] = x = 1.2 \times 10^{-8}\ M$$

(b) Dissolution of Ag_2SO_4 occurs as

$$Ag_2SO_4(s) \rightleftharpoons 2\ Ag^+(aq) + SO_4^{2-}(aq)$$

Let x equal the molar solubility of Ag_2SO_4. Dissolving Ag_2SO_4 produces 2 Ag^+ and 1 SO_4^{2-}, so the concentrations in terms of x are

$$[Ag^+] = 2x$$

$$[SO_4^{2-}] = x$$

$$K_{sp} = [Ag^+]^2[SO_4^{2-}] = 1.18 \times 10^{-5}$$

$$(2x)^2(x) = 4x^3 = 1.18 \times 10^{-5}$$

$$x^3 = 2.95 \times 10^{-6}$$

$$x = 1.4 \times 10^{-2}$$

$$[Ag^+] = 2.8 \times 10^{-2}\ M; \quad [SO_4^{2-}] = 1.4 \times 10^{-2}\ M$$

(d) Dissolution of $Sr(OH)_2 \cdot 8H_2O$ occurs as

$$Sr(OH)_2 \cdot 8H_2O \rightleftharpoons Sr^{2+}(aq) + 2OH^-(aq) + 8H_2O$$

Let x equal the molar solubility of $Sr(OH)_2 \cdot 8H_2O$. Dissolving 1 mol of $Sr(OH)_2 \cdot 8H_2O$ produces 1 mol of Sr^{2+} and 2 mol of OH^-.

$$K_{sp} = [Sr^{2+}][OH^-]^2 = 3.2 \times 10^{-4}$$

The ion concentrations in terms of x are

$$x = [Sr^{2+}]; \quad [OH^-] = 2x$$

$$(x)(2x)^2 = 3.2 \times 10^{-4}$$

$$4x^3 = 3.2 \times 10^{-4}$$

$$x^3 = 8.0 \times 10^{-5}$$

$$[Sr^{2+}] = x = 0.043 = 4.3 \times 10^{-2}\ M$$

$$[OH^-] = 8.6 \times 10^{-2}\ M$$

17. Calculate the concentration of Sr^{2+} when SrF_2 ($K_{sp} = 3.7 \times 10^{-12}$) starts to precipitate from a solution that is 0.0025 M in F^-.

Solution

Precipitation of SrF_2 will begin when the ion product of the ions Sr^{2+} and F^- exceed the K_{sp} of SrF_2.

$$SrF_2 \rightleftharpoons Sr^{2+}(aq) + 2F^-(aq)$$

$$K_{sp} = [Sr^{2+}][F^-]^2 = 3.7 \times 10^{-12}$$

The maximum concentration of Sr^{2+} that will remain in solution is

$$[Sr^{2+}] = \frac{3.7 \times 10^{-12}}{(0.0025)^2} = 5.9 \times 10^{-7}\ M$$

21. The solubility product of $CaSO_4 \cdot 2H_2O$ is 2.4×10^{-5}. What mass of this salt will dissolve in 1.0 L of 0.010 *M* K_2SO_4?

Solution

The amount of $CaSO_4 \cdot 2H_2O$ that dissolves is limited by the presence of a substantial amount of SO_4^{2-} already in solution from the K_2SO_4. This is a common-ion problem. Let x = concentration of Ca^{2+} and of SO_4^{2-} that dissociates.

$$CaSO_4(s) \rightleftharpoons Ca^{2+}(aq) + SO_4^{2-}(aq)$$

$$K_{sp} = [Ca^{2+}][SO_4^{2-}] = 2.4 \times 10^{-5}$$

Substitution of 0.010 *M* SO_4^{2-} generated from the complete dissociation of 0.010 *M* K_2SO_4, gives

$$[x][x + 0.010] = 2.4 \times 10^{-5}$$

Here, x cannot be neglected in comparison with 0.010 *M*; the quadratic equation must be used. In standard form,

$$x^2 + 0.010x - 2.4 \times 10^{-5} = 0$$

$$x = \frac{-0.01 \pm \sqrt{1 \times 10^{-4} + 9.6 \times 10^{-5}}}{2} = \frac{-0.01 \pm 1.4 \times 10^{-2}}{2}$$

Only the positive value will give a meaningful answer.

$$x = 2.0 \times 10^{-3} = [Ca^{2+}]$$

This is also the concentration of $CaSO_4 \cdot 2H_2O$ that has dissolved. The mass of the salt in 1 L is

$$\text{Mass } CaSO_4 \cdot 2H_2O = 2.0 \times 10^{-3} \text{ mol/L} \times 172.16 \text{ g/mol} = 0.34 \text{ g}$$

Note that the presence of the common ion, SO_4^{2-}, has caused a decrease in the concentration of Ca^{2+} that otherwise would be in solution

$$\sqrt{2.4 \times 10^{-5}} = 4.9 \times 10^{-3}\ M$$

26. A solution of 0.075 *M* $CoBr_2$ is saturated with H_2S ($[H_2S] = 0.10\ M$). What is the minimum pH at which CoS ($K_{sp} = 5.9 \times 10^{-21}$) will precipitate?

Solution

Two equilibria are in competition for the ions and must be considered simultaneously. Precipitation of CoS will occur when the concentration of S^{2-} in conjunction with 0.075 *M* Co^{2+} exceeds the K_{sp} of CoS. But the $[S^{2-}]$ must come from the ionization of H_2S as defined by the equilibrium:

$$H_2S(aq) \rightleftharpoons 2H^+(aq) + S^{2-}(aq)$$

$$\frac{[H^+]^2[S^{2-}]}{[H_2S]} = K_1K_2(H_2S) = 1.3 \times 10^{-20}$$

Since a saturated solution of H_2S is 0.10 *M*, this expression becomes

$$[H^+]^2[S^{2-}] = 1.3 \times 10^{-21}$$

From the equilibrium of CoS, the minimum concentration of S^{2-} required to cause precipitation is calculated as

$$CoS(s) \rightleftharpoons Co^{2+}(aq) + S^{2-}(aq)$$

$$K_{sp} = [Co^{2+}][S^{2-}] = 5.9 \times 10^{-21}$$

$$[S^{2-}] = \frac{5.9 \times 10^{-21}}{0.075} = 7.87 \times 10^{-20}$$

This amount of S^{2-} will exist in solution at a pH defined by the H_2S equilibrium.

$$[H^+]^2(7.87 \times 10^{-20}) = 1.3 \times 10^{-21}$$

$$[H^+]^2 = 1.65 \times 10^{-2}$$

$$[H^+] = 1.29 \times 10^{-1}$$

$$pH = -\log 1.29 \times 10^{-1} = 0.89$$

27. (a) What are the concentrations of Ca^{2+} and $CO_3^{\ 2-}$ in a saturated solution of $CaCO_3$ ($K_{sp} = 4.8 \times 10^{-9}$)?

Solution

(a) The dissolution of $CaCO_3$ is highly dependent on the pH of the aqueous medium. Assume that association of $CO_3^{\ 2-}$ with H_3O^+ from water is insignificant; this means that the $[Ca^{2+}]$ at equilibrium equals $[CO_3^{\ 2-}]$.

$$CaCO_3(s) \rightleftharpoons Ca^{2+}(aq) + CO_3^{\ 2-}(aq)$$

$$K_{sp} = 4.8 \times 10^{-9} = [Ca^{2+}][CO_3^{\ 2-}]$$

Let x equal the number of moles of $CaCO_3$ dissolving.

$$x = [Ca^{2+}] = [CO_3^{\ 2-}]$$

$$K_{sp} = (x)(x) = 4.8 \times 10^{-9}$$

$$x = 6.9 \times 10^{-5}\ M = [Ca^{2+}] = [CO_3^{\ 2-}]$$

(b) What are the concentrations of Ca^{2+} and $CO_3^{\ 2-}$ in a buffer solution with a pH of 4.55 in contact with an excess of $CaCO_3$?

Solution

(b) The dissolution of $CaCO_3$ produces one ion each of Ca^{2+} and CO_3^{2-}. The carbonic acid is a weak acid and CO_3^{2-} will associate with H_3O^+ according to the following reactions:

$$CO_3^{2-} + H_3O^+ \rightleftharpoons HCO_3^- + H_2O \quad (1)$$

$$HCO_3^- + H_3O^+ \rightleftharpoons H_2CO_3 + H_2O \quad (2)$$

The concentration of Ca^{2+} at equilibrium equals the total concentration of the species containing carbonate.

$$\text{Solubility} = [Ca^{2+}] = [CO_3^{2-}] + [HCO_3^-] + [H_2CO_3] \quad (3)$$

$$K_{sp}(CaCO_3) = [Ca^{2+}][CO_3^{2-}] = 4.8 \times 10^{-9} \quad (4)$$

The solubility expression involves four unknowns, so four simultaneous equations are needed for solving the expression for $[Ca^{2+}]$. Equations (1) and (2) are merely the reverse of the successive ionizations of H_2CO_3, and the K values for H_2CO_3 can be used.

$$H_2CO_3(aq) \rightleftharpoons H^+(aq) + HCO_3^-(aq) \qquad K_1 = 4.3 \times 10^{-7} \quad (5)$$

$$HCO_3^-(aq) \rightleftharpoons H^+(aq) + CO_3^{2-}(aq) \qquad K_2 = 7.0 \times 10^{-11} \quad (6)$$

Equations (5) and (6) can be rearranged and expressed in terms of $[CO_3^{2-}]$. At a pH of 4.55 the $[H^+]$ is 2.8×10^{-5}.

$$\frac{[H^+][CO_3^{2-}]}{[HCO_3^-]} = 7.0 \times 10^{-11}$$

$$[HCO_3^-] = \frac{2.8 \times 10^{-5}}{7.0 \times 10^{-11}}[CO_3^{2-}] = 4.0 \times 10^5[CO_3^{2-}]$$

and

$$\frac{[H^+][HCO_3^-]}{[H_2CO_3]} = 4.3 \times 10^{-7}$$

$$[H_2CO_3] = \frac{2.8 \times 10^{-5}}{4.3 \times 10^{-7}}[HCO_3^-]$$

$$= 6.5 \times 10^1[HCO_3^-] = 6.5 \times 10^1(4.0 \times 10^5)[CO_3^{2-}]$$

$$= 2.6 \times 10^7[CO_3^{2-}]$$

These values for $[HCO_3^-]$, $[H_2CO_3]$, and $[CO_3^{2-}]$ can be substituted into Equation (3), which yields

$$[Ca^{2+}] = [CO_3^{2-}] + 4.0 \times 10^5[CO_3^{2-}] + 2.6 \times 10^7[CO_3^{2-}]$$

$$[Ca^{2+}] = 2.6 \times 10^7[CO_3^{2-}]$$

or

$$[CO_3^{2-}] = \frac{[Ca^{2+}]}{2.6 \times 10^7}$$

Substitution into the expression for K_{sp} yields

$$K_{sp} = [Ca^{2+}][CO_3^{2-}] = [Ca^{2+}]\frac{[Ca^{2+}]}{2.6 \times 10^7} = 4.8 \times 10^{-9}$$

$$[Ca^{2+}]^2 = 1.25 \times 10^{-1}$$

$$[Ca^{2+}] = 0.353 \text{ mol/L} = 0.35\ M$$

$$[CO_3^{2-}] = \frac{[Ca^{2+}]}{2.6 \times 10^7} = \frac{0.353}{2.6 \times 10^7} = 1.4 \times 10^{-8}\ M$$

33. A volume of 50 mL of 1.8 *M* NH_3 is mixed with an equal volume of a solution containing 0.95 g of $MgCl_2$. What mass of NH_4Cl must be added to the resulting solution to prevent the precipitation of $Mg(OH)_2$?

Solution

The hydroxide ion concentration in solution depends on two simultaneous equilibria. The maximum allowable $[OH^-]$ can be calculated from the K_{sp} of $Mg(OH)_2$ based on the $[Mg^{2+}]$.

$$Mg(OH)_2 \rightleftharpoons Mg^{2+}(aq) + 2OH^-(aq) \qquad K_{sp} = 1.5 \times 10^{-11}$$

$$[Mg^{2+}] = [MgCl_2]$$

$$[Mg^{2+}] = \frac{0.95 \text{ g } MgCl_2 \times \frac{1 \text{ mol}}{95.211 \text{ g}}}{0.10 \text{ L}} = 0.0998\ M$$

$$[Mg^{2+}][OH^-]^2 = 1.5 \times 10^{-11}$$

$$[OH^-]^2 = \frac{1.5 \times 10^{-11}}{0.0998} = 1.50 \times 10^{-10}$$

$$[OH^-] = 1.22 \times 10^{-5}\ M$$

The $[OH^-]$ produced from NH_3 must be suppressed to 1.2×10^{-5} *M* by buffering the solution through the addition of NH_4Cl. The required $[NH_4^+]$ can be calculated from the equilibrium constant expression for ammonia.

$$NH_3(aq) + H_2O(\ell) \rightleftharpoons NH_4^+(aq) + OH^-(aq)$$

$$K_b = 1.8 \times 10^{-5} = \frac{[NH_4^+][OH^-]}{[NH_3]}$$

At equilibrium, the $[NH_3]$ approximately equals 1.8 *M*/2 = 0.90 *M* since 1.22×10^{-5} is small with respect to 0.90. Therefore, the $[NH_4^+]$ is

$$1.8 \times 10^{-5} = \frac{[NH_4^+](1.22 \times 10^{-5})}{0.90}$$

$$[NH_4^+] = \frac{(0.90)(1.8 \times 10^{-5})}{1.22 \times 10^{-5}} = 1.33\ M$$

$$\text{Mass } NH_4Cl \text{ required} = (1.33 \text{ mol/L})(53.5 \text{ g/mol})(0.10 \text{ L}) = 7.1 \text{ g}$$

37. A solution is 0.15 *M* in both Pb^{2+} and Ag^+. If Cl^- is added to this solution, what is $[Ag^+]$ when $PbCl_2$ begins to precipitate?

Solution

Lead and silver both form slightly soluble chlorides. The ion with the least molar solubility will precipitate first. This precipitation will continue until the chloride ion increases sufficiently to exceed the K_{sp} of the other chloride. Since both Ag^+ and Pb^{2+} must exist simultaneously in solution, both K_{sp} equilibria must be satisfied.

$$PbCl_2(s) \rightleftharpoons Pb^{2+}(aq) + 2Cl^-(aq) \qquad K_{sp} = 1.7 \times 10^{-5}$$

$$AgCl(s) \rightleftharpoons Ag^+(aq) + Cl^-(aq) \qquad K_{sp} = 1.8 \times 10^{-10}$$

First calculate the maximum chloride ion concentration that can be present without causing the precipitation of Pb^{2+}. Given that $[Pb^{2+}]$ equals 0.15 *M*, the $[Cl^-]$ maximum is

$$[Pb^{2+}][Cl^-]^2 = 1.7 \times 10^{-5}$$

$$[Cl^-]^2 = \frac{1.7 \times 10^{-5}}{0.15} = 1.133 \times 10^{-4}$$

$$[Cl^-] = 1.06 \times 10^{-2}$$

Precipitation of AgCl will continue until the chloride ion concentration equals 1.06×10^{-2} *M*. At this point the $[Ag^+]$ is controlled by the K_{sp} of AgCl. Substituting this $[Cl^-]$ into K_{sp} of AgCl gives

$$[Ag^+][1.06 \times 10^{-2}] = 1.8 \times 10^{-10}$$

$$[Ag^+] = \frac{1.8 \times 10^{-10}}{1.06 \times 10^{-2}} = 1.7 \times 10^{-8} \; M$$

41. Calculate the cadmium ion concentration $[Cd^{2+}]$, in a solution prepared by mixing 0.100 L of 0.0100 *M* $Cd(NO_3)$: with 0.150 L of 0.100 *M* $NH_3(aq)$.

Solution

Cadmium ions associate with ammonia molecules in solution to form the complex ion, $[Cd(NH_3)_4]^{2+}$, which is defined by the following equilibrium:

$$Cd^{2+}(aq) + 4NH_3(aq) \rightleftharpoons [Cd(NH_3)_4]^{2+}(aq) \qquad K_f = 4.0 \times 10^6$$

The formation of the complex requires 4 mol of NH_3 for each mole of Cd^{2+}. First calculate the initial amounts of Cd^{2+} and of NH_3 available for association.

$$[Cd^{2+}] = \frac{(0.100 \text{ L})(0.0100 \text{ mol/L})}{0.250 \text{ L}} = 4.00 \times 10^{-3} \; M$$

$$[NH_3] = \frac{(0.150\ L)(0.100\ mol/L)}{0.250\ L} = 6.00 \times 10^{-2}\ M$$

For the reaction, 4.00×10^{-3} mol/L of Cd^{2+} would require $4(4.00 \times 10^{-3}$ mol/L) of NH_3 or 1.6×10^{-2} mol. Due to the large value of K_f and the substantial excess of NH_3, it can be assumed that the reaction goes to completion with only a small amount of the complex dissociating to form the ions. After reaction, concentrations of the species in solution are

$$[NH_3] = 6.00 \times 10^{-2}\ mol/L - 1.6 \times 10^{-2}\ mol/L = 4.4 \times 10^{-2}\ M$$

$[Cd^{2+}]$ = equilibrium concentration from the dissociation of the complex

$$[Cd(NH_3)_4]^{2+} \approx 4.00 \times 10^{-3}\ M$$

At equilibrium: Let x = moles of $Cd(NH_3)_4^{2+}$ that dissociate.

$$x = [Cd^{2+}]$$

$$[NH_3] = 4.4 \times 10^{-2} + x$$

$$K_f = 4.0 \times 10^6 = \frac{[Cd(NH_3)_4^{2+}]}{[Cd^{2+}][NH_3]^4}$$

$$4.0 \times 10^6 = \frac{(4.00 \times 10^{-3} - x)}{(x)(4.4 \times 10^{-2} + x)^4}$$

Since x is small, this expression can be simplified to

$$4.0 \times 10^6 = \frac{(4.00 \times 10^{-3})}{(x)(4.4 \times 10^{-2})^4}$$

Rearranging and solving for x yields

$$1.50 \times 10^1\ x = 4.00 \times 10^{-3}$$

$$x = 2.7 \times 10^{-4}\ M = [Cd^{2+}]$$

49. (a) What mass of AgCl will dissolve in 1.0 L of 1.0 M NH_3?

Solution

Silver ion from dissolved AgCl complexes with ammonia to form a diamine-silver complex as follows:

$$AgCl(s) + 2NH_3(aq) \rightleftharpoons Ag(NH_3)_2^+(aq) + Cl^-(aq) \quad (1)$$

The formation constant for this equilibrium is

$$K_f = \frac{[Ag(NH_3)_2^+]}{[Ag^+][NH_3]^2} = 1.6 \times 10^7 \quad (2)$$

A small amount of AgCl dissolves according to the equilibrium

$$AgCl(s) \rightleftharpoons Ag^+(aq) + Cl^-(aq) \qquad K_{sp} = 1.8 \times 10^{-10} \qquad (3)$$

At equilibrium both equilibria, $Ag(NH_3)_2^+$ and AgCl, must be satisfied simultaneously. The Ag^+ in solution applies to both equilibria when AgCl and $Ag(NH_3)_2^+$ are present. At equilibrium, let x equal the number of moles of AgCl that dissolve. According to Equation (1), x equals the $[Cl^-]$ in solution and the amount of Ag^+ available for producing the complex, the $[Ag(NH_3)_2^+]$. By using Equation (3), the $[Ag^+]$ in terms of K_{sp} is

$$[Ag^+](x) = 1.8 \times 10^{-10}$$

$$[Ag^+] = \frac{1.8 \times 10^{-10}}{x}$$

The formation constant for the complex is very large; hence essentially all the available Ag^+ is complexed with only a small amount remaining in solution. The formation of the complex requires 2 mol of NH_3 for each Ag^+, so the concentration of NH_3 at equilibrium is $1.0 - 2x$. Substitution into Equation (2) gives

$$\frac{x}{\left(\frac{1.8 \times 10^{-10}}{x}\right)(1.0 - 2x)^2} = 1.6 \times 10^7$$

Clearing the fraction and rearranging to standard quadratic form gives

$$x^2 = 2.88 \times 10^{-3}(1 - 4x + 4x^2)$$

$$0.988x^2 + 1.15 \times 10^{-2}x - 2.88 \times 10^{-3} = 0$$

Substitution into the quadratic formula gives

$$x = \frac{-1.15 \times 10^{-2} \pm 0.107}{1.98}$$

The positive root of x is 4.8×10^{-2} mol.

$$\text{Mass AgCl dissolved} = 4.8 \times 10^{-2} \text{ mol} \times 143.321 \text{ g/mol} = 6.9 \text{ g}$$

59. The calcium ions in human blood serum are necessary for coagulation. In order to prevent coagulation when a blood sample is drawn for laboratory tests, an anticoagulant is added to the sample. Potassium oxalate, $K_2C_2O_4$, can be used as an anticoagulant because it removes the calcium as a precipitate of $CaC_2O_4 \cdot H_2O$. In order to prevent coagulation, it is necessary to remove all but 1.0% of the Ca^{2+} in serum. If normal blood serum with a buffered pH of 7.40 contains 9.5 mg of Ca^{2+} per 100 mL, what mass of $K_2C_2O_4$ is required to prevent the coagulation of a 10-mL blood sample that is 55% serum by volume? [All volumes are accurate to two significant figures. Note that the volume of fluid (serum) in a 10-mL blood sample is 5.5 mL. Assume that the K_{sp} value for CaC_2O_4 in serum is the same as in water.]

Solution

Although oxalic acid is a weak acid, the oxalate ions, $C_2O_4^{2-}$, do not associate significantly with H_3O^+ at the pH of blood. The amount of $K_2C_2O_4$ required equals the equivalent of 99% of the available Ca^{2+} plus the amount needed to maintain the equilibrium defined by the solubility product.

$$K_{sp}(CaC_2O_4) = [Ca^{2+}][C_2O_4^{2-}] = 2.27 \times 10^{-9}$$

The amount of available Ca^{2+} in 5.5 mL of serum is

$$\text{Moles } Ca^{2+} = 5.5 \text{ mL} \times \frac{9.5 \text{ mg}}{100 \text{ mL}} \times \frac{1.0 \text{ g}}{1000 \text{ mg}} \times \frac{1 \text{ mol}}{40.08 \text{ g}}$$

$$= 1.304 \times 10^{-5} \text{ mol}$$

$$\text{Moles } Ca^{2+} \text{ to be precipitated}$$

$$= (0.99)(1.304 \times 10^{-5} \text{ mol}) = 1.291 \times 10^{-5} \text{ mol}$$

At equilibrium the $[Ca^{2+}]$ is 1% of the original:

$$[Ca^{2+}] = \frac{(0.01)(1.304 \times 10^{-5} \text{ mol})}{0.0055 \text{ L}} = 2.371 \times 10^{-5} \text{ mol/L}$$

From the K_{sp} expression,

$$[C_2O_4^{2-}] = \frac{2.27 \times 10^{-9}}{2.371 \times 10^{-5}} = 9.574 \times 10^{-5} \text{ mol/L}$$

$$\text{Mol } C_2O_4^{2-} \text{ in 5.5 mL} = 9.574 \times 10^{-5} \text{ mol/L} \times 0.0055 \text{ L}$$

$$= 5.27 \times 10^{-7} \text{ mol}$$

$$\text{Total moles } C_2O_4^{2-} \text{ required} = 5.266 \times 10^{-7} \text{ mol} + 1.291 \times 10^{-5} \text{ mol}$$

$$= 1.34 \times 10^{-5} \text{ mol}$$

$$\text{Mass } K_2C_2O_4 = 1.34 \times 10^{-5} \text{ mol} \times 166.2162 \text{ g/mol} = 2.2 \times 10^{-3} \text{ g}$$

60. All urinary calculi (kidney stones) consist of about 50% calcium phosphate, $Ca_3(PO_4)_2$. The normal midrange calcium content excreted in the urine is 0.10 g of Ca^{2+} per day. The normal midrange amount of urine passed may be taken as 1.4 L per day. What is the maximum concentration of phosphate ion possible in urine before a calculus begins to form?

Solution

The dissolution of $Ca_3(PO_4)_2$ yields

$$Ca_3(PO_4)_2(s) \rightleftharpoons 3Ca^{2+}(aq) + 2PO_4^{3-}(aq)$$

Given the concentration of Ca^{2+} in solution, the maximum $[PO_4^{3-}]$ can be calculated by using the K_{sp} expression for $Ca_3(PO_4)_2$.

$$K_{sp} = 1 \times 10^{-25} = [Ca^{2+}][PO_4^{3-}]^2$$

$$[Ca^{2+}]_{urine} = \frac{0.10\ g\ \frac{1\ mol}{40.08\ g}}{1.4\ L} = 1.8 \times 10^{-3}\ M$$

$$[PO_4^{3-}]^2 = \frac{1 \times 10^{-25}}{(1.8 \times 10^{-3})^3} = 1.7 \times 10^{-17}$$

$$[PO_4^{3-}] = 4 \times 10^{-9}\ M$$

61. The pH of a normal urine sample is 6.30, and the total phosphate concentration

$$[PO_4^{3-}] + [HPO_4^{2-}] + [H_2PO_4^-] + [H_3PO_4] = 0.020\ M$$

What is the minimum concentration of Ca^{2+} necessary to induce calculus formation? (See Exercise 60 for additional information.)

Solution

The concentration of Ca^{2+} depends on the concentration of PO_4^{3-} in solution. But the $[PO_4^{3-}]$ is dependent on pH and the H_3PO_4 equilibrium. Because the total phosphate concentration is 0.020 M, the $[PO_4^{3-}]$ must be calculated from the three expressions involving H_3PO_4. These expressions, solved in terms of $[PO_4^{3-}]$, yield a ratio of $[PO_4^{3-}]$ to the other components.

$$H_3PO_4 \rightleftharpoons H^+ + H_2PO_4^- \qquad K_1 = 7.5 \times 10^{-3} \qquad (1)$$

$$H_2PO_4^- \rightleftharpoons H^+ + HPO_4^{2-} \qquad K_2 = 6.2 \times 10^{-8} \qquad (2)$$

$$HPO_4^{2-} \rightleftharpoons H^+ + PO_4^{3-} \qquad K_3 = 3.6 \times 10^{-13} \qquad (3)$$

At a pH of 6.3, the $[H^+] = 5.01 \times 10^{-7}$. For Equation (3),

$$\frac{[H^+][PO_4^{3-}]}{[HPO_4^{2-}]} = 3.6 \times 10^{-13}$$

$$[HPO_4^{2-}] = \frac{5.01 \times 10^{-7}}{3.6 \times 10^{-13}} [PO_4^{3-}] = 1.39 \times 10^6 [PO_4^{3-}]$$

For Equation (2),

$$\frac{[H^+][HPO_4^{2-}]}{[H_2PO_4^-]} = 6.2 \times 10^{-8}$$

$$[H_2PO_4^-] = \frac{5.01 \times 10^{-7}}{6.2 \times 10^{-8}} [HPO_4^{2-}]$$

$$= 8.08(1.39 \times 10^6)[PO_4^{3-}] = 1.12 \times 10^7 [PO_4^{3-}]$$

For equation (1),

$$\frac{[H^+][H_2PO_4^-]}{[H_3PO_4]} = 7.5 \times 10^{-3}$$

$$[H_3PO_4] = \frac{5.01 \times 10^{-7}}{7.5 \times 10^{-3}}[H_2PO_4^-]$$

$$= 6.7 \times 10^{-5}(1.12 \times 10^{7})[PO_4^{3-}] = 750[PO_4^{3-}]$$

$$[PO_4^{3-}] : [HPO_4^{2-}] : [H_2PO_4^-] : [H_3PO_4]$$

$$1 : 1.39 \times 10^{6} : 1.12 \times 10^{7} : 750$$

The fraction of PO_4^{3-} in solution is

$$\frac{[PO_4^{3-}]}{[PO_4^{3-}] + [HPO_4^{2-}] + [H_2PO_4^-] + [H_3PO_4]} = \frac{1}{1.26 \times 10^{7}} = 7.94 \times 10^{-8}$$

The concentration of PO_4^{3-} in urine at pH 6.3 is

$$[PO_4^{3-}] = 7.94 \times 10^{-8}(0.020) = 1.59 \times 10^{-9}$$

Use the K_{sp} of $Ca_3(PO_4)_2$ to calculate the minimum $[Ca^{2+}]$ present. For the reaction $Ca_3(PO_4)_2 \rightleftharpoons 3Ca^{2+} + 2PO_4^{3-}$,

$$K_{sp} = [Ca^{2+}]^3[PO_4^{3-}]^2$$

$$[Ca^{2+}]^3 = \frac{K_{sp}}{[PO_4^{3-}]^2} = \frac{1 \times 10^{-25}}{(1.59 \times 10^{-9})^2} = 3.96 \times 10^{-8}\ M$$

$$[Ca^{2+}] = (3.96 \times 10^{-8})^{1/3} = 3.4 \times 10^{-3}\ M$$

RELATED EXERCISES

1. Calculate the concentration of silver ion required to initiate precipitation of AgI from a 0.010 *M* KI solution.

   Answer: $[Ag^+] = 1.5 \times 10^{-14}\ M$

2. Silver plating can be accomplished through the electrolysis of a solution containing silver cyanide. The equilibrium of AgCN is represented as

$$2AgCN(s) \longrightarrow Ag^+ + Ag(CN)_2^- \qquad K_{sp} = 4.0 \times 10^{-12}$$

   Calculate the concentration of free Ag^+ at equilibrium in a saturated solution.

   Answer: $[Ag^+] = 2.0 \times 10^{-6}\ M$

3. Sea water contains a substantial amount of dissolved magnesium salts. Calculate the concentration of hydroxide ion necessary to remove 60% of the Mg^{2+} from sea water that is 0.050 *M* in Mg^{2+}.

Answer: $[OH^-] = 2.7 \times 10^{-5}$ *M*

4. Calculate the mass of lead(II) fluoride required to prepare 2.0 L of saturated solution.

Answer: *1.0 g*

5. The inside walls of cooking ware often become encrusted with a variety of slightly soluble compounds. Calculate the volume of water that would be required to dissolve 10.0 g of crust composed of pure $CaCO_3$. Assume that $CaCO_3$ dissolves and does not undergo hydrolysis.

Answer: 1.4×10^3 *L*

UNIT 13

CHAPTER 18: CHEMICAL THERMODYNAMICS

INTRODUCTION

The beginnings of modern thermodynamics can be traced to Benjamin Thompson, better known as Count Rumford of the Holy Roman Empire. His observations of the boring of cannon barrels in the early 1800s convinced him that the heat liberated in the machining resulted from the dissipation of mechanical work. Later, in the 1840s, James Joule made quantitative measurements of the conversion of mechanical energy into heat. His value was close to the modern value of the equivalency of heat and work: 1 calorie of heat = 4.184 joules of work.

This interconvertibility of heat and work makes it possible to define energy as the ability to produce heat or to do work. The First Law of Thermodynamics interrelates these two quantities and is basically an outgrowth of the human experience that energy can neither be created nor destroyed. Mathematically stated, the First Law is

$$\Delta E = q - w$$

where ΔE is the change in internal energy, q is the heat absorbed by the system, and w is the work done *by* the system

This concept is important in chemistry because chemists generally are interested in three basic questions that ultimately deal with energy: (1) Will two or more substances react? (2) If a reaction does occur, what is the associated energy change? (3) If the reaction occurs, what will be the equilibrium concentrations of the reactants and products?

The application of the First Law of Thermodynamics to chemical systems is immediately evident if a reaction is considered involving a phase change, where the work done may be ignored. Thus ice at 0°C may be converted to water at 0°C solely by the addition of heat q. In this reaction the energy content of ice has increased from that initially present by the amount of heat necessary to make this transition. This can be written

$$\Delta E = q - w = q - 0 = q$$

The heat change in this reaction is therefore the change in the internal energy of the system.

In general, during the course of *any* chemical reaction, the internal energy of the system will change by a specific amount because of a change in heat or work. The enthalpy (H), a function related to the internal energy, is introduced to simplify the study of the energy changes that occur in chemical reactions. The study of these heat effects is called *thermochemistry* and constitutes one important aspect of thermodynamics. This field answers the second question above.

In order to answer the other two questions, two more functions must be introduced. These are also the result of the study of fundamental laws in nature, namely: (1) systems tend to attain a state of minimum potential energy, and (2) systems tend toward a state of maximum disorder.

The first of these functions, entropy (S), is historically an outgrowth of the study of steam engines by the Frenchman Sadi Carnot. Carnot's study dates from 1824, but it was Clausius who first introduced the term *entropy* in 1840. Entropy, a measure of the disorder in a system, when properly combined with the enthalpy under the two constraints above, leads to a new function that conveniently allows us to determine whether a reaction will occur. The American J. Willard Gibbs introduced this new function, which was originally known as the *free energy*, but it is now also called the *Gibbs energy* (G).

The free energy or Gibbs energy is an extremely powerful tool, for it not only allows us to determine whether a reaction will occur based only on the knowledge of the state of the system but it also allows us to predict the concentrations of the reacting substances at equilibrium through the expression

$$\log K = \frac{\Delta G^\circ}{-2.303\ RT}$$

where K is the equilibrium constant. Thus, the three questions of interest to chemists posed earlier can be answered through the application of several simple concepts.

The problems in this chapter deal with the calculation of internal energy, enthalpy, entropy, and free energy or Gibbs energy changes as well as the determination of the equilibrium constant.

FORMULAS AND DEFINITIONS

Chemical thermodynamics That branch of chemistry that studies the energy transformations and transfers that accompany chemical and physical changes.

System That part of the universe on which we focus our attention and with whose properties we are concerned.

Surroundings All of the universe except the system we are studying.

State The condition of the system, defined by n, P, V, and T.

State function A function that depends only on the particular state of a system and not on how the system got to that state.

First Law of Thermodynamics A statement of the Law of Conservation of Energy: The total amount of energy in the universe is constant. Mathematically, $\Delta E = E_2 - E_1 = q - w$. Here ΔE is the internal energy change of the system due to a change in state, q is heat, and w is work.

Heat (q) A form of energy. A positive sign indicates a heat increase in the system; a negative sign corresponds to loss of heat from the system.

Work (w) A positive sign indicates work done by the system; a negative sign corresponds to work done on the system. Work has a pressure-volume equivalent defined as $w = P(V_2 - V_1)$, where P is the pressure restraining the system, V_1 is the initial volume, and V_2 is the final volume.

Enthalpy (H) The heat content or enthalpy of the system. The change in enthalpy ΔH is the quantity of heat absorbed or liberated by the system when a reaction takes place at constant pressure; therefore, $\Delta H = q$. If a reaction is endothermic, q is positive; if exothermic, q is negative. By definition, $\Delta H = \Delta E + \Delta(PV)$, or $\Delta H = \Delta E + P\Delta V$ for a constant-pressure process.

Standard state An agreed upon specific set of conditions designed to facilitate the handling of data. The standard state of a pure substance is taken as 25°C (298.15 K) and 1 atm pressure.

ΔH°_{f298} Standard molar enthalpy of formation. This is the change in enthalpy when one mole of a pure substance is formed from the free elements in their most stable state under standard conditions. For any free element in its most stable form, the value of the standard molar enthalpy is zero.

Hess's law For any process that can be considered the sum of several stepwise processes, the enthalpy change for the total process must equal the sum of the enthalpy changes for the various steps.

Bond energy A measure of the strength of the chemical bonds in a compound. It is determined through the summation of the heats of dissociation of all the individual chemical bonds in a compound.

Entropy (S) The entropy is a measure of the order or randomness of a system. The smaller the value of the entropy, the more ordered is the system; the larger the entropy, the greater the disorder or randomness of the system. The entropy is also a function of the temperature; it decreases as the temperature decreases and increases as the temperature increases. The importance of entropy lies in our ability to predict the direction of a chemical process if both the entropy of the system and the entropy of the surroundings are known.

Second Law of Thermodynamics In any spontaneous change the entropy of the universe increases.

Third Law of Thermodynamics The entropy of any pure, perfectly crystalline substance at the absolute zero of temperature (0 K) equals zero. Basically, this law allows the establishment of a beginning point, or zero point, for entropy measurements.

Free energy change (Gibbs energy change) (ΔG) Perhaps the most useful function of thermodynamics. It is the maximum amount of useful work that can be accomplished by a reaction at constant temperature and pressure. The free energy can also be used to predict the direction of a chemical process using only information about the system. These predictions are possible because reactions tend to proceed to a state of maximum disorder (positive ΔS) and minimum energy (negative ΔH). The sign that accompanies ΔG derived from the free energy expression, $\Delta G = \Delta H - T\,\Delta S$, is used to determine the spontaneity of the reaction: A negative sign indicates a spontaneous reaction as written, a positive value indicates a nonspontaneous reaction, and a value of zero indicates a reaction at equilibrium.

Relation of the free energy change (ΔG) *to the equilibrium constant* K_e The relation between the standard free energy and the equilibrium constant is given by the equation

$$\Delta G° = -RT \ln K_e$$

where the term $\ln K_e$ is the natural or Naperian logarithm of K_e. Natural logarithms ($\ln$) occur in calculus and are defined in terms of the base e, where $e = 2.71828$. If one is interested in the $\ln$ of the number x, x can be expressed as e raised to some power a. Thus,

$$x = e^a$$

Then a is called the natural logarithm of x and

$$a = \ln x$$

Tables are available in handbooks to calculate a. However, it is more customary to work with numbers to the base 10. These are called common or Briggsian logarithms (log) and written

$$x = 10^b$$

where b is called the common logarithm of x. This is expressed as $b = \log x$. (If you are unfamiliar with the fundamental rules for using logs, you should work through the programmed instruction section on logs in Part Two of this manual.)

A relation between these two bases can be derived so that equations written in terms of natural logarithms can be expressed in terms of common logarithms. This is done in the following manner. Since

$$x = e^a \quad \text{and} \quad x = 10^b$$

$$e^a = 10^b$$

Taking logarithms to the base e of both sides gives

$$\ln e^a = \ln 10^b$$

because the $\ln$ of e raised to any power is simply that power

$$a = b \ln 10$$

However, as has already been seen, $a = \ln x$; therefore substitution is again possible since $b = \log x$. Consequently,

$$\ln x = \log x \ln 10$$

or

$$\ln x = 2.303 \log x$$

The free energy expression can, therefore, be written as

$$\Delta G° = -2.303\ RT \log K_e$$

Note that the *standard* free energy change for the reaction is used. The free energy change by itself will not allow the calculation of the equilibrium constant.

EXERCISES

2. Calculate the missing value of ΔE, q, or w for a system given the following data:

   (a) $\Delta E = -7500$ J; $w = -4500$ J

   (d) The system absorbs 1 kJ of heat energy and does 650 J of work on the surroundings.

   Solution

   (b) The internal energy change ΔE of a system is a balance between the heat and work. This reaction is given by the First Law of Thermodynamics, $\Delta E = q - w$. Solving for the unknown q,

   $$q = \Delta E + w = -7500 \text{ J} + (-4500 \text{ J}) = -12{,}000 \text{ J}$$

   (d) Heat absorbed by the system is considered to be positive and work done by the system is considered positive. From the First Law,

   $$\Delta E = q - w = 1000 \text{ J} - (+650 \text{ J}) = 350 \text{ J}$$

4. Calculate the work involved in compressing a system consisting of 1 mol of H_2O as it changes from a gas at 373 K (volume = 30.6 L) to a liquid at 373 K (volume = 18.9 mL) under a constant pressure of 1 atm. Does this work increase or decrease the internal energy of the system?

   Solution

   For pressure-volume work done at constant pressure, the work is given by $P(V_2 - V_1)$. In order to have the work in joules, the pressure must be expressed in pascals and the volume in (meter)^3. Recall that 1 atm = 101,325 Pa, and 1 L = 10^{-3} m^3, so the two volumes may be expressed as

   $$V_1 = 30.6 \text{ L} \times \frac{10^{-3} \text{ m}^3}{\text{L}} = 3.06 \times 10^{-2} \text{ m}^3$$

   and

   $$V_2 = 18.9 \text{ mL} \times \frac{1 \text{ L}}{10^3 \text{ mL}} \times \frac{10^{-3} \text{ m}^3}{\text{L}} = 1.89 \times 10^{-5} \text{ m}^3$$

   From $P(V_2 - V_1)$ the work is

   $$w = P(V_2 - V_1) = 101{,}325 \text{ Pa}(1.89 \times 10^{-5} - 3.06 \times 10^{-2})\text{m}^3$$

   $$= -3.10 \times 10^3 \text{ Pa m}^3 = -3.10 \times 10^3 \text{ J} = -3.10 \text{ kJ}$$

Hence 3.10 kJ of work would be involved. The work done on the system has a negative value. From the First Law, $\Delta E = q - w$ and since w enters the equation as a negative quantity, the negative sign in the equation causes an overall increase in the internal energy.

9. Does $\Delta H^\circ_{f_{H_2O(g)}}$ differ from ΔH°_{298} for the reaction $2H_2(g) + O_2(g) \longrightarrow 2H_2O(g)$? If so, how?

Solution

Heats of formation, such as $\Delta H^\circ_{f_{H_2O(g)}}$, refer to the formation of one mole of the substance with each reactant and product in its standard state. Here the $H_2O(g)$ is not its standard state. The value of ΔH°_{298} refers to the reaction as written; in this case, the reaction is the formation of 2 mol of $H_2O(g)$, not 1 mole of $H_2O(\ell)$ as would be required for its value to be the same as $\Delta H^\circ_{f_{H_2O(g)}}$

11. (a) Using the data in Appendix J, calculate the standard enthalpy change, ΔH°, for each of the following reactions:

(1) $Fe_2O_3(s) + 13CO(g) \longrightarrow 2Fe(CO)_5(g) + 3CO_2(g)$

(3) $CH_4(g) + N_2(g) \longrightarrow HCN(g) + NH_3(g)$

Solution

(1) $\Delta H = \Delta H(\text{products}) - \Delta H(\text{reactants})$ (ΔH° values in kJ)

$$\Delta H^\circ = 2\Delta H^\circ_{Fe(CO)_5(g)} + 3\Delta H^\circ_{CO_2(g)} - \Delta H^\circ_{Fe_2O_3(s)} - 13\Delta H^\circ_{CO(g)}$$

$$= 2(-733.9) + 3(-393.5) - (-824.2) - 13(-110.5) = -387.6 \text{ kJ}$$

(3) $$\Delta H^\circ = \Delta H^\circ_{HCN(g)} + \Delta H^\circ_{NH_3(g)} - \Delta H^\circ_{CH_4(g)} - \Delta H^\circ_{N_2(g)}$$

$$= (135) + (-46.11) - (-74.81) - (0) = 164 \text{ kJ}$$

14. How many kilojoules of heat energy will be liberated when 49.70 g of manganese are burned to form $Mn_3O_4(s)$ at standard state conditions? ΔH°_{f298} of Mn_3O_4 is equal to $-1388 \text{ kJ mol}^{-1}$.

Solution

First write the overall reaction

$$3Mn(s) + 2O_2(g) \longrightarrow Mn_3O_4(s)$$

Next calculate the amount of $Mn_3O_4(s)$ in moles that will be formed by the reaction of 49.70 g of Mn. Based on the balanced equation,

$$3 \text{ mol Mn} \longrightarrow 1 \text{ mol } Mn_3O_4$$

Therefore,

$$\text{Amount of } Mn_3O_4 = \frac{\text{no. mol Mn}}{3} = \frac{\text{Mass Mn}}{3(\text{at.wt Mn})}$$

$$= \frac{49.70 \text{ g}}{3(54.9380 \text{ g/mol})} = 0.3016 \text{ mol}$$

Multiply the moles of M_3O_4 produced by the heat released per mole to obtain the heat liberated. Since $\Delta H^\circ_{f298_{Mn_3O_4}} = -1388 \text{ kJ mol}^{-1}$,

$$\text{Heat} = 0.3015 \text{ mol} \times -1388 \text{ kJ mol}^{-1} = 418.5 \text{ kJ}$$

15. The heat of formation of $OsO_4(s)$, $\Delta H^\circ_{f_{OsO_4(s)}}$, is -391 kJ mol^{-1} at 298 K and the heat of sublimation is 56.4 kJ mol^{-1}. What is ΔH°_{298} for the process $Os(s) + 2O_2(g) \longrightarrow OsO_4(g)$ under standard state conditions?

Solution

Sublimation is the process of the direct conversion of a solid to a gas. The sublimation process is endothermic and requires heat, as shown by the positive sign for the heat of sublimation. On the other hand, the negative sign for the heat of formation of $OsO_4(s)$ indicates that the formation is exothermic. For the overall reaction, heat is liberated by the formation of $OsO_4(s)$ and absorbed by its sublimation. Thus,

$$Os(s) + 2O_2(g) \longrightarrow OsO_4(s) \qquad \Delta H^\circ_f = -391 \text{ kJ}$$

followed by

$$OsO_4(s) \longrightarrow OsO_4(g) \qquad \Delta H^\circ_{sub} = 56.4 \text{ kJ}$$

The sum of the two reactions gives the heat of formation of $OsO_4(g)$.

$$Os(s) + 2O_2(g) \longrightarrow OsO_4(g)$$

$$\Delta H^\circ_{f_{OsO_4(g)}} = \Delta H^\circ_{f_{OsO_4(s)}} + \Delta H^\circ_{sub}$$

$$= -391 \text{ kJ} + 56.4 \text{ kJ} = -335 \text{ kJ}$$

18. Calculate the standard molar enthalpy of formation of $NO(g)$ from the following data:

$$N_2(g) + 2O_2(g) \longrightarrow 2NO_2(g) \qquad \Delta H^\circ = 66.4 \text{ kJ}$$

$$2NO(g) + O_2(g) \longrightarrow 2NO_2(g) \qquad \Delta H^\circ = -114.1 \text{ kJ}$$

Solution

Hess's Law can be applied to the two equations above by reversing the sense of the second equation. The first equation is a formation reaction and is so indicated by writing ΔH°_{f298}.

$$N_2(g) + 2O_2(g) \longrightarrow 2NO_2(g) \qquad \Delta H^\circ_{f298} = 66.4 \text{ kJ}$$

$$2NO_2(g) \longrightarrow 2NO(g) + O_2(g) \qquad \Delta H^\circ = 114.1 \text{ kJ}$$

Adding,

$$N_2(g) + 1O_2(g) \longrightarrow 2NO(g) \qquad \Delta H^\circ = 180.5 \text{ kJ}$$

This is the heat of formation of 2 mol of NO. For 1 mol,

$$\Delta H^\circ_{f298} = \frac{180.5 \text{ kJ}}{2} = 90.3 \text{ kJ mol}^{-1} \text{ of NO}$$

20. The heat of combustion of a hydrocarbon (a compound of carbon and hydrogen) is the standard state enthalpy change for the reaction of the compound with oxygen to give $CO_2(g)$ and $H_2O(\ell)$. Determine the heats of combustion of (a) octane (C_8H_{18}, $\Delta H^\circ_f = -208.4$ kJ mol^{-1}), a component of gasoline and (b) methane ($CH_4(g)$, $\Delta H^\circ_f = -74.81$ kJ mol^{-1}), the major component of natural gas. Which has the higher heat content per gram?

Solution

(a) The reaction for the combustion of octane is given with the ΔH°_f values written as products minus reactants below in kJ mol^{-1}.

$$C_8H_{18}(\ell) + \frac{25}{2} O_2(g) \longrightarrow 8CO_2(g) + 9H_2O(\ell)$$

$$\Delta H^\circ_{comb} = 8\Delta H^\circ_{f298_{CO_2}} + 9\Delta H^\circ_{f298_{H_2O}} - \Delta H^\circ_{f298_{C_8H_{18}}} - \Delta H^\circ_{f298_{O_2}}$$

$$\Delta H^\circ_{comb} = 8(-393.5) + 9(-285.8) - (-208.4) - (0)$$

$$= -3148 - 2572.2 + 208.4 = -5512 \text{ kJ mol}^{-1}$$

(b) For methane, the combustion process is

$$CH_4(g) + 2O_2(g) \longrightarrow CO_2(g) + 2H_2O(\ell)$$

$$\Delta H^\circ_{comb} = \Delta H^\circ_{f298_{CO_2}} + 2\Delta H^\circ_{f298_{H_2O}} - \Delta H^\circ_{f298_{CH_4}} - 2\Delta H^\circ_{f298_{O_2}}$$

$$= -393.5 + 2(-285.8) - (-74.81) - 2(0)$$

$$= -393.5 - 571.6 + 74.81 = -890.3 \text{ kJ mol}^{-1}$$

The substance with the larger amount of heat released per gram is calculated by dividing the individual heats released through combustion by the molar mass (molecular weight) of the respective organic compound.

For octane: $\dfrac{5512 \text{ kJ mol}^{-1}}{114.230 \text{ g mol}^{-1}} = 48.25 \text{ kJ g}^{-1}$

For methane: $\dfrac{890.3 \text{ kJ mol}^{-1}}{16.043 \text{ g mol}^{-1}} = 55.49 \text{ kJ g}^{-1}$

Thus, methane has a higher heat content per unit mass.

21. Using the data in Appendix J, calculate the bond energies of F_2, Cl_2, and FCl. All are gases in their most stable form at standard state conditions.

Solution

The bond energy, *D*, for a diatomic molecule equals the change in standard enthalpy for the dissociation of the molecule. In general, the reaction can be expressed as

$$XY(g) \longrightarrow X(g) + Y(g) \qquad \Delta H°(\text{reaction}) = D$$

For F—F: $F_2(g) \longrightarrow F(g) + F(g)$

$$D = \Delta H° = 2\Delta H°_{f298_{F(g)}} - \Delta H°_{f298_{F_2(g)}} = 2(78.99 \text{ kJ}) - 0 \text{ kJ}$$

$$= 158.0 \text{ kJ per mole of bonds}$$

For Cl—Cl: $Cl_2(g) \longrightarrow Cl(g) + Cl(g)$

$$D = \Delta H° = 2\Delta H°_{f298_{Cl(g)}} - \Delta H°_{f298_{Cl_2(g)}} = 2(121.68 \text{ kJ}) - 0 \text{ kJ}$$

$$= 243.36 \text{ kJ per mole of bonds}$$

For F—Cl: $FCl(g) \longrightarrow F(g) + Cl(g)$

$$D = \Delta H° = \Delta H°_{f298_{F(g)}} + \Delta H°_{f298_{Cl(g)}} - \Delta H°_{f298_{FCl}}$$

$$= 78.99 + 121.68 - (-54.48) \text{ (units of kJ per mole of bonds)}$$

$$= 255.15 \text{ kJ per mole of bonds}$$

24. (a) Using the bond energies given in Table 18-2 in Section 18.7, determine the approximate enthalpy change for the formation of ethylene from ethane, which is described in Section 15.14:

$$C_2H_6(g) \longrightarrow C_2H_4(g) + H_2(g)$$

(b) Compare this with the standard state enthalpy change.

Solution

(a) Heats of reaction are calculated from bond energies by using the same procedure that is used with enthalpies:

$$\Delta \text{ bond energy} = \text{bond energy of products} - \text{bond energy of reactants}$$

Each bond in every compound involved in the reaction must be considered. First write the structural formula of the compounds to better determine the bonds involved.

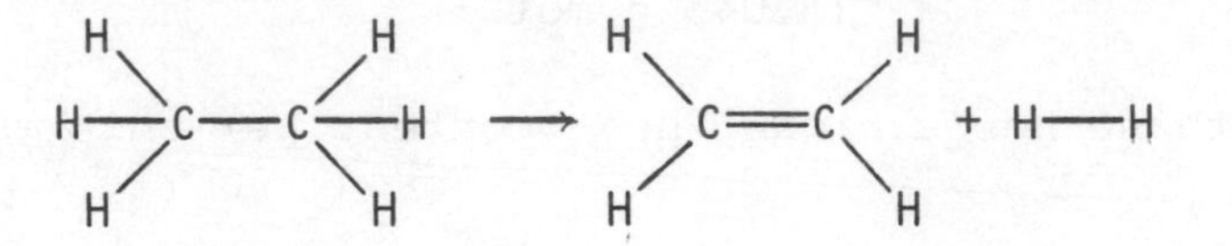

and then write the number of bond types:

Ethane	Ethylene + Hydrogen	
6 C—H	4 C—H	1 H—H
1 C—C	1 C=C	

The change in bond energies is obtained by taking the value of the bond energies of the products minus the values of those of the reactants. Because the change in bond energies is negative, more energy is required to break the bonds in ethane than is released on the formation of C_2H_4. Consequently, the heat of reaction has the same numerical value but is positive.

$$\Delta \text{ bond energy} = \underbrace{[4(415) + 611] + [436]}_{\text{Products}} - \underbrace{[6(415) + 345]}_{\text{Reactants}}$$

$$\Delta D = \Delta \text{ bond energy} = -128 \text{ kJ mol}^{-1}$$

The heat of reaction is $-\Delta D = +128 \text{ kJ mol}^{-1}$.

(b) The standard enthalpy change is calculated from

$$\Delta H = \Delta H^\circ_{f298_{\text{ethylene}}} + \Delta H^\circ_{f298_{H_2}} - \Delta H^\circ_{f298_{\text{ethane}}}$$

$$= 52.26 + 0 - (-84.68) = 136.9 \text{ kJ mol}^{-1}$$

32. (a) Using the data in Appendix J, calculate the standard entropy change for each reaction in part (a) of Exercise 11.

(b) For which of the reactions in Exercise 11 are the entropy changes favorable for the reaction to proceed spontaneously?

Solution

This problem requires that ΔS be found for each reaction as a difference of the entropies of the products minus the entropies of the reactants.

(a) (1) $Fe_2O_3(s) + 13CO(g) \longrightarrow 2Fe(CO)_5(g) + 3CO_2(g)$

$$\Delta S^\circ = \Delta S^\circ(\text{products}) - \Delta S^\circ(\text{reactants}) \quad (\Delta S^\circ \text{ values in J K}^{-1} \text{ mol}^{-1})$$

$$\Delta S^\circ = 2\Delta S^\circ_{Fe(CO)_5(g)} + 3\Delta S^\circ_{CO_2(g)} - \Delta S^\circ_{Fe_2O_3(s)} - 13\Delta S^\circ_{CO(g)}$$

$$= 2(445.2) + 3(213.6) - (87.40) - 13(197.56) = -1124.5 \text{ J/K}$$

(3) $CH_4(g) + N_2(g) \longrightarrow HCN(g) + NH_3(g)$

$$\Delta S° = \Delta S°_{HCN(g)} + \Delta S°_{NH_3(g)} - \Delta S°_{CH_4(g)} - \Delta S°_{N_2(g)}$$

$$= [(201.7) + (192.3)] - [(186.15) + (191.5)]$$

$$= 16.4 \text{ J/K}$$

(b) Of the two reactions above, (3) is favorable for the reaction to proceed spontaneously because the sign of ΔS is positive. However, ΔS alone cannot be used to determine whether a reaction is spontaneous. Indeed, the values for ΔG as found in Exercise 41 [(1) -52.4 kJ and (3) 159.0 kJ] show that reaction (1) is spontaneous and (3) is not.

33. What is the entropy change accompanying the evaporation of 1 mol of chloroform, $CHCl_3(\ell) \longrightarrow CHCl_3(g)$, under standard state conditions?

Solution

For this process, which occurs at 298 K,

$$\Delta S° = \text{absolute entropy for } CHCl_3(g) \text{ (final state)}$$

$$- \text{ absolute entropy for } CHCl_3(\ell) \text{ (initial state)}$$

$$= S°_{CHCl_3(g)_{298}} - S°_{CHCl_3(\ell)_{298}}$$

$$= 295.6 - 202 = 94 \text{ J K}^{-1} \text{ mol}^{-1}$$

41. (a) Using the data in Appendix J, calculate the standard free energy changes for the reactions given in Exercise 11.
(b) Which of those reactions are spontaneous? Why?

Solution

(a) (1) $Fe_2O_3(s) + 13CO(g) \longrightarrow 2Fe(CO)_5(g) + 3CO_2(g)$

$$\Delta G° = \Delta G°(\text{products}) - \Delta G°(\text{reactants}) \quad (\Delta G° \text{ values in kJ mol}^{-1})$$

$$\Delta G° = 2\Delta G°_{Fe(CO)_5(g)} + 3\Delta G°_{CO_2(g)} - \Delta G°_{Fe_2O_3(s)} - 13\Delta G°_{CO(g)}$$

$$= 2(-697.26) + 3(-394.36) - (-742.2) - 13(-137.15)$$

$$= -52.4 \text{ kJ}$$

(3) $CH_4(g) + N_2(g) \longrightarrow HCN(g) + NH_3(g)$

$$\Delta G° = \Delta G°_{HCN(g)} + \Delta G°_{NH_3(g)} - \Delta G°_{CH_4(g)} - \Delta G°_{N_2(g)}$$

$$= +124.7 + (-16.5) - (-50.75) - (0)$$

$$= 159 \text{ kJ}$$

(b) Of the two reactions above, (1) is spontaneous because the sign of ΔG is negative.

42. The standard enthalpies of formation of $NO(g)$, $NO_2(g)$, and $N_2O_3(g)$ are 90.25 kJ mol^{-1}, 33.2 kJ mol^{-1}, and 83.72 kJ mol^{-1}, respectively. Their standard entropies are 210.65 J mol^{-1} K^{-1}, 239.9 J mol^{-1} K^{-1}, and 312.2 J mol^{-1} K^{-1}, respectively.

(a) Use the data above to calculate the free energy change for the following reaction at 25.0°C.

$$N_2O_3(g) \longrightarrow NO(g) + NO_2(g)$$

Solution

We apply the equation

$$\Delta G° = \Delta H° - T\Delta S°$$

$$\Delta H° = \Delta H°_{f\,NO(g)} + \Delta H°_{f\,NO_2(g)} - \Delta H°_{f\,N_2O_3(g)} \qquad (\Delta H°_f \text{ values in kJ mol}^{-1})$$

$$= 90.25 + 33.2 - 83.72 = 39.7 \text{ kJ}$$

$$\Delta S° = \Delta S°_{NO(g)} + \Delta S°_{NO_2(g)} - \Delta S°_{N_2O_3(g)} \qquad (\Delta S° \text{ values in J mol}^{-1}\text{ K}^{-1})$$

$$= 210.65 + 239.9 - 312.2 = 138.4 \text{ J K}^{-1}$$

Then,

$$\Delta G° = 39{,}700 \text{ J} - 298.2(138.4) \text{ J}$$

$$= -1570 \text{ J} = -1.6 \text{ kJ}$$

48. Consider the reaction

$$I_2(g) + Cl_2(g) \longrightarrow 2ICl(g)$$

(a) For this reaction $\Delta H° = -26.9$ kJ and $\Delta S° = 11.3$ J K^{-1}. Calculate $\Delta G°$ for the reaction.

(b) Calculate the equilibrium constant for this reaction at 25.0°C.

Solution

(a)

$$\Delta G°_{298} = \Delta H°_{298} - T\Delta S°_{298}$$

$$= -26.9 \text{ kJ} - 298.2(11.3 \text{ J K}^{-1})$$

$$= -26{,}900 - 3370 = -30{,}270 \text{ J} = -30.3 \text{ kJ}$$

(b) The equilibrium constant for the reaction at 25°C is related to the free energy (Gibbs energy) by the equation

$$\Delta G° = -2.303\, RT \log K$$

Thus, the value of the standard state free energy (Gibbs energy) for a reaction can be used to determine the equilibrium constant K for the reaction. The value of $\Delta G°$ is -30.3 kJ; therefore,

$$-30{,}300 \text{ J} = (-2.303)(8.314 \text{ J K}^{-1})(298.2 \text{ K}) \log K$$

$$\log K = \frac{30,300}{(2.303)(8.314)(298.2)} = 5.307$$

$$K = 2.03 \times 10^5$$

51. For the decomposition of $CaCO_3(s)$ into $CaO(s)$ and $CO_2(g)$ at 1 atm,

(a) estimate the minimum temperature at which you would conduct the reaction.

Solution

Write the balanced chemical reaction:

$$CaCO_3(s) \longrightarrow CaO(s) + CO_2(g)$$

Decomposition will occur spontaneously when the system is at equilibrium: ΔG for the reaction is zero. In order to determine the equilibrium temperature, first calculate ΔH° and ΔS° for the reaction.

$$\Delta H^\circ = \Delta H^\circ_{298}{}_{CaO(s)} + \Delta H^\circ_{298}{}_{CO_2(g)} - \Delta H^\circ_{298}{}_{CaCO_3(s)}$$

$$= -635.5 + (-393.5) - (-1206.9) \quad (\Delta H^\circ \text{ in kJ mol}^{-1})$$

$$= 177.9 \text{ kJ}$$

$$\Delta S^\circ = \Delta S^\circ_{298}{}_{CaO(s)} + \Delta S^\circ_{298}{}_{CO_2(g)} - \Delta S^\circ_{298}{}_{CaCO_3(s)}$$

$$= 40. + 213.6 - 92.9 \quad (\Delta S^\circ \text{ in J mol}^{-1} \text{ K}^{-1})$$

$$= 160.7 \text{ J K}^{-1}$$

Then, from $\Delta G = \Delta H - T\Delta S = 0$

on the assumption that $\Delta H = \Delta H^\circ$ and $\Delta S = \Delta S^\circ$, we have

$$T = \frac{\Delta H^\circ}{\Delta S^\circ} = \frac{177,900 \text{ J}}{160.7 \text{ J K}^{-1}} = 1107 \text{ K} = 834^\circ\text{C}$$

This calculation is based on the assumption that ΔH and ΔS are independent of temperature over the range from 298 K to the temperature of dissociation. This is not strictly true, and the actual values of ΔH and ΔS near the calculated temperature should be used to obtain a more accurate value.

52. If the enthalpy of vaporization of CH_2Cl_2 is 29.0 kJ mol^{-1} at 25.0°C and the entropy of vaporization is 92.5 J mol^{-1} K^{-1}, calculate a value for the normal boiling temperature of CH_2Cl_2.

Solution

The wording of this problem seems quite different from the previous problem. However, the same basic idea applies: ΔG is equal to zero at equilibrium (the boiling point in this case). Thus,

$$\Delta G = \Delta H - T\Delta S = 0$$

$$T = \frac{\Delta H}{\Delta S} = \frac{29{,}000\ \text{J}}{92.5\ \text{J K}^{-1}} = 313.5\ \text{K or } 40.4°\text{C}$$

53. The equilibrium constant, K_p, for the reaction $N_2O_4(g) \rightleftharpoons 2NO_2(g)$ is 0.142 at 298 K. What is $\Delta G°$ for the reaction?

Solution

$$\Delta G° = -2.303\ RT \log K_p$$

$$= -2.303(8.314\ \text{J K}^{-1}\ \text{mol}^{-1})(298.2\ \text{K})(\log 0.142)$$

$$= -5709.7\ \text{J mol}^{-1}(-0.8477)$$

$$= 4840\ \text{J mol}^{-1} = 4.84\ \text{kJ mol}^{-1}$$

58. For the vaporization of bromine liquid to bromine gas
   (a) calculate the change in enthalpy and the change in entropy at standard state conditions.
   (c) estimate the value of $\Delta G°$ for the vaporization of bromine from the data in Appendix J.
   (d) state what you can about the spontaneity of the process from the value you obtained for $\Delta G°$ in part (c)
   (e) estimate the temperature at which liquid and gaseous bromine are in equilibrium with each other at 1 atm (assume $\Delta H°$ and $\Delta S°$ are independent of temperature).

Solution

(a) For $Br_2(\ell) \longrightarrow Br_2(g)$,

$$\Delta H°_{298} = \Delta H°_{f298_{Br_2(g)}} - \Delta H°_{f298_{Br_2(\ell)}}$$

$$= 30.91\ \text{kJ} - 0\ \text{kJ} = 30.91\ \text{kJ mol}^{-1}$$

$$\Delta S°_{298} = \Delta S°_{298_{Br_2(g)}} - \Delta S°_{298_{Br_2(\ell)}}$$

$$= 245.35 - 152.23 = 93.12\ \text{J mol}^{-1}\ \text{K}^{-1}$$

(c) For vaporization, $\Delta G°_{298} = \Delta H°_{298} - T\Delta S°_{298}$

$$\Delta G°_{298} = 30{,}910\ \text{J} - 298.2(93.12)\text{J} = 3142\ \text{J} = 3.142\ \text{kJ mol}^{-1}$$

(d) The vaporization should not be a spontaneous process at 1 atm of pressure and 298 K.

(e) For equilibrium to occur, $\Delta G°_T$ must equal zero. Therefore,

$$\Delta G°_T = 0 = \Delta H° - T\Delta S°$$

or

$$\Delta H° = T\Delta S°$$

Assuming that ΔH° and ΔS° do not change significantly with a change in temperature,

$$T = \frac{\Delta H^\circ_{298}}{\Delta S^\circ_{298}} = \frac{30910}{93.12} = 331.9 \text{ K} \quad \text{or} \quad 58.7^\circ\text{C}$$

62. Acetic acid, CH_3CO_2H, can form a dimer, $(CH_3CO_2H)_2$, in the gas phase.

$$2CH_3CO_2H(g) \rightleftharpoons (CH_3CO_2H)_2(g)$$

The dimer is held together by two hydrogen bonds

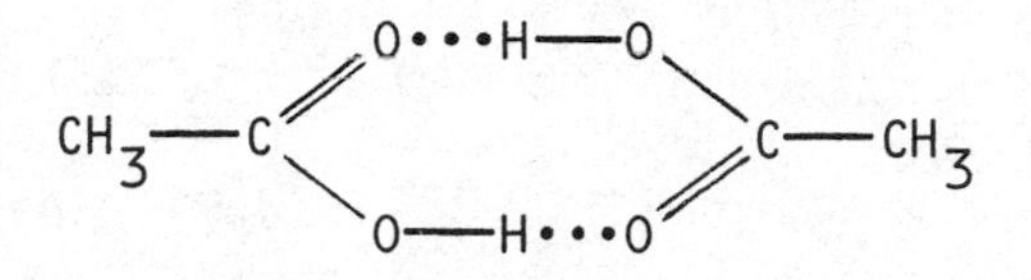

with a total strength of 66.5 kJ per mole of dimer. At 25°C the equilibrium constant for the dimerization is 1.3×10^3 (pressure in atmospheres). What is ΔS° for the reaction at 25°C?

Solution

Two items of information are given to use in this problem. The equilibrium constant allows the calculation of ΔG°_{298}. The strength of the bond (66.5 kJ) means that it requires 66.5 kJ to pull one mole of bonds apart. In other words, $\Delta H^\circ_{f298} = -66.5$ kJ. The values of ΔG°_{298} and ΔH°_{f298} allow us to calculate ΔS°_{298} by use of the equation

$$\Delta G^\circ_{298} = \Delta H^\circ_{f298} - T\Delta S^\circ_{298}$$

First calculate ΔG°_{298}:

$$\Delta G^\circ_{298} = -2.303\, RT \log K$$
$$= -2.303(8.314 \text{ J K}^{-1} \text{ mol}^{-1})(298.2 \text{ K})(\log 1.3 \times 10^3)$$
$$= -17.78 \text{ kJ mol}^{-1}$$

$$\Delta S^\circ_{298} = -\frac{\Delta G^\circ_{298} - \Delta H^\circ_{298}}{T}$$

Then, $$\Delta S^\circ_{298} = -\frac{(-17{,}780) - (-66{,}500)}{298.2} = -163 \text{ J K}^{-1}$$

63. At 1000 K the equilibrium constant for the reaction $Br_2(g) \rightleftharpoons 2Br$ is 2.8×10^4 (pressure in atmospheres). What is ΔG° for the reaction? Assume that the bond energy of Br_2 does not change between 298 K and 1000 K and calculate ΔS° for the reaction at 1000 K.

Solution

$$\Delta G^\circ = -2.303\ RT \log K$$

$$\Delta G^\circ_{1000} = -2.303(8.314)(1000)(\log 2.8 \times 10^4)$$
$$= -19{,}147(4.447) = -85 \text{ kJ mol}^{-1}$$

The value of the change in enthalpy of the reaction $Br_2(g) \rightleftharpoons 2Br(g)$ may be calculated from the ΔH_f's at 298 K on the assumption that the value of ΔH will not change with temperature. Over such a large temperature range, this is a first approximation. Techniques are developed in a course in physical chemistry to treat this situation. As an approximation, then,

$$\Delta H = 2\Delta H^\circ_{f_{Br(g)}} - \Delta H^\circ_{f_{Br_2(g)}}$$

$$= 2(111.88) - 30.91 \quad (\Delta H^\circ_f \text{ in kJ})$$

$$= 192.85 \text{ kJ}$$

Then, from $\Delta G = \Delta H - T\Delta S$

or $\Delta S = -\dfrac{\Delta G - \Delta H}{T}$

$$= -\frac{-85{,}000 - 192{,}850}{1000 \text{ K}} = \frac{277{,}860 \text{ J}}{1000 \text{ K}}$$

$$= 278 \text{ J K}^{-1}$$

RELATED EXERCISES

1. Boron compounds have been investigated extensively in recent years because of peculiarities that sometimes occur in their bonding. Diborane, B_2H_6, exhibits such bonding in which the two boron atoms are joined through two hydrogen atoms, one hydrogen located on each side of the center line between the two boron atoms. Calculate the standard heat of formation of diborane if the combustion of diborane at constant pressure is

$$B_2H_6(g) + 3O_2(g) \longrightarrow B_2O_3(s) + 3H_2O(g)$$

The combustion proceeds with the liberation of 2020 kJ per mole. Combustion of elemental boron produces B_2O_3 with the liberation of 1264 kJ mol^{-1}. $(\Delta H^\circ_{f298_{H_2O(g)}} = -241.8 \text{ kJ mol}^{-1})$

Answer: +30.6 kJ mol^{-1}

2. The enthalpies of individual ions in solution may be developed using the heat of formation of hydrogen ion as an arbitrary reference set at 0. All other heats are then compared to this standard. If the standard enthalpy of formation of the fumarate ions is -777.4 kJ mol^{-1}, calculate the enthalpy of formation of the succinate ion. The standard enthalpy change of the reaction

$$fumarate^{2-}(aq) + H_2 \longrightarrow succinate^{2-}(aq)$$

is -131.4 kJ. *Answer: -908.8 kJ mol⁻¹*

3. It is estimated that the human brain consumes the equivalent of 10 g of glucose, $C_6H_{12}O_6$, per hour. The metabolism of glucose produces CO_2 and H_2O and proceeds in aqueous solution, where the standard state is taken at a concentration of 1 molal at 25°C. References indicate that

$$\Delta H^\circ_{f298_{(glucose)}}(aq) = -1263.1 \text{ kJ mol}^{-1}$$

and

$$\Delta H^\circ_{f298_{(CO_2)}}(aq) = -413.8 \text{ kJ mol}^{-1}$$

Estimate (a) the energy utilized per hour and (b) the power output of the brain in watts [1 watt (W) = 1 J s^{-1}].

Answer: (a) 163.0 kJ/hr; (b) 45 W

4. Calculate the enthalpy change accompanying the conversion of 1 mol of glucose into maltose according to the reaction

$$2C_6H_{12}O_6(s) \longrightarrow C_{12}H_{22}O_{11}(s) + H_2O(l)$$

The heats of combustion are (ΔH°_{comb}/kJ mol^{-1})

α-D-glucose, $C_6H_{12}O_6(s)$: -2809

maltose, $C_{12}H_{22}O_{11}(s)$: -5645

Answer: 13.6 kJ mol⁻¹ (glucose)

5. The heat of sublimation of graphite to carbon atoms [$C_{graphite} \longrightarrow C(g)$] is estimated as 715 kJ mol^{-1}. The dissociation of molecular hydrogen into atoms [$H_2(g) \longrightarrow 2H(g)$] has $\Delta H^\circ = 436$ kJ mol^{-1}. Using $\Delta H^\circ_{f_{CH_4}} = -74.9$ kJ mol^{-1}, estimate ΔH° for the reaction

$$C(g) + 4H(g) \longrightarrow CH_4(g)$$

One-fourth of this value is a measure of the C—H bond strength in methane.

Answer: -1662 kJ or -415.5 kJ per C—H bond

UNIT 14

CHAPTER 20: ELECTROCHEMISTRY AND OXIDATION-REDUCTION

INTRODUCTION

Historically, the terms *oxidation* and *reduction* came from the field of metallurgy. Oxidation referred to reactions involving the addition of oxygen to a metal. Reduction, on the other hand, referred to the removal of oxygen from a reacting metal oxide, often by the passage of hydrogen over or through it. Today, these terms are used in chemistry in a much broader sense to describe reactions involving the transfer of electrons from one substance to another.

Reactions involving oxidation and reduction are called *oxidation-reduction* or *redox* reactions. Such reactions are quite common. For example, the energy that is required for you to assimilate your thoughts while you read this page is derived from foods that have undergone a variety of complex biochemical oxidation-reduction reactions. Another example may be drawn from industry. The extraction of aluminum from bauxite ore is accomplished by a redox reaction in which an electric current is passed through a molten mixture of the ore. These examples by no means exhaust the diverse forms possible in redox reactions.

The first evidence of a relationship between chemical reactions and electricity was the result of work done by Luigi Galvani on the twitching of frog legs caused by an electrical shock. The results of further work by Galvani set the stage in 1800 for the development by Allesandro Volta of a practical battery based on a silver-zinc couple. This discovery made it possible for the first time in history to obtain a continuous source of electric current. In 1807, Sir Humphry Davy discovered the elements sodium and potassium by electrolyzing fused mixtures of their solid hydroxides. Bear in mind that all this work was accomplished prior to the discovery of subatomic particles — protons, electrons, and neutrons.

The development of this chapter treats the quantitative relationships between chemical change and electrical phenomena. The fundamental laws of electrochemical work were discovered by Michael Faraday during the years 1832 and 1833. Faraday's work showed that the decomposition that occurs in a quantity of substance brought about by electrical means is proportional to the electrical current passing through that substance and to the time the current is applied. He further extended his research to show that the masses of substances deposited at the electrodes in electrolysis cells are proportional to their chemical equivalent weights. Faraday's Laws are now widely applied in research and industry.

The exercises included in this chapter involve application of electrochemical principles to computing cell potentials or voltages, equilibrium constants, free energies (Gibbs energies), and the quantities appearing in Faraday's Laws.

FORMULAS AND DEFINITIONS

Oxidation Oxidation involves the removal or loss of electrons from a substance. The element(s) undergoing oxidation are determined from changes in oxidation states or numbers in the following way. The number 0 (called an oxidation number) is assigned to neutral species. For each electron in excess of the number in the neutral species, the state is decreased by -1, and for each electron less than that occurring in the neutral species, the state is increased by +1. Thus Cl^{-1} indicates a gain of one electron; its oxidation number is -1. On the other hand, Cu^{2+} indicates a loss of two electrons; its oxidation number is +2. By use of this convention, the substance undergoing oxidation shows a net increase in oxidation number. Examples include

$$Sn^{2+} - 2e^- \longrightarrow Sn^{4+}$$

$$Fe^0 - 2e^- \longrightarrow Fe^{2+}$$

$$O^{2-} - 2e^- \longrightarrow O^0$$

Reduction The reduction of a substance involves the absorption or gain of electrons. By the convention used to assign oxidation numbers to reactants, the substance undergoing reduction shows a net decrease in oxidation number. Examples include

$$Fe^{+3} + e^- \longrightarrow Fe^{2+}$$

$$Cl^0 + e^- \longrightarrow Cl^-$$

$$Ag^+ + e^- \longrightarrow Ag^0$$

Oxidizing agent The substance in a reaction causing the oxidation of a substance.

Reducing agent The substance in a reaction causing the reduction of a substance.

In the following reaction,

$$Fe^0 + Sn^{2+} \longrightarrow Fe^{2+} + Sn^0$$

iron (0) is oxidized to iron (2+) and tin (2+) is reduced to tin (0). Tin (2+) is the oxidizing agent and iron (0) is the reducing agent.

Anode The electrode toward which negatively charged ions are attracted; electrons are withdrawn from the electrolytic liquid causing oxidation.

Anions Negatively charged ions.

Cathode The electrode toward which positively charged ions are attracted; electrons enter the electrolytic liquid causing reduction.

Cations Positively charged ions.

Electrolytic cell A chemical reaction system in which electrical energy is consumed to bring about desired chemical changes. Such chemical changes are by definition nonspontaneous. In the process, electrons are forced from an outside source on to the cathode, making it negatively charged, and electrons are withdrawn from the anode, making it positively charged.

Electrolysis An oxidation-reduction reaction taking place in an electrolytic cell.

Electromotive series An ordering of the elements according to their tendency to form positive ions. In terms of reduction potentials, potassium has the largest negative value, -2.925 volts for the reaction $K^+ + e^- \longrightarrow K$.

emf An acronym for electromotive force, a force or potential causing an electron flow. It is normally measured in volts.

Faraday's Law During electrolysis, 96,487 coulombs (1 faraday) of electricity reduce one gram-equivalent of the oxidizing agent and oxidize one gram-equivalent of the reducing agent. In other words, the amount of substance reacted at each electrode during electrolysis is directly proportional to the quantity of electricity passed through the electrolytic cell.

Nernst equation The equation is defined for reactions and for half-reactions having the general form

$$aA + bB \rightleftharpoons cC + dD \qquad (1)$$

and

$$xM + ne^- \rightleftharpoons yN \qquad (2)$$

as

$$E = E^\circ - \frac{0.05915}{n} \log Q$$

where

$$Q = \frac{[C]^c[D]^d}{[A]^a[B]^b}$$

For Equation (1) and

$$Q = \frac{[N]^y}{[M]^x}$$

for Equation (2).

E = emf for the reaction or half-reaction.

E° = standard electrode potential for the cell reaction or the half-reaction.

n = number of electrons required in the redox transfer process according to the balanced equation or half-reactions.

Standard hydrogen electrode Prepared by bubbling hydrogen gas at 25°C and a pressure of 1 atm around a platinized platinum electrode immersed in a solution in which hydrogen ions are at unit activity (approximately 1 *M*).

Standard potential Potential of the electrode measured at 25°C when the concentration of the ions in the solution are at unit activity ($\approx$ 1 *M*) and the pressure of any gas involved is 1 atm.

Thermodynamic functions and their relation to E° Several relations are possible:

$$\Delta G^\circ = -nFe^\circ$$

$$E^\circ = \frac{RT}{nF} \ln K_e = \frac{0.05915}{n} \log K_e$$

where n is the number of electrons transferred and F is the faraday, a constant in units of kilojoules per volt (F = 96.487 kJ/V) and the temperature is defined at 298.15 K.

Voltaic cells Commonly thought of as batteries, voltaic cells have as their negative terminal the anode, where oxidation occurs. Reduction occurs at the cathode, but in contrast to the situation in electrolytic cells, the cathode is the positive terminal.

BALANCING EQUATIONS: TECHNIQUES FOR OXIDATION-REDUCTION REACTIONS

Several methods are commonly used to balance equations involving oxidation and reduction. Regardless of the method used to balance equations, electrons lost through oxidation must be absorbed through reduction; no net gain or net loss of electrons can occur in a reaction. Further, for an equation to be balanced, the total of the ion charges on both sides of the equation must be equal. The two methods most used for balancing equations are presented here for your study.

Change in oxidation number method: This is based on the concept that in a redox reaction the total increase in units of positive oxidation number must equal the total decrease in units of negative oxidation number. Consider the following reaction as an example:

$$MnO_4^- + C_2O_4^{2-} + H^+ \longrightarrow Mn^{2+} + CO_2 + H_2O$$

Assign oxidation numbers to each element, determine those elements that undergo oxidation or reduction, and then write half-reactions that indicate the oxidation number changes (sometimes these are half-reactions for the overall reaction). In this case, the half-reactions are

$$C_2^{3+} \longrightarrow 2C^{4+} + 2e^- \uparrow 2 \text{ (gain in positive oxidation number, 2)}$$

$$Mn^{7+} \longrightarrow Mn^{2+} - 5e^- \downarrow 5 \text{ (decrease in positive oxidation number, 5)}$$

To balance the equation, the proper number of manganese-containing and carbon-containing ions must be selected, so that the total increase of electrons will equal the total decrease. This can be done by choosing the smallest factor common to both 2 and 5 and multiplying each half-reaction by the multiple of each in the factor. For these half-reactions, 10 is the common factor, and the half-reactions must be multiplied by 5 and 2, respectively.

$$5[C_2^{3+} \longrightarrow 2C^{4+} + 2e^-] \quad \uparrow 2 \quad 5 \times 2e^- = 10e^-$$

$$2[Mn^{7+} \longrightarrow Mn^{2+} - 5e^-] \quad \downarrow 5 \quad 2 \times 5e^- = 10e^-$$

The proper coefficients for those substances changing oxidation state become

$$2MnO_4^- + 5C_2O_4^{2-} + ?H^+ \longrightarrow 2Mn^{2+} + 10CO_2 + ?H_2O$$

The remaining factors in the equation can be balanced by inspection, yielding

$$2MnO_4^- + 5C_2O_4^{2-} + 16H^+ \longrightarrow 2Mn^{2+} + 10CO_2 + 8H_2O$$

Ion-electron method: Based on combining half-reactions, one-half of which represents the oxidation step and the other, the reduction step. This method is very useful in electrochemical cells, for which, generally, the half-reactions are known. The same reaction used in the change in oxidation number method will be used to illustrate this method.

The oxalate ion is oxidized to carbon dioxide

$$C_2O_4^{2-} \longrightarrow 2CO_2 + 2e^- \qquad (1)$$

and is balanced by inspection. The oxidizing agent in this reaction, permanganate, is reduced in acid solution to manganous ion and water:

$$MnO_4^- + 8H^+ + 5e^- \longrightarrow Mn^{2+} + 4H_2O \qquad (2)$$

Again, the balancing is done by inspection.

It is now obvious that multiplying Equation (1) by 5 and Equation (2) by 2 will make the number of electrons lost by oxalate the same as the number gained by permanganate. The two equations are added to give the following:

$$5C_2O_4^{2-} \longrightarrow 10CO_2 + 10e^-$$

$$2MnO_4^- + 16H^+ + 10e^- \longrightarrow 2Mn^{2+} + 8H_2O$$

$$\overline{2MnO_4^- + 5C_2O_4^{2-} + 16H^+ \longrightarrow 2Mn^{2+} + 10CO_2 + 8H_2O}$$

Both methods give the same result, as should be expected.

EXERCISES

4. Complete and balance the following half-reactions. (In each case indicate whether oxidation or reduction occurs.)

   (e) $O_2 \longrightarrow OH^-$ (in base) (g) $MnO_4^- \longrightarrow Mn^{2+}$ (in acid)

   Solution

   (e) Oxygen in O_2 has an oxidation state of 0 and in OH^- it has a value of -2. The gain of electrons is a reduction. To balance the equation, water and electrons are required on the left-hand side.

   $$O_2 + 2H_2O + 4e^- \longrightarrow 4OH^-$$

   (g) Manganese goes from an oxidation state of +7 in MnO_4^- to +2 in Mn^{2+}. This is a gain of electrons or a reduction. Acid as H^+ is added to the left-hand side along with 5 electrons. Water is added to the right-hand side to balance the O and H on the left.

$$MnO_4^- + 8H^+ + 5e^- \longrightarrow Mn^{2+} + 4H_2O$$

7. Using a diagram of the type

anode | anode solution || cathode solution | cathode

diagram the electrolytic cell for each of the following cell reactions:

(b) $2NaCl(aq) + 2H_2O \longrightarrow 2NaOH(aq) + H_2 + Cl_2$ (using inert electrodes of specially treated titanium, Ti)

Solution

The anodic reaction is the oxidation of chloride ions:

$$2Cl^- \longrightarrow Cl_2 + 2e^-$$

The cathodic reaction is the reduction of water:

$$2H_2O + 2e^- \longrightarrow H_2 + 2OH^-$$

giving an overall reaction of

$$2H_2O(\ell) + 2Cl^- \longrightarrow H_2(g) + Cl_2(g) + 2OH^-$$

Using the convention of the anode on the left, we have

$$(Ti)Cl_2 \mid Cl^- \parallel H_2O, OH^- \mid H_2(Ti)$$

8. Write the anode half-reaction, the cathode half-reaction, and the cell reaction for the following electrolytic cells:

$$\xrightarrow{e^-}$$

(a) $(C)NaCl(\ell) \mid Cl_2 \parallel NaCl(\ell) \mid Na(\ell)(Fe)$

Solution

The anodic half-reaction is

$$2Cl^- \longrightarrow Cl_2 + 2e^- \quad \text{(anodic oxidation)}$$

The cathodic half-reaction is

$$Na^+ + e^- \longrightarrow Na \quad \text{(cathodic reduction)}$$

The overall cell reaction is obtained by adding the electrode reactions.

$$\begin{array}{lrcl} \textit{Anode:} & 2Cl^- & \longrightarrow & Cl_2 + 2e^- \\ \textit{Cathode:} & 2[Na^+ + e^- & \longrightarrow & Na] \\ \hline & 2Na^+ + 2Cl^- & \longrightarrow & Cl_2 + 2Na \end{array}$$

(d) $\overrightarrow{e^-}$ $(Pb)PbSO_4, PbO_2 \mid H_2SO_4(aq) \parallel H_2SO_4(aq) \mid PbSO_4, (Pb)$

Solution

When the cell is delivering a current, the anode is the electrode at which oxidation occurs

$$Pb \longrightarrow Pb^{2+} + 2e^- \quad \text{(anodic oxidation)}$$

The electrons flow from the negatively charged lead electrode through the external circuit to the lead dioxide electrode where PbO_2 is reduced in the presence of hydrogen ion to Pb^{2+}.

$$PbO_2 + 4H^+ + 2e^- \longrightarrow Pb^{2+} + 2H_2O \quad \text{(cathodic reduction)}$$

In both cases the lead ions form an insoluble precipitate with the sulfate ions present. The two reactions are

$$Pb + SO_4^{2-} \longrightarrow PbSO_4(s) + 2e^- \quad \text{(at lead anode)}$$

$$PbO_2 + 4H^+ + SO_4^{2-} + 2e^- \longrightarrow PbSO_4(s) + 2H_2O \quad \text{(at lead dioxide cathode)}$$

The net cell reaction is found by adding these two reactions.

$$Pb + PbO_2 + 4H^+ + 2SO_4^{2-} \longrightarrow 2PbSO_4(s) + 2H_2O$$

12. Calculate the value of the Faraday constant, F, from the charge on a single electron, 1.6021×10^{-19} C.

Solution

The Faraday constant is the charge on one mole of electrons.

$$F = 1.6021 \times 10^{-19} \frac{C}{e^-} \times \frac{6.022 \times 10^{23}\ e^-}{mol} = 9.648 \times 10^4\ C$$

13. How many moles of electrons are involved in the following electrochemical changes?

(a) 0.800 mol of I_2 is converted to I^-

(c) 27.6 g of SO_3 is converted to SO_3^{2-}

(g) 15.80 mL of 0.1145 M MnO_4^- is converted to Mn^{2+}

Solution

(a) The reduction of 1 mol of I_2 molecules to 2 mol of I^- ions requires 2 mol of electrons.

$$I_2 + 2e^- \longrightarrow 2I^-$$

$$\text{Moles of } e^- = \frac{2 \text{ mol } e^-}{\text{mol } I_2} \times 0.800 \text{ mol} = 1.60 \text{ mol}$$

(c) The reduction of 1 mol of SO_3 to SO_3^{2-} requires 2 mol of electrons.

$$SO_3 + 2e^- \longrightarrow SO_3^{2-}$$

$$\text{Moles of } e^- = 27.6 \text{ g } SO_3 \times \frac{1 \text{ mol}}{80.1 \text{ g}} \times \frac{2 \text{ mol } e^-}{\text{mol } SO_3} = 0.690 \text{ mol}$$

(g) The reduction of 1 mol of MnO_4^- to Mn^{2+} requires 5 mol of electrons.

$$8H^+ + MnO_4^- + 5e^- \longrightarrow Mn^{2+} + 4H_2O$$

$$\text{Mol } MnO_4^- = (0.01580 \text{ L})(0.1145 \text{ mol/L}) = 1.8091 \times 10^{-3} \text{ mol}$$

$$\text{Moles of } e^- = 1.8091 \times 10^{-3} \text{ mol } MnO_4^- \times \frac{5 \text{ mol } e^-}{\text{mol } MnO_4^-} = 9.046 \times 10^{-3} \text{ mol}$$

14. How many faradays of electricity are involved in the electrochemical changes described in Exercise 13?

Solution

One faraday of charge provides 1 mol of electrons for a reduction process.

(a) $$\text{No. faradays} = 1.60 \text{ mol } e^- \times \frac{1 \text{ faraday}}{\text{mol } e^-} = 1.60$$

(c) $$\text{No. faradays} = 0.690 \text{ mol } e^- \times \frac{1 \text{ faraday}}{\text{mol } e^-} = 0.690$$

(g) $$\text{No. faradays} = 9.046 \times 10^{-1} \text{ mol } e^- \times \frac{1 \text{ faraday}}{\text{mol } e^-} = 9.046 \times 10^{-3}$$

15. How many coulombs of electricity are involved in the electrochemical changes described in Exercise 13?

Solution

One faraday of charge is equivalent to 96,487 coulombs.

(a) $$\text{No. coulombs} = 1.60 \text{ mol } e^- \times \frac{1 \text{ faraday}}{\text{mol } e^-} \times \frac{96{,}487 \text{ C}}{\text{faraday}} = 1.54 \times 10^5$$

(c) $$\text{No. coulombs} = 0.690 \text{ mol } e^- \times \frac{1 \text{ faraday}}{\text{mol } e^-} \times \frac{96{,}487 \text{ C}}{\text{faraday}} = 6.66 \times 10^4$$

(g) $$\text{No. coulombs} = 9.046 \times 10^{-3} \text{ mol } e^- \times \frac{1 \text{ faraday}}{\text{mol } e^-} \times \frac{96{,}487 \text{ C}}{\text{faraday}} = 8.728 \times 10^2$$

18. Ammonium perchlorate, used in the solid fuel in the booster rockets on the space shuttle, is prepared from sodium perchlorate, $NaClO_4$, which is produced commercially by the electrolysis of a hot, stirred solution of sodium chloride.

$$NaCl + 4H_2O \longrightarrow NaClO_4 + 4H_2$$

How many moles of electrons are required to produce 1.00 kg of sodium perchlorate? How many faradays? How many coulombs?

Solution

The reaction at the anode is an oxidation

$$Cl^- \longrightarrow Cl^{7+} + 8e^-$$

This equation tells us that 1 mol of Cl^{7+} is produced for each 8 mol of electrons passing through the cell. There are

$$1000 \text{ g} \times \frac{1 \text{ mol}}{122.440 \text{ g}} = 8.17 \text{ mol } NaClO_4$$

in 1.00 kg. This requires

$$8.17 \text{ mol } NaClO_4 \times \frac{8 \text{ mol } e^-}{\text{mol } NaClO_4} = 65.3 \text{ mol}$$

$$65.3 \text{ mol } e^- \times \frac{1 \text{ faraday}}{1 \text{ mol } e^-} = 65.3 \text{ faraday}$$

$$65.3 \text{ faradays} \times \frac{96{,}487 \text{ coulombs (C)}}{1 \text{ faraday}} = 6.30 \times 10^6 \text{ C}$$

20. How many grams of zinc will be deposited from a solution of zinc(II) sulfate by 3.40 faraday of electricity?

Solution

The reduction of 1 mol of zinc(II) ions requires 2 mol of electrons.

$$Zn^{2+} + 2e^- \longrightarrow Zn$$

And 1 faraday of charge furnishes 1 mol of electrons for the reaction. Hence, 2 faraday of charge reduce 1 mol of zinc(II) ions.

$$\text{Mass Zn} = 3.40 \text{ faraday} \times \frac{1 \text{ mol Zn}}{2 \text{ faraday}} \times \frac{65.38 \text{ g}}{\text{mol}} = 111 \text{ g}$$

23. How many grams of cobalt will be deposited from a solution of cobalt(II) chloride electrolyzed with a current of 20.0 A for 54.5 min?

Solution

The reduction of Co^{2+} to Co requires two faradays of charge per mole of Co^{2+} ions reduced.

$$Co^{2+} + 2e^- \longrightarrow Co$$

$$\text{No. coulombs} = (20\ A)(54.5\ min \times 60\ s/min) = 6.54 \times 10^4\ C$$

$$\text{No. faradays} = 6.54 \times 10^4\ C \times \frac{1\ \text{faraday}}{96{,}487\ C} = 0.678$$

$$\text{Mass Co} = 0.678\ \text{faraday} \times \frac{1\ \text{mol Co}}{2\ \text{faraday}} \times \frac{58.9332\ g}{mol} = 20.0\ g$$

29. Diagram voltaic cells having the following net reactions:

(a) $Mn + 2Ag^+ \longrightarrow Mn^{2+} + Ag$

Solution

By convention the anode is placed on the left. Single vertical lines represent a phase separation, and the double vertical lines represent a salt bridge or separation of the two electrolytic solutions. The oxidized form of the anode is placed adjacent to and to the right of the anode substance. The cathode always appears to the extreme right with its oxidized form to its left separated by a single vertical line

$$Mn \mid Mn^{2+} \parallel Ag^+ \mid Ag$$

34. Calculate the emf at standard conditions of the cell based on each of the following reactions:

(a) $Zn + I_2 \longrightarrow Zn^{2+} + 2I^-$

Solution

The two half-cell reactions are written along with their respective $E°$ values from Table 20-1.

$$Zn^{2+} + 2e^- \rightleftharpoons Zn \qquad E° = -0.763\ V$$

$$I_2 + 2e^- \rightleftharpoons 2I^- \qquad E° = +0.5355$$

The Zn^{2+} reaction lies higher in the electromotive series than does the I_2 reaction. Because the reduced form of any element will reduce the oxidized form of any element below it in the electromotive series, Zn will reduce I_2. The first reaction is reversed along with the sign of its $E°$ and the two reactions are added.

$$\begin{array}{rll} Zn \rightleftharpoons Zn^{2+} + 2e^- & E° = +0.763\ V \\ I_2 + 2e^- \rightleftharpoons 2I^- & E° = +0.5355\ V \\ \hline Zn + I_2 \rightleftharpoons Zn^{2+} + 2I^- & E° = +1.299\ V \end{array}$$

(e) $MnO_4^- + 8H^+ + 5Au \longrightarrow Mn^{2+} + 4H_2O + 5Au^+$

Solution

From Tables 20-1 and 20-2,

$$Au^+ + e^- \rightleftharpoons Au \qquad E° = +1.68 \text{ V}$$

$$MnO_4^- + 8H^+ + 5e^- \rightleftharpoons Mn^{2+} + 4H_2O \qquad E° = +1.51 \text{ V}$$

If the two tables were combined, the MnO_4^- reaction would lie above the Au^+ reaction, meaning that the reduced form of MnO_4^-, Mn^{2+}, would reduce Au^+, the oxidized form of Au. This is the reverse of the way in which the reaction is written. A negative voltage should be expected, which would indicate that the reaction is not spontaneous. In order to calculate the potential in this case, the gold reaction is reversed and multiplied by 5 to allow the electrons to cancel when the equations are added. The voltages, however, are not changed except for the change in sign.

$$5Au \rightleftharpoons 5Au^+ + 5e^- \qquad E° = -1.68 \text{ V}$$

$$MnO_4^- + 8H^+ + 5e^- \rightleftharpoons Mn^{2+} + 4H_2O \qquad E° = +1.51 \text{ V}$$

$$MnO_4^- + 8H^+ + 5Au \longrightarrow Mn^{2+} + 4H_2O + 5Au^+ \qquad E° = -0.17 \text{ V}$$

35. Determine the standard emf for each of the following cells:

(a) $Co \mid Co^{2+}, M = 1 \overset{e^-}{\parallel} Cr^{3+}, M = 1 \mid Cr$

Solution

In this cell Co is oxidized to Co^{2+}, and Cr^{3+} is reduced to Cr — both reactions occurring in 1 *M* solutions. Cr lies higher than Co in the electromotive series and will reduce cobalt ion. However, in this cell Co is oxidized to Co^{2+}, and Cr^{3+} is reduced to Cr. This reversal from normal conditions would indicate that an overall negative sign is expected. To obtain the net cell reaction, the half-reactions are added as follows:

$$Co \rightleftharpoons Co^{2+} + 2e^- \qquad E° = +0.277 \text{ V}$$

$$Cr^{3+} + 3e^- \rightleftharpoons Cr \qquad E° = -0.74 \text{ V}$$

$$3Co + 2Cr^{3+} \rightleftharpoons 3Co^{2+} + 2Cr \qquad E° = -0.46 \text{ V}$$

36. Write the cell reaction for a voltaic cell based on each of the following pairs of half-reactions, and calculate the emf of the cell under standard conditions:

(a) $Sc^{3+} + 3e^- \longrightarrow Sc$

$Ag^+ + e^- \longrightarrow Ag$

(e) $HClO_2 + 2H^+ + 2e^- \longrightarrow HClO + H_2O$

$ClO_3^- + 3H^+ + 2e^- \longrightarrow HClO_2 + H_2O$

Solution

(a) The E° values are

$$Sc^{3+} + 3e^- \rightleftharpoons Sc \quad E^\circ = -2.08\ V$$
$$Ag^+ + e^- \rightleftharpoons Ag \quad E^\circ = +0.7991\ V$$

A voltaic cell requires a positive voltage output. The anode is the electrode at which oxidation occurs. From a comparison of the sign and size of the two reduction potentials, the oxidation reaction $Sc \longrightarrow Sc^{3+} + 3e^-$ with its larger positive sign is favored over the $Ag \longrightarrow Ag^+ + e^-$ reaction. The half-cell reactions are added after multiplying the Ag reaction by 3 so that the electrons will cancel.

Anode:	$Sc \longrightarrow Sc^{3+} + 3e^-$	$E^\circ = +2.08\ V$
Cathode:	$3(Ag^+ + e^- \longrightarrow Ag)$	$E^\circ = +0.7991\ V$
Cell reaction:	$Sc + 3Ag^+ \longrightarrow 3Ag + Sc^{3+}$	$E^\circ = +2.88\ V$

(e) The E° values are

$$HClO_2 + 2H^+ + 2e^- \rightleftharpoons HClO + H_2O \quad E^\circ = 1.64\ V$$
$$ClO_3^- + 3H^+ + 2e^- \rightleftharpoons HClO_2 + H_2O \quad E^\circ = 1.21\ V$$

The first reaction is more favored to occur in the direction written than the second one because of the larger positive sign. Therefore, reverse the sense of the second equation for the anode reaction and add.

Anode:	$HClO_2 + H_2O \longrightarrow ClO_3^- + 3H^+ + 2e^-$	$E^\circ = -1.21\ V$
Cathode:	$HClO_2 + 2H^+ + 2e^- \longrightarrow HClO + H_2O$	$E^\circ = +1.64\ V$
Cell reaction:	$2HClO_2 \longrightarrow HClO + ClO_3^- + H^+$	$E^\circ = +0.43\ V$

37. Rechargeable nickel-cadmium cells are used in calculators and other battery-powered devices. The Telstar communication satellite also uses these cells. The cell reaction is

$$NiO_2 + Cd + 2H_2O \longrightarrow Ni(OH)_2 + Cd(OH)_2$$

Calculate the emf of such a cell using the following half-cell potentials:

$$NiO_2 + 2H_2O + 2e^- \longrightarrow Ni(OH)_2 + 2OH^- \quad E^\circ = +0.49\ V$$
$$Cd(OH)_2 + 2e^- \longrightarrow Cd + 2OH^- \quad E^\circ = -0.81\ V$$

Solution

The tendency of Cd to go to $Cd(OH)_2$ is greater than that of NiO_2 to go to $Ni(OH)_2$. Write the $Cd(OH)_2$ half-cell in the reverse sense and change the sign of its $E°$. The two half-cells are

$$NiO_2 + 2H_2O + 2e^- \longrightarrow Ni(OH)_2 + 2OH^- \quad E° = +0.49 \text{ V}$$

$$Cd + 2OH^- \longrightarrow Cd(OH)_2 + 2e^- \quad E° = +0.81 \text{ V}$$

The reactions are added and gives an $E°$ of 0.49 V + 0.81 V = +1.30 V for the cell reaction.

44. Calculate the emf for each of the following half-reactions:

(a) $Sn^{2+}(0.0100\ M) + 2e^- \longrightarrow Sn$

(c) $O_2(0.0010 \text{ atm}) + 4H^+(0.100\ M) + 4e^- \longrightarrow 2H_2O(\ell)$

(d) $Cr_2O_7^{2-}(0.150\ M) + 14H^+(0.100\ M) + 6e^- \longrightarrow 2Cr^{3+}(0.000100\ M) + 7H_2O(\ell)$

Solution

(a) $Sn^{2+}(0.0100\ M) + 2e^- \longrightarrow Sn$.

The emf for this half-cell is calculated from the Nernst equation by assuming that the activity or concentration of the free metal has a value of 1. At unit concentration

$$E = E° - \frac{0.05915}{n} \log Q$$

the emf, $E°$, is -0.136 V, but at a lower concentration of Sn^{2+}, 0.0100 M, the cell potential is

$$E = -0.136 \text{ V} - \frac{0.05915}{2} \log \frac{[Sn]}{[Sn^{2+}]}$$

$$= -0.136 \text{ V} - \frac{0.05915}{2} \text{ V}\left(\log \frac{1}{.01}\right)$$

$$= -0.136 \text{ V} - \frac{0.05915}{2} (2) \text{ V} = -0.195 \text{ V}$$

(c) $O_2(0.0010 \text{ atm}) + 4H^+(0.100\ M) + 4e^- \longrightarrow 2H_2O(\ell)$

$$E = E° - \frac{0.05915}{n} \log Q$$

$$= 1.23 \text{ V} - \frac{0.05915}{4} \log \frac{[H_2O]^2}{pO_2[H^+]^4}$$

The water produced is pure and its concentration or activity is considered unity.

$$E = 1.23\text{ V} - \frac{0.05915}{4}\log\frac{(1)^2}{(10^{-3})(10^{-1})^4}$$

$$= 1.23\text{ V} - \frac{0.05915}{4}(7)\text{ V} = 1.13\text{ V}$$

(d) $Cr_2O_7^{2-}(0.150\ M) + 14H^+(0.100\ M) + 6e^- \longrightarrow 2Cr^{3+}(0.000100\ M) + 7H_2O(\ell)$

$$E = E° - \frac{0.05915}{n}\text{ V}(\log Q)$$

$$= 1.33\text{ V} - \frac{0.05915}{6}\text{ V}\left(\log\frac{[Cr^{3+}]^2}{[Cr_2O_7^{2-}][H^+]^{14}}\right)$$

$$= 1.33\text{ V} - \frac{0.05915}{6}\text{ V}\left(\log\frac{(1.00\times10^{-4})^2}{(0.150)(10^{-1})^{14}}\right)$$

$$= 1.33\text{ V} - \frac{0.05915}{6}\text{ V}(\log 6.67\times10^6)$$

$$= 1.33\text{ V} - 0.0673\text{ V} = 1.26\text{ V}$$

45. Hypochlorous acid, HOCl, is a stronger oxidizing agent in acidic solution than in neutral solution. Calculate the potential for the reduction of HOCl to Cl^- in a solution with a pH of 7.00 in which [HOCl] and $[Cl^-]$ both equal 1.00 *M*.

Solution

In neutral solution

$$HClO + H^+ + 2e^- \longrightarrow Cl^- + H_2O \qquad E° = +1.49\text{ V}$$

From the Nernst equation

$$E = E° - \frac{0.05915}{n}\log Q$$

$$= 1.49\text{ V} - \frac{0.05915}{2}\text{ V}\left(\log\frac{[Cl^-]}{[HClO][H^+]}\right) \quad (1)$$

$$= 1.49\text{ V} - \frac{0.05915}{2}\text{ V}\left(\log\frac{1.00}{(1.00)(10^{-7})}\right) \quad (2)$$

$$= 1.49\text{ V} - \frac{0.05915}{2}\log 10^7\text{ V} \quad (3)$$

$$= 1.49\text{ V} - \frac{0.05915}{2}(7)\text{ V} \quad (4)$$

$$= 1.49\text{ V} - 0.20702\text{ V} = 1.28\text{ V} \quad (5)$$

At stage (2):

Keystrokes: 1 ÷ 1 [EE] [+/−] 7 = [2nd] [log] × .05915 ÷ 2 = ans. 2.07025 − 01

48. Calculate the voltage produced by each of the following cells:

(a) Zn | Zn^{2+}, $M = 0.0100$ || Cu^{2+}, $M = 1.00$ | Cu

(c) (Pt)$Br_2(\ell)$ | Br^-, $M = 0.450$ || Cl^-, $M = 0.0500$ | $Cl_2(g)$, 0.900 atm (Pt)

Solution

(a) First calculate $E°$ for the reaction. The half-cells are from Table 20-1.

$$Zn^{2+} + 2e^- \rightleftharpoons Zn \quad E° = -0.763 \text{ V}$$

Cathode half-reaction:

$$Cu^{2+} + 2e^- \rightleftharpoons Cu \quad E° = +0.337 \text{ V}$$

The first of these is reversed along with the sign of its $E°$.

Anode half-reaction:

$$Zn \rightleftharpoons Zn^{2+} + 2e^- \quad E° = +0.763$$

This and the cathode reaction are added.

Net cell reaction:

$$Zn + Cu^{2+} \longrightarrow Zn^{2+} + Cu \quad E° = 1.100 \text{ V}$$

This value can now be used in the Nernst equation

$$E = E° - \frac{0.05915}{n} \log Q$$

$$E = 1.100 \text{ V} - \frac{0.05915 \text{ V}}{2} \log \frac{[Zn^{2+}]}{[Cu^{2+}]}$$

where $n = 2$ since two electrons are transferred in each reaction.

$$E = 1.100 - 0.029575 \log \left(\frac{0.01}{1.00}\right) = 1.100 + 0.05915 = 1.16 \text{ V}$$

(c) The $E°$ is determined from the half-cells in Table 20-1, where the bromine reaction is reversed from the way it is written in the table.

Anode half-reaction: $\quad 2Br^- \longrightarrow Br_2(\ell) + 2e^- \quad E° = -1.0652$

Cathode half-reaction: $\quad Cl_2(g) + 2e^- \longrightarrow 2Cl^- \quad E° = +1.3595$

Net reaction: $\quad 2Br^- + Cl_2(g) \longrightarrow Br_2(\ell) + 2Cl^- \quad E° = +0.2943$

The Nernst equation is

$$E = E^\circ - \frac{0.05915}{n} \log Q = E^\circ - \frac{0.05915}{n} \log \frac{[Cl^-]^2}{[Br^{-2}][0.9]}$$

$$= 0.2943 - \frac{0.05915}{2} \log \frac{(0.0500)^2}{(0.450)^2(0.9)}$$

$$= 0.2943 - \frac{0.05915}{2} \log \left(\frac{(0.0025)}{(0.2025)(0.9)} \right)$$

$$= 0.2943 + 0.0551 = +0.3494 \text{ V}$$

51. For a cell based on each of the following reactions run at standard conditions, calculate the emf of the cell, the standard free energy change of the reaction, and the equilibrium constant of the reaction:

(a) $Mn(s) + Cd^{2+}(aq) \longrightarrow Mn^{2+}(aq) + Cd(s)$

(c) $2Br^-(aq) + I_2(s) \longrightarrow Br_2(\ell) + 2I^-(aq)$

Solution

(a) The reaction represents the oxidation of Mn to Mn^{2+} and reduction of Cd^{2+} to Cd. Under standard conditions the emf of the cell is the sum of the half-cell potentials.

Anode half-cell:	$Mn \longrightarrow Mn^{2+} + 2e^-$	$E^\circ = +1.18$ V
Cathode half-cell:	$Cd^{2+} + 2e^- \longrightarrow Cd$	$E^\circ = -0.40$ V
Net cell reaction:	$Mn + Cd^{2+} \longrightarrow Mn^{2+} + Cd$	$E^\circ = +0.78$ V

The change in free energy for this reaction at standard conditions is calculated from

$$\Delta G^\circ = -nFE^\circ$$

in which 2 mol of electrons are transferred from Mn to Cd per mole of reactant

$$\Delta G^\circ = -(2)\left(96.487 \frac{\text{kJ}}{\text{V}}\right)(0.78 \text{ V}) = -150 \text{ kJ}$$

The equilibrium constant can be calculated directly from ΔG° or from E° equated to K. This latter method is preferred if a value for E° is available; the reason for this is that less chance for error owing to rounding is involved. In this equation,

$$E^\circ = \frac{0.05915 \text{ V}}{n} \log K$$

the unit of E° is the volt, but the unit is conventionally deleted and the equation left as

$$E^\circ = \frac{0.05915}{n} \log K$$

with volt understood as the unit of measurement. In this problem the value of K is calculated as

$$0.78 = \frac{0.05915}{2} \log K$$

$$\log K = \frac{2(0.78)}{0.05915} = 26.3736$$

$$K = 10^{26.3736} = 2.4 \times 10^{26}$$

Keystrokes: 26.3736 [2nd] [10^x] = ans. 2.3638852 26

(c) The emf of the cell is the sum of the half-cell potentials.

Anode half-cell:	$2Br^- \longrightarrow Br_2(\ell) + 2e^-$	$E° = -1.0652$ V
Cathode half-cell:	$I_2 + 2e^- \longrightarrow 2I^-$	$E° = +0.5355$ V
Net cell reaction:	$2Br^- + I_2 \longrightarrow Br_2 + 2I^-$	$E° = -0.5297$ V

$$\Delta G° = -nFE° = -(2)\left(96.487\ \frac{kJ}{V}\right)(-0.5297\ V) = 102.2\ kJ$$

And,

$$E° = \frac{0.05915}{n} \log K$$

$$-0.5297 = \frac{0.05915}{2} \log K$$

$$\log K = -17.9104$$

$$K = 10^{-17.9104} = 1.229 \times 10^{-18}$$

52. Calculate the free energy change and equilibrium constant for the reaction

$$2Br^- + F_2 \longrightarrow 2F^- + Br_2$$

Solution

The change in free energy for this reaction can be calculated from the emf of the cell:

$$\Delta G° = -nFE°$$

Anode half cell:	$2Br^- - 2e^- \longrightarrow Br_2$	$E° = -1.0652$ V
Cathode half cell:	$F_2 + 2e^- \longrightarrow 2F^-$	$E° = +2.87$ V
Net cell reaction:	$2Br^- + F_2 \longrightarrow 2F^- + Br_2$	$E°_{cell} = 1.805$ V

$$\Delta G° = -2\left(96.487\ \frac{kJ}{V}\right)(1.805\ V) = 348\ kJ$$

and
$$E^\circ = \frac{0.05915}{n} \log K$$

$$1.805 = \frac{0.05915}{2} \log K$$

$$\log K = 61.03$$

$$K = 10^{61.03} = 1.06 \times 10^{61}$$

54. Copper(I) salts disproportionate in water to form copper(II) salts and copper metal:

$$2Cu^+ \longrightarrow Cu^{2+} + Cu$$

What concentration of Cu^+ remains at equilibrium in 1.00 L of a solution prepared from 1.00 mol of Cu_2SO_4?

Solution

The concentration of Cu^+ remaining at equilibrium is determined from the equilibrium constant which can be found from the emf. The half-reactions and calculations of the emf of the cell are

$$Cu^{2+} + 2e^- \longrightarrow Cu \qquad E_1^\circ = 0.337 \text{ V} \qquad (1)$$

$$Cu^{2+} + e^- \longrightarrow Cu^+ \qquad E_2^\circ = 0.153 \text{ V} \qquad (2)$$

If 2 × (2) is subtracted from reaction (1),

$$2Cu^+ \longrightarrow Cu^{2+} + Cu \quad E^\circ = E_1^\circ - E_2^\circ = 0.184 \text{ V}$$

then
$$\log K = \frac{n\,E^\circ}{0.05916} = \frac{2 \times 0.1841}{0.05916} = 6.22$$

$$K = 1.66 \times 10^6$$

Let x = the moles of Cu^+ which remain/L. Then, $(1.00 - x)$ is the moles of Cu^+ that have reacted (x is small compared to 1.00). At equilibrium:

$$[Cu^{2+}] = (1/2)(1.00 - x) \approx 0.500 \text{ mol/L}$$

and
$$[Cu^+] = x \text{ mol/L}$$

$$K = \frac{[Cu^{2+}]}{[Cu^+]^2} = 1.66 \times 10^6 = \frac{0.500}{x^2}$$

$$x = 5.48 \times 10^{-4} \text{ mol/L} = 5.48 \times 10^{-4}\ M$$

57. Balance the following redox equations:

(b) $IF_5 + Fe \longrightarrow FeF_3 + IF_3$

Solution

On assignment of oxidation numbers, it is evident that iodine is reduced from +5 to +3 and that iron is oxidized from 0 to +3.

$$Fe^0 + I^{5+} \longrightarrow Fe^{3+} + I^{3+}$$

Balance the equation with regard to the atoms that change in oxidation numbers.

$$2[Fe^0 \longrightarrow Fe^{3+} + 3e^-] \quad \uparrow 3 \quad 3 \times 2 = 6$$

$$3[I^{5+} \longrightarrow I^{3+} - 2e^-] \quad \downarrow 3 \quad 2 \times 3 = 6$$

Two Fe atoms will increase by 6 units of oxidation numbers in going to $2Fe^{3+}$ ions. Three I^{5+} ions will decrease by 6 units of oxidation numbers.

$$3I^{5+} + 2Fe^0 \longrightarrow 3I^{3+} + 2Fe^{3+}$$

$$2Fe + 3IF_5 \longrightarrow 2FeF_3 + 3IF_3$$

(d) $H_2S + Hg_2^{2+} \longrightarrow Hg + S + H^+$

Solution

$$S^{2-} \longrightarrow S^0 + 2e^- \quad \uparrow 2 \quad 2 \times 1 = 2$$

$$Hg_2^{2+} \longrightarrow 2Hg^0 - 2e^- \quad \downarrow 2 \quad 2 \times 1 = 2$$

$$Hg_2^{2+} + S^{2-} \longrightarrow 2Hg^0 + S^0$$

Then, by inspection,

$$H_2S + Hg_2^{2+} \longrightarrow 2Hg + S + 2H^+$$

58. Balance the following redox equations:

(a) $Zn + BrO_4^- + OH^- + H_2O \longrightarrow [Zn(OH)_4]^{2-} + Br^-$

Solution

$$4[Zn^0 \longrightarrow Zn^{2+} + 2e^-] \quad \uparrow 2 \quad 2 \times 4 = 8$$

$$Br^{7+} \longrightarrow Br^- - 8e^- \quad \downarrow 8 \quad 8 \times 1 = 8$$

$$4Zn^0 + Br^{7+} \longrightarrow 4Zn^{2+} + Br^-$$

Then, by inspection,

$$4Zn + BrO_4^- + OH^- + H_2O \longrightarrow 4Zn(OH)_4^{2-} + Br^-$$

Charge balance is important and convenient to use here; the right side shows 9- and is fixed, while BrO_4^- is the only fixed negative charge on the left side. Hence, $8OH^-$ are needed to balance the charge.

$$4Zn + BrO_4^- + 8OH^- + 4H_2O \longrightarrow 4Zn(OH)_4^{2-} + Br^-$$

(c) $MnO_4^- + S^{2-} + H_2O \longrightarrow MnO_2 + S + OH^-$

Solution

$$\begin{array}{lll} 3[S^{2-} \longrightarrow S^0 + 2e^-] & \uparrow 2 & 2 \times 3 = 6 \\ 2[Mn^{7+} \longrightarrow Mn^{4+} - 3e^-] & \downarrow 3 & 3 \times 2 = 6 \\ \hline 3S^{2-} + 2Mn^{7+} \longrightarrow 3S^0 + 2Mn^{4+} & & \end{array}$$

$$2MnO_4^- + 3S^{2-} + 4H_2O \longrightarrow 2MnO_2 + 3S + 8OH^-$$

(e) $Cu + H^+ + NO_3^- \longrightarrow Cu^{2+} + NO_2 + H_2O$

Solution

$$\begin{array}{lll} [Cu^0 \longrightarrow Cu^{2+} + 2e^-] & \uparrow 2 & 2 \times 1 = 2 \\ 2[N^{5+} \longrightarrow N^{4+} - 1e^-] & \downarrow 1 & 1 \times 2 = 2 \\ \hline Cu^0 + 2N^{5+} \longrightarrow Cu^{2+} + 2N^{4+} & & \end{array}$$

$$Cu + 4H^+ + 2NO_3^- \longrightarrow Cu^{2+} + 2NO_2 + 2H_2O$$

(g) $Cu + H^+ + NO_3^- \longrightarrow Cu^{2+} + NO + H_2O$

Solution

$$\begin{array}{lll} 3[Cu^0 \longrightarrow Cu^{2+} + 2e^-] & \uparrow 2 & 2 \times 3 = 6 \\ 2[N^{5+} \longrightarrow N^{2+} - 3e^-] & \downarrow 3 & 3 \times 2 = 6 \\ \hline 3Cu^0 + 2N^{5+} \longrightarrow 3Cu^{2+} + 2N^{2+} & & \end{array}$$

$$3Cu + 8H^+ + 2NO_3^- \longrightarrow 3Cu^{2+} + 2NO + 4H_2O$$

(i) $MnO_4^- + NO_2^- + H_2O \longrightarrow MnO_2 + NO_3^- + OH^-$

Solution

$$\begin{array}{lll} 3[N^{3+} \longrightarrow N^{5+} + 2e^-] & \uparrow 2 & 2 \times 3 = 6 \\ 2[Mn^{7+} \longrightarrow Mn^{4+} - 3e^-] & \downarrow 3 & 3 \times 2 = 6 \\ \hline 2Mn^{7+} + 3N^{3+} \longrightarrow 2Mn^{4+} + 3N^{5+} & & \end{array}$$

$$2MnO_4^- + 3NO_2^- + H_2O \longrightarrow 2MnO_2 + 3NO_3^- + 2OH^-$$

(k) $Br_2 + SO_2 + H_2O \longrightarrow H^+ + Br^- + SO_4{}^{2-}$

Solution

$$S^{4+} \longrightarrow S^{6+} + 2e^- \quad \uparrow 2 \quad 2 \times 1 = 2$$

$$Br_2{}^0 \longrightarrow 2Br^- - 2e^- \quad \downarrow 2 \quad 2 \times 1 = 2$$

$$Br_2 + S^{4+} \longrightarrow 2Br^- + S^{6+}$$

$$Br_2 + SO_2 + 2H_2O \longrightarrow 4H^+ + 2Br^- + SO_4{}^{2-}$$

59. Balance the following redox equations:

(a) $Al + [Sn(OH)_4]^{2-} \longrightarrow [Al(OH)_4]^- + Sn + OH^-$

Solution

$$2[Al^0 \longrightarrow Al^{3+} + 3e^-] \quad \uparrow 3 \quad 3 \times 2 = 6$$

$$3[Sn^{2+} \longrightarrow Sn^0 - 2e^-] \quad \downarrow 2 \quad 2 \times 3 = 6$$

$$2Al^0 + 3Sn^{2+} \longrightarrow 2Al^{3+} + 3Sn^0$$

$$2Al + 3[Sn(OH)_4]^{2-} \longrightarrow 2[Al(OH)_4]^- + 3Sn + 4OH^-$$

(e) $H_2S + H_2O_2 \longrightarrow S + H_2O$

Solution

$$S^{2-} \longrightarrow S^0 + 2e^- \quad \uparrow 2 \quad 2 \times 1 = 2$$

$$O_2{}^{2-} \longrightarrow 2O^{2-} - 2e^- \quad \downarrow 2 \quad 2 \times 1 = 2$$

$$S^{2-} + O_2{}^{2-} \longrightarrow S^0 + 2O^{2-}$$

$$H_2S + H_2O_2 \longrightarrow S + 2H_2O$$

(f) $OH^- + NO_2 \longrightarrow NO_3{}^- + NO_2{}^- + H_2O$

Solution

$$N^{4+} \longrightarrow N^{5+} + e^- \quad \uparrow 1 \quad 1 \times 1 = 1$$

$$N^{4+} \longrightarrow N^{3+} - e^- \quad \downarrow 1 \quad 1 \times 1 = 1$$

$$2N^{4+} \longrightarrow N^{5+} + N^{3+}$$

$$2OH^- + 2NO_2 \longrightarrow NO_3{}^- + NO_2{}^- + H_2O$$

(i) $C + HNO_3 \longrightarrow NO_2 + H_2O + CO_2$

Solution

$$
\begin{array}{rcll}
C^0 & \longrightarrow & C^{4+} + 4e^- & \uparrow 4 \quad 4 \times 1 = 4 \\
4[N^{5+} & \longrightarrow & N^{4+} - 1e^-] & \downarrow 1 \quad 1 \times 4 = 4 \\
\hline
C^0 + 4N^{5+} & \longrightarrow & C^{4+} + 4N^{4+} &
\end{array}
$$

$$C + 4HNO_3 \longrightarrow 4NO_2 + 2H_2O + CO_2$$

60. Complete and balance the following equations. (Remember that when a reaction occurs in acidic solution, H^+ and/or H_2O may be added on either side of the equation, as necessary, to balance the equation properly; when a reaction occurs in basic solution, OH^- and/or H_2O may be added, as necessary, on either side of the equation. No indication of the acidity of the solution is given if neither H^+ nor OH^- is involved as a reactant or product.)

(a) $Zn + NO_3^- \longrightarrow Zn^{2+} + N_2$ (acidic solution)

Solution

$$
\begin{array}{rcll}
5[Zn^0 & \longrightarrow & Zn^{2+} + 2e^-] & \uparrow 2 \quad 2 \times 5 = 10 \\
2N^{5+} & \longrightarrow & N_2^{\,0} - 10e^- & \downarrow 5 \quad 10 \times 1 = 10 \\
\hline
5Zn^0 + 2N^{5+} & \longrightarrow & 5Zn^{2+} + N_2^{\,0} &
\end{array}
$$

$$5Zn + 2NO_3^- + 12H^+ \longrightarrow 5Zn^{2+} + N_2 + 6H_2O$$

(c) $CuS + NO_3^- \longrightarrow Cu^{2+} + S + NO$ (acidic solution)

Solution

$$
\begin{array}{rcll}
3[S^{2-} & \longrightarrow & S^0 + 2e^-] & \uparrow 2 \quad 2 \times 3 = 6 \\
2[N^{5+} & \longrightarrow & N^{2+} - 3e^-] & \downarrow 3 \quad 3 \times 2 = 6 \\
\hline
3S^{2-} + 2N^{5+} & \longrightarrow & 3S^0 + 2N^{2+} &
\end{array}
$$

$$8H^+ + 3CuS + 2NO_3^- \longrightarrow 3Cu^{2+} + 3S + 2NO + 4H_2O$$

(f) $Cl_2 + OH^- \longrightarrow Cl^- + ClO_3^-$ (basic solution)

Solution

$$
\begin{array}{rcll}
Cl^0 & \longrightarrow & Cl^{5+} + 5e^- & \uparrow 5 \quad 5 \times 1 = 5 \\
5[Cl^0 & \longrightarrow & Cl^- - e^-] & \downarrow 1 \quad 1 \times 5 = 5 \\
\hline
6Cl^0 & \longrightarrow & 5Cl^- + Cl^{5+} &
\end{array}
$$

$$3Cl_2 + 6OH^- \longrightarrow 5Cl^- + ClO_3^- + 3H_2O$$

(h) $NO_2 \longrightarrow NO_3^- + NO_2^-$ (basic solution)

Solution

$$\begin{array}{rcll} N^{4+} & \longrightarrow & N^{5+} + e^- & \uparrow 1 \quad 1 \times 1 = 1 \\ N^{4+} & \longrightarrow & N^{3+} - e^- & \downarrow 1 \quad 1 \times 1 = 1 \\ \hline 2N^{4+} & \longrightarrow & N^{5+} + N^{3+} & \end{array}$$

$$2NO_2 + 2OH^- \longrightarrow NO_3^- + NO_2^- + H_2O$$

(j) $Fe^{3+} + I^- \longrightarrow Fe^{2+} + I_2$

Solution

$$\begin{array}{rcll} 2I^- & \longrightarrow & I_2{}^0 + 2e^- & \uparrow 2 \quad 2 \times 1 = 2 \\ 2[Fe^{3+} & \longrightarrow & Fe^{2+} - e^-] & \downarrow 1 \quad 1 \times 2 = 2 \\ \hline 2Fe^{3+} + 2I^- & \longrightarrow & 2Fe^{2+} + I_2 & \end{array}$$

$$2Fe^{3+} + 2I^- \longrightarrow 2Fe^{2+} + I_2$$

(l) $P_4 \longrightarrow PH_3 + HPO_3^{2-}$ (acidic solution)

Solution

$$\begin{array}{rcll} P^0 & \longrightarrow & P^{3+} + 3e^- & \uparrow 3 \quad 3 \times 1 = 3 \\ P^0 & \longrightarrow & P^{3-} - 3e^- & \downarrow 3 \quad 3 \times 1 = 3 \\ \hline 2[2P^0 & \longrightarrow & P^{3+} + P^{3-}] & \end{array}$$

$$4P^0 \longrightarrow 2P^{3+} + 2P^{3-}$$

P exists as a unit of four atoms, so the 4P becomes P_4.

$$P_4 + 6H_2O \longrightarrow 2PH_3 + 2HPO_3^{2-} + 4H^+$$

67. A current of 10.0 A is applied for 1.0 h to 1.0 L of a solution containing 1.0 mol of HCl. Calculate the pH of the solution at the end of this time.

Solution

The reduction of H^+ to produce H_2 at the cathode will produce a change in pH as the reaction proceeds.

Cathode half reaction: $2H^+ + 2e^- \longrightarrow H_2$

Anode half reaction: $2Cl^- \longrightarrow Cl_2 + 2e^-$

The number of moles of H^+ consumed by the reaction is

No. mol H^+ reduced = no. faradays passed

No. faradays = (10.0 A)(1.0 h x 3600 s/h) $\frac{1.0 \text{ faraday}}{96,487 \text{ C}}$ = 0.37

No. mol H^+ reduced = 0.37 mol

$[H^+]$ after reduction = 1.0 mol/L − 0.37 mol/L = 0.63 mol/L

pH = −log 0.63 = 0.20

71. The standard reduction potentials for the reactions

$$Ag^+ + e^- \longrightarrow Ag \quad \text{and} \quad AgCl + e^- \longrightarrow Ag + Cl^-$$

are +0.7991 V and +0.222 V, respectively. From these data and the Nernst equation, calculate a value for the solubility product (K_{sp}) for AgCl. Compare your answer with the value given in Appendix E.

Solution

The solubility product constant for the dissolution of AgCl, is

$$AgCl(s) \longrightarrow Ag^+(aq) + Cl^-(aq)$$
$$K_{sp} = [Ag^+][Cl^-]$$

The two half-cells given in the problem can be rearranged to give the desired equilibrium.

$$\begin{array}{lll} Ag \longrightarrow Ag^+ + e^- & E° = & -0.7991 \text{ V} \\ \underline{AgCl + e^- \longrightarrow Ag + Cl^-} & \underline{E° =} & \underline{0.222 \text{ V}} \\ AgCl \longrightarrow Ag^+ + Cl^- & E°_{cell} = & -0.557 \text{ V} \end{array}$$

This value can be substituted into the equation relating $E°$ to log K. We solve that equation to obtain

$$E° = \frac{0.05915}{n} \log K$$

$$-0.577 = \frac{0.05915}{1} \log K$$

$$\log K = \frac{-0.577}{0.05915} = -9.755$$

$$K = K_{sp} = 10^{-9.755} = 1.76 \times 10^{-10}$$

This is compared to 1.8×10^{-10} as given in Appendix E.

RELATED EXERCISES

1. Calculate the number of faradays required to deposit 2.94 g of nickel at a cathode.

   Answer: 0.10 faraday

2. The potential for the cell

   $$Zn \mid Zn^{2+}(1.70\ M) \parallel Cu^{2+}(M = ?) \mid Cu$$

   is 1.09 V. Calculate the concentration of Cu^{2+} in this cell.

   Answer: $[Cu^{2+}] = 0.779\ M$

3. Given the following two half-cells, calculate the equilibrium constant for the cell combination that reacts spontaneously at standard conditions.

   $$Ni^{2+} + 2e^- \longrightarrow Ni \quad E° = -0.25\ V$$

   $$Cu^{2+} + 2e^- \longrightarrow Cu \quad E° = 0.34\ V$$

   Answer: $K = 8.9 \times 10^{19}$

4. The storage battery used for starting an automobile engine is designed to deliver a high current for a short period of time. Given that a current of 160 A is delivered for 2.5 s, calculate the amount of lead that would be converted to lead(II) sulfate at each anode in a six-cell battery (cells are connected in series).

   Answer: 0.43 g/cell

5. Given the following cell,

   $$Zn \mid Zn^{2+}(0.10\ M) \parallel Fe^{2+}(0.01\ M) \mid Fe$$

   calculate the expected potential and the Gibbs energy for the cell.

   Answer: $E = 0.29\ V$; $\Delta G = 56\ kJ$

UNIT 15

CHAPTER 28: NUCLEAR CHEMISTRY

INTRODUCTION

In 1896, Henri Becquerel observed that ores containing the compound potassium uranyl sulfate, $K_2SO_4 \cdot UO_2SO_4 \cdot 2H_2O$, emitted high-energy rays similar to X rays. One of his students, Marie Sklodowska Curie, named this phenomenon *radioactivity*. Two years later, Marie and her husband Pierre were able to identify two new elements, polonium (Po), named after her native Poland, and radium (Ra), both of which were radioactive. The Nobel Prize was awarded to the Curies and Becquerel in 1903 for their work. Their efforts marked the beginning of the nuclear era.

Their discovery led to the understanding that certain atomic nuclei are unstable and spontaneously disintegrate, with the release of energy and penetrating radiations to produce new nuclei. These may be either nuclei of an element different from the original element or an isotope of the original element. About three-fourths of the more than 1000 known isotopes are unstable or radioactive and can exhibit the phenomena described. All isotopes with atomic number Z greater than 83 are radioactive, and all elements beyond uranium, $Z = 92$, in the Periodic Table are artificially created as well as radioactive.

Atoms of stable isotopes can be made radioactive through bombardment of their nuclei with nuclear particles. Some of these nuclides so produced are useful in medicine, research, and in the generation of electrical energy. As an example, cobalt-60, an isotope widely used in cancer therapy, is prepared by bombarding stable cobalt-59 with neutrons according to the reaction

$$^{59}_{27}\text{Co} + ^{1}_{0}\text{n} \longrightarrow ^{60}_{27}\text{Co}$$

But cobalt-60 is unstable and decays to stable nickel-60 in the following way:

$$^{60}_{27}\text{Co} \longrightarrow ^{60}_{28}\text{Ni} + ^{0}_{-1}\beta + \text{energy}$$

This process requires 5.29 years for one-half of the original amount of cobalt to decompose. This time is known as the *half-life*. Some larger radioactive nuclei will split into two or more smaller nuclear fragments upon addition of another particle. This process, called *fission*, is exemplified by the splitting of uranium-235, which is used as an energy source for generating electricity. The reaction shown here is only one of many that have been observed in the fission of uranium-235.

$$^{235}_{92}U + ^{1}_{0}n \longrightarrow ^{236}_{92}U \longrightarrow ^{90}_{36}Kr + ^{144}_{56}Ba + 2\,^{1}_{0}n$$

The purpose of this chapter is to acquaint you with phenomena related to nuclear charge. Items specifically treated are characteristics of nuclear change, including natural and artificial nuclear reactions, nuclear stability, methods for detecting and accelerating particles, rates of nuclear reactions, energies associated with nuclear change, and some research and technological applications of radioactive isotopes.

FORMULAS AND DEFINITIONS

Binding energy (B) The mass of a nucleus is always less than the combined mass of its constituent particles. The mass difference, or mass defect, is related to energy through the equation $E = mc^2$. This energy, called the binding energy of the nucleus, is the energy required to break up the nucleus into its constituent particles. To calculate the binding energy, calculate the mass difference between the mass of a nuclide and the mass of its components. This difference, the mass defect, is converted to energy through substitution into $E = mc^2$.

Energy equivalences Energy produced by conversion of a mass equivalent to 1 amu, or $1.6605655 \times 10^{-27}$ kg, to energy through the Einstein equation, $E = mc^2$, is

$$E = (1.6605 \times 10^{-27}\ \text{kg})(2.99792468 \times 10^{8}\ \text{m/s})^2 = 1.492442 \times 10^{-10}\ \text{J}$$

$$1\ \text{joule} = 6.24146 \times 10^{18}\ \text{electron volts} = 6.24146 \times 10^{12}\ \text{MeV}$$

or

$$1\ \text{MeV} = 1.602189 \times 10^{-13}\ \text{J}$$

$$E = (1.49244 \times 10^{-10}\ \text{J})(6.24146 \times 10^{12}\ \text{MeV/J}) = 931.450\ \text{MeV}$$

Half-life ($t_{1/2}$) The amount of time required for one-half the number of nuclei in a radioactive sample to decay to new nuclei. Half-life values range from microseconds to billions of years.

Istopic mass The experimentally measured mass of an isotope (nuclide) relative to the mass of carbon-12.

Mass number (A) Sum of the number of neutrons and protons in the nucleus of a specific nuclide.

Nuclear reaction rate The rate of a nuclear reaction is kinetically of the first order. That is, the number of nuclei of a specific nuclide remaining in a sample after an elapsed time, t, is a function of the initial number of nuclei, N_0, and the half-life of the nuclide. The value N_t, or the number of nuclei at time t, is related to N_0 by the equation $N_t = N_0 e^{-kt}$, where k is a proportionality constant called the *decay constant*. This equation is conveniently rearranged as

$$\log \frac{N_0}{N_t} = \frac{kt}{2.303} \quad \text{and} \quad k = \frac{0.693}{t_{1/2}}$$

EXERCISES

Atomic Particle Masses
Hydrogen atom (proton + electron) = 1.007825 amu = 1.673559×10^{-27} kg
Neutron = 1.008665 amu = $1.6749543 \times 10^{-27}$ kg
Proton = 1.0072764 amu = $1.6726485 \times 10^{-27}$ kg
1 amu = $1.6605655 \times 10^{-27}$ kg

6. The mass of the isotope $^{27}_{13}Al$ is 26.98154.

   (a) Calculate its binding energy per atom in millions of electron-volts.
   (b) Calculate its binding energy per nucleon. (See Appendix D.)

   Solution

   (a) The binding energy per atom of $^{27}_{13}Al$ is calculated from the difference between the actual mass of the nuclide and its theoretical mass and by converting this mass to energy. An atom of Al-27 contains 14 neutrons and the equivalent of 13 hydrogen atoms (13 protons + 13 electrons in atomic orbitals).

   Theoretical mass:

$$14 \text{ neutrons} = 14 \times 1.008665 \text{ amu} = 14.121310 \text{ amu}$$

$$13 \text{ H atoms} = 13 \times 1.007825 \text{ amu} = \underline{13.101725 \text{ amu}}$$

$$\text{total} = 27.223035 \text{ amu}$$

$$\text{Mass defect} = (27.223035 - 26.98154) \text{ amu} = 0.24150 \text{ amu}$$

$$\text{Binding energy} = 0.24150 \text{ amu} \times 931.50 \frac{\text{MeV}}{\text{amu}} = 224.96 \text{ MeV}$$

   (b) The binding energy per nucleon is calculated by dividing the binding energy per atom by the number of nucleons; in this case,

$$\text{Binding energy/nucleon} = 224.96 \text{ MeV}/27 = 8.332 \text{ MeV}$$

8. The mass of deuteron $\left(^{2}_{1}H\right)$ is 2.01355 amu; that of an α particle, 4.00150 amu. How much energy per mole of $^{4}_{2}He$ produced is released by the following reaction?

$$^{2}_{1}H + ^{2}_{1}H \longrightarrow ^{4}_{2}He$$

Solution

Calculate the mass change that occurs; then convert this mass to energy.

$$\text{Mass difference per } ^{4}He = [2(2.01355) - 4.00150] \text{ amu} = 0.02560 \text{ amu}$$

$$E(^{4}He) = 0.02560 \text{ amu} \times 931.50 \text{ MeV/amu} = 23.846 \text{ MeV}$$

$$E(^{4}He) = 23.846 \text{ MeV} \times 1.602189 \times 10^{-13} \text{ J/MeV}$$
$$= 3.821 \times 10^{-12} \text{ J}$$

$$E(\text{mol } ^{4}He) = 3.821 \times 10^{-12} \text{ J/atom} \times 6.022 \times 10^{23} \text{ atoms/mol}$$
$$= 2.301 \times 10^{12} \text{ J} = 2.301 \times 10^{9} \text{ kJ}$$

10. What percentage of $^{212}_{82}Pb$ remains of a 1.00-g sample, 1.0 min after it is formed (half-life of 10.6 s)? 10 min after it is formed?

Solution

The radioactivity of a sample is a function of the number of nuclei in the sample and the time. The number of nuclei in a sample determines its mass, its concentration, and its activity. Therefore, the formula

$$\log \frac{N_0}{N_t} = \frac{kt}{2.303}$$

is equally valid for concentration, activity, and mass,

$$\log \frac{C_0}{C_t} = \log \frac{A_0}{A_t} = \log \frac{M_0}{M_t} = \frac{kt}{2.303}$$

The time reference for t and k in the equation must be consistent. In this case, the half-life is given in seconds and the experimental time period in minutes. The same unit must be used for both measurements to satisfy the above equation:

$$k = \frac{0.693}{10.6 \text{ sec}} = \frac{0.0654}{\text{sec}} = 0.0654 \text{ s}^{-1}$$

$$\log \frac{M_0}{M_t} = \frac{0.0654 \text{ s}^{-1} t}{2.303}$$

$$= \frac{0.0654 \text{ s}^{-1}(1 \text{ min} \times 60 \text{ s/min})}{2.303}$$

$$= 1.704$$

$$\log M_0 - \log M_t = 1.704$$

Because $M_0 = 1.0$ g, the log $M_0 = 0$, and

$$\log M_t = -1.704$$

$$M_t = 10^{-1.704} = 0.02 \text{ g at } 1.0 \text{ min}$$

$$\% = \frac{0.020 \text{ g}}{1.0 \text{ g}} \times 100 = 2.0\%$$

11. The isotope ^{208}Tl undergoes β decay with a half-life of 3.1 min.

(a) What isotope is the product of the decay?

(b) Is ^{208}Tl more stable or less stable than an isotope with a half-life of 54.5 s?

(c) How long will it take for 99.0% of a sample of pure ^{208}Tl to decay?

(d) What percentage of a sample of pure ^{208}Tl will remain undecayed after an hour?

Solution

(a) $^{208}_{81}Tl - ^{0}_{-1}\beta \longrightarrow ^{208}_{82}Pb$

(b) Tl-208 with a half-life of 3.1 min is considered to be more stable than a nuclide with a 54.5-s half-life.

(c) The percentage of a sample remaining after a period of time is independent of mass. If 99% of the sample has decayed, 1%, or a fraction of 0.01 of the original remains (that is, $M_t = 0.01$ when $M_0 = 1$). The time is calculated from the first-order rate law.

$$k = \frac{0.693}{3.1 \text{ min}} = 0.224 \text{ min}^{-1}$$

$$\log \frac{M_0}{M_t} = \frac{kt}{2.303} = \frac{0.224 \text{ min}^{-1}}{2.303} t = 0.0973 \text{ min}^{-1} t$$

$$t = 10.3 \text{ min}\left(\log \frac{1}{0.01}\right) = 10.3(-\log 0.01)$$

$$= (10.3 \text{ min})(2) = 20.6 \text{ min}$$

(d) Let $M_0 = 1.00$ and then on substitution

$$\log \frac{M_0}{M_t} = \frac{kt}{2.303} \qquad \log \frac{1.00}{M_t} = \frac{0.224 \text{ min}^{-1} \times 60 \text{ min}}{2.303}$$

$$\log M_t = -5.835 \quad \text{or} \quad M_t = 1.46 \times 10^{-6}$$

This is the amount remaining from one unit. Converting this to percentage is done by dividing the part by the whole and multiplication by 100%. Thus,

$$\frac{1.46 \times 10^{-6}}{1} \times 100\% = 1.5 \times 10^{-4}\%$$

12. Calculate the time required for 99.999% of each of the following radioactive isotopes to decay:

(a) $^{226}_{88}Ra$ (half-life, 1590 yr).

Solution

Consider the initial amount of Ra-226 to be 1.00. Therefore, 0.001% or a fraction of 0.00001, remains. Calculate k and t as in Exercise 11c.

$$k = \frac{0.693}{1590\ yr} = 4.36 \times 10^{-4}\ yr^{-1}$$

$$\log \frac{1}{0.00001} = \frac{4.36 \times 10^{-4}}{2.303}\ yr^{-1}\ t$$

$$= 1.89 \times 10^{-4}\ yr^{-1}\ t$$

$$t = (5.28 \times 10^{3}\ yr)(-\log 0.00001)$$

$$= (5.28 \times 10^{3}\ yr)(5)$$

$$= 26{,}400\ yr$$

13. The isotope $^{90}_{38}Sr$ is an extremely hazardous species in the fallout from a nuclear fission explosion. A 0.500-g sample diminishes to 0.393 g in 10.0 yr. Calculate the half-life.

Solution

Sufficient information is given in the problem to calculate the value of k, which in turn can be used to calculate the half-life of Sr-90.

$$\log \frac{M_0}{M_t} = \frac{kt}{2.303}$$

$$\log \frac{0.500\ g}{0.393\ g} = \frac{k(10\ yr)}{2.303} = k(4.3422\ yr)$$

$$\log 1.2723 = 4.3422\ yr(k)$$

$$k = \frac{0.10459}{4.3422\ yr} = 0.02409\ yr^{-1}$$

$$\text{Half-Life} = \frac{0.693}{0.02409\ yr^{-1}} = 28.8\ yr$$

22. Complete the following equations:

(a) $^{27}_{13}Al + ^{2}_{1}H \longrightarrow ? + ^{4}_{2}He$

(b) $^{7}_{3}Li + ? \longrightarrow 2\ ^{4}_{2}He$

(c) $^{9}_{4}Be + ^{4}_{2}He \longrightarrow ^{12}_{6}C + ?$

Solution

(a) In (a) think of the arrow as an equals sign. Then the sum of the mass numbers on both sides must be equal; that is, $27 + 2 = Z + 4$ and $Z = 25$. In the same way the sum of the number of protons must be equal; that is, $13 + 1 = p + 2$ and $p = 12$. All that remains is to identify the element whose atomic number is 12. This is magnesium.

$$^{27}_{13}Al + ^{2}_{1}H \longrightarrow ^{25}_{12}Mg + ^{4}_{2}He$$

(b) The mass and charge totals of both products and reactants must be equal in a correctly balanced equation.

$$^{7}_{3}Li + ^{1}_{1}H \longrightarrow 2\ ^{4}_{2}He$$

(c) $$^{9}_{4}Be + ^{4}_{2}He \longrightarrow ^{12}_{6}C + ^{1}_{0}n$$

23. Fill in the atomic number of the initial nucleus and write out the complete nuclear symbol for the product of each of the following nuclear reactions:

(a) $^{65}Cu(n,2n)$ (b) $^{54}Fe(\alpha,2p)$ (c) $^{33}S(n,p)$ (g) $^{14}N(p,\gamma)$

Solution

(a) $^{65}Cu(n,2n)$ $$^{65}_{29}Cu + ^{1}_{0}n \longrightarrow ^{64}_{29}Cu + 2\ ^{1}_{0}n$$

(b) $^{54}Fe(\alpha,2p)$ $$^{54}_{26}Fe + ^{4}_{2}He \longrightarrow ^{56}_{26}Fe + 2\ ^{1}_{1}H$$

(c) $^{33}S(n,p)$ $$^{33}_{16}S + ^{1}_{0}n \longrightarrow ^{33}_{15}P + ^{1}_{1}H$$

(g) $^{14}N(p,\gamma)$ $$^{14}_{7}N + ^{1}_{1}H \longrightarrow ^{15}_{8}O + \gamma$$

24. Complete the following notations by filling in the missing parts:

(a) $^{2}H(d,n)$ (b) $(\alpha,n)\ ^{30}P$ (e) $^{232}Th(\ ,n)\ ^{235}U$

Solution

(a) $^{2}H(d,n)$ $$^{2}_{1}H + ^{2}_{1}H \longrightarrow ^{3}_{2}He + ^{1}_{0}n$$

(b) $(\alpha,n)\ ^{30}P$ $$^{27}_{13}Al + ^{4}_{2}He \longrightarrow ^{30}_{15}P + ^{1}_{0}n$$

(e) $^{232}Th(\ ,n)\ ^{235}U$ $$^{232}_{90}Th + ^{4}_{2}He \longrightarrow ^{235}_{92}U + ^{1}_{0}n$$

RELATED EXERCISES

1. Which one of the following nuclear equations is incorrectly written?

(a) ${}^{226}_{88}Ra \longrightarrow {}^{222}_{86}Rn + {}^{4}_{2}He$

(b) ${}^{14}_{7}N + {}^{4}_{2}He \longrightarrow {}^{17}_{8}O + {}^{1}_{0}n$

(c) ${}^{10}_{5}B + {}^{4}_{2}He \longrightarrow {}^{13}_{7}N + {}^{1}_{0}n$

(d) ${}^{235}_{92}U + {}^{1}_{0}n \longrightarrow {}^{142}_{56}Ba + {}^{91}_{36}Kr + 3\,{}^{1}_{0}n$

Answer: (b)

2. The activity of carbon-12 in the ash remains of an ancient civilization is found to be 1/8 its original activity. Based on an archeological carbon-12 half-life of 5730 yr, calculate the time elapsed since the demise of the civilization.

Answer: 17,200 yr

3. The mass loss incurred as 1 mol of radium-226 decays to radon-222 is 2.4×10^{21} amu. Calculate the energy produced by conversion of this mass to energy. Convert this mass loss from *amu* to *kg*. Given that the complete combustion of 1 mol of methane releases 882 kJ, calculate the mass of methane required to produce an energy equivalent to the decay of 1 mol of radium.

Answer: 3.6×10^{11} J; 4×10^{-6} kg; 6.5×10^{3} kg

4. The initial activity of a cobalt-60 source is 9.0×10^{6} counts per second (cps). Given that the half-life of this nuclide is 5.2 years, calculate the activity that would remain 8.0 yr hence.

Answer: 3.1×10^{6} cps

5. The Group I metals, potassium and sodium, are essential for the growth and maintenance of healthy plants and animals. Cesium is in the same periodic group, so one would expect it to be incorporated in living tissue in much the same manner as potassium and sodium. Cesium-137 is radioactive, has a half-life of 30 years, and is a significant radioactive waste. Given that the initial activity in 1984 of a 1.0-g sample of cesium-137 is 3.2×10^{12} cps, calculate the activity of the sample in the year 2500.

Answer: 2.1×10^{7} cps

PART 2

EXPONENTIAL NOTATION AND LOGARITHMS

Part Two is a programmed unit designed as a comprehensive review of algebraic operations involving exponential numbers and logarithms. After a careful study of this unit, you should be able to

1. Write numbers in scientific notation or exponential form.
2. Multiply and divide numbers written in scientific notation.
3. Raise a number to a specified power.
4. Find a specified root of a number.
5. Perform logarithmic operations to calculate products, quotients, powers, and roots.

Make certain you are able to answer *correctly* all parts of the question in a frame before proceeding to the next frame. Answers to questions in this part begin on page 271.

EXPONENTIAL NOTATION

[1] Scientific study involves writing and using numbers in the base 10 system. Exponential notation makes it possible to write very large and very small numbers with base 10 in a convenient manner. For example, 0.0000000007 can be written 7×10^{-10} and 7,000,000,000,000,000 can be written 7×10^{15}. Operations with exponential numbers are easily performed with practice. This study of exponential notation will begin with some moderate-sized numbers as examples. The number 20 can be written 2×10^{1}; 40 can be written $4 \times$ _____; 400 can be written $4 \times$ _____.

[2] The number 100 can be thought of as 10×10 or as $10^{1} \times 10^{1}$, both of which equal 10^{2}. So the number 600 can be written 6×10^{2}. The number 750 can be written $7.5 \times$ _____. The number 625 can be written _____ $\times$ _____.

[3] Just as 100 equals 10^2, 1000 equals $10 \times 10 \times 10$, or $10^1 \times 10^1 \times 10^1$, or 10^3. The number 4000 can be written 4×10^3. Rewrite the following numbers:

(a) 7700 is $7.7 \times 10^?$.

(b) 8250 is $8.25 \times 10^?$.

(c) 6255 is _____ × _____.

[4] The number 100 is 10^2, 1000 is 10^3, 10,000 is 10^4, and 1,000,000 is 10^6. These numbers have a common factor. That is, 100 has two zeros and the exponent of 10 is 2; 100,000 has five zeros and the exponent of 10 is 5. It follows that the exponent of 10 for the number 1,000,000,000 should be _____.

[5] Any number or expression with zero as an exponent has a value of one. For example: $10^0 = 1$; $a^0 = 1$; $(ab)^0 = 1$; etc. Calculate the values for the following exercises:

(a) $5^0 =$ _____

(b) $100^0 =$ _____

(c) $5^0 \times 10^0 =$ _____

(d) $a^0 \times 10^0 \times b^0 =$ _____

[6] In general, a number written in scientific or exponential form should be written with one nonzero digit to the left of the decimal point and all other significant figures to the right of the decimal point. Experimentally determined digits are referred to as significant figures. For example: 2254. contains 4 digits that have been experimentally determined and can be written exponentially as 2.254×10^3. The number 254,000 would be written exponentially as 2.54×10^5; the 3 zeros are merely place holders. Complete the following exercises:

(a) $7655 = 7.655 \times 10^?$.

(b) $2{,}225{,}000{,}000 = 2.225 \times 10^?$.

(c) $786{,}000 = 7.86 \times 10^?$.

[7] Write the following numbers in exponential form:

(a) 2250 __________

(b) 60,200,000,000 __________

(c) 5,575,000,000 __________

[8] Basically the same procedure as that outlined in frame 6 can be used to rewrite very small numbers in exponential form. Examine the accompanying group of numbers and note that the exponent of 10 has a negative sign and represents the number of digits from the decimal point to the right of the first nonzero digit.

$$0.1 = 1 \times 10^{-1}$$
$$0.01 = 1 \times 10^{-2}$$
$$0.001 = 1 \times 10^{-3}$$
$$0.0001 = 1 \times 10^{-4}$$

With this concept in mind, study the following examples and rewrite the exercises:

$$0.00000001 = 1 \times 10^{-8}$$
$$0.00025 = 2.5 \times 10^{-4}$$

(a) $0.000007 = 7 \times 10^?$.

(b) $0.0000425 = 4.25 \times 10^?$.

(c) $0.0000608 =$ __________

[9] By the same reasoning as in frame 8, the exponent of 10 for 0.000007 is ______________ and for 0.0000425 is ______________.

[10] The number 0.00000235 is correctly written 2.35×10^{-6}, and 506,000 as 5.06×10^{5}. That is, the exponential form of each number contains all the significant figures in the number with one nonzero digit written to the left of the decimal point. In general, the exponent of 10 equals the number of digits from the decimal point of the number to the right of the first nonzero digit. The sign of the exponent is "+" if the number is greater than 1 and "-" if the number is less than 1. For the number 0.00000235, there are <u>6</u> digits from the decimal point to the right of 2, and for 506,000 there are <u>5</u> digits from the right of 5 to the last digit. (No answer required.)

[11] Write the following numbers in scientific notation:

(a) 0.003 ______________ (c) 0.0002354 ______________

(b) 0.0075 ______________ (d) 0.0000072 ______________

[12] Sometimes it is necessary to change the form a number is written in; care must be taken, however, not to change the magnitude of the number. The change in the value of the exponent must reflect the change in position of the decimal point. For base 10 numbers, if the decimal point is moved three places to the left, the exponent of 10 must be increased by the same amount; it is decreased when the decimal point is moved to the right. Consider the following examples.

The number 2.25×10^{2} can be written 22.5×10^{1} or 225×10^{0} without changing the value of the original number. Also, 0.00075 can be written 7.5×10^{-4} or 75×10^{-5}. Using this concept, rewrite the following numbers as indicated:

(a) $235{,}000 = 23.5 \times 10^{?} = 2.35 \times 10^{?}$

(b) $0.00068 = 6.8 \times 10^{?} = 68 \times 10^{?}$

[13] The implication of the concept discussed in frame 12 is that all real base 10 numbers can be written as the product of a rational number and an exponential term. Therefore, it follows that common arithmetical operations, such as multiplication and division, can be performed with numbers written in exponential form. First consider the rules for multiplying exponential numbers. In multiplying expressions having the same number base system (in this case, base 10), the exponent of the product is merely the sum of the exponents of the multipliers. For example,

$$10^{2} \cdot 10^{3} = 10^{2+3} = 10^{5} = \underline{100{,}000}$$

Determine the following products:

(a) $10^{1} \cdot 10^{5}$ = __________ (c) $10^{4} \cdot 10^{10}$ = __________

(b) $10^{3} \cdot 10^{9}$ = __________

[14] The same rule applies when the exponents are negative. For example,

$$10^{-2} \cdot 10^{-3} = 10^{-5}$$

Determine the following products:

(a) $10^{-8} \cdot 10^{-10}$ = __________ (c) $10^{-5} \cdot 10^{-5}$ = __________

(b) $10^{-5} \cdot 10^{-3}$ = __________

[15] When both positive and negative exponents are present in the same expression, the same rule applies. Merely add the exponents algebraically. For example,

$$10^{-3} \cdot 10^{5} = 10^{2}$$

(a) $10^{-15} \cdot 10^{12}$ = __________ (c) $10^{14} \cdot 10^{-12} \cdot 10^{3}$ = __________

(b) $10^{-2} \cdot 10^{-3} \cdot 10^{6}$ = __________

[16] Now consider the possibility of multiplying expressions written in scientific notation form. To find the product, add the exponents of the exponential terms and multiply the rational terms. For example,

$$(2 \times 10^{5})(4 \times 10^{6}) = (2 \times 4)(10^{5} \times 10^{6}) = 8 \times 10^{11}$$

Calculate the following:

(a) $(3.2 \times 10^{4})(2.0 \times 10^{5})$ = __________

(b) $(4.0 \times 10^{6})(8.0 \times 10^{-5})$ = __________

(c) $(8.0 \times 10^{6})(9.1 \times 10^{-2})$ = __________

(d) $(2.0 \times 10^{3})(3.0 \times 10^{-5})(4.0 \times 10^{2})$ = __________

[17] The operations involved in dividing exponential expressions are just as simple as multiplication operations. With base 10 numbers (or any numbers having the same base), the exponent of the quotient is the algebraic difference between the exponent of the numerator and the exponent of the denominator. Study the following examples:

$$\frac{a^x}{a^y} = a^{x-y}$$

$$\frac{a^x}{a^{-y}} = a^{x-(-y)} = a^{x+y}$$

Now with numbers:

$$\frac{10^3}{10^2} = 10^{3-2} = 10^1$$

$$\frac{10^5}{10^{-2}} = 10^{5-(-2)} = 10^{5+2} = 10^7$$

(No answer required.)

[18] Divide as indicated:

(a) $\frac{10^4}{10^1}$ = ________ (c) $\frac{10^5}{10^{-5}}$ = ________

(b) $\frac{10^6}{10^2}$ = ________ (d) $\frac{10^{1.2}}{10^{2.4}}$ = ________

[19] The procedure for dividing two expressions written in scientific notation is similar to multiplication. Consider dividing 4000 by 200. The numbers may be written as 4×10^3 and 2×10^2. Set up the division in the following manner:

$$\frac{4 \times 10^3}{2 \times 10^2}$$

Two division operations are required to obtain the quotient: first, division of the rational terms, and second, division of the exponential terms. The quotient should be left in scientific notation.

$$\frac{4}{2} = 2$$

$$\frac{10^3}{10^2} = 10^{3-2} = 10^1$$

The quotient is 2×10^1, or 20. Determine the quotient for each of the following examples:

(a) $\frac{5.0 \times 10^4}{2.0 \times 10^2}$ = ________ (c) $\frac{9.6 \times 10^{13}}{2.4 \times 10^{-12}}$ = ________

(b) $\frac{7.5 \times 10^{-1}}{3.8 \times 10^3}$ = ________

[20] Now consider the division of two numbers written in scientific notation in which the rational term in the numerator is smaller than the rational term in the denominator. The procedure for determining the quotient of the following expression is the same as in frame 19:

$$\frac{2.4 \times 10^1}{4.8 \times 10^2}$$

The quotient of 2.4/4.8 is 0.5 and of $10^1/10^2$ is 10^{-1}, which can be combined as 0.5×10^{-1}. This answer is correct but it is not conventionally left in this form. The value should be rewritten as 5×10^{-2}. This type of change was considered briefly in frame 12. The expression 5×10^{-2} is the product of two numbers, so there is no change in the value of the expression when one factor is multiplied by a number and the other divided by the same number. To obtain 5×10^{-2} from 0.5×10^{-1}, multiply and divide both terms by 10 as follows:

$$0.5 \times 10^{-1} \equiv (0.5)(10^1) \times \frac{10^{-1}}{10^1} \equiv 5 \times 10^{-1-(1)} = 5 \times 10^{-2}$$

Using the same logic, calculate the quotient for each of the following expressions and leave the quotient in the proper form for scientific notation:

(a) $\dfrac{2.2 \times 10^{-2}}{6.6 \times 10^{-4}} =$ ______________

(c) $\dfrac{1.42 \times 10^{-3}}{96.2 \times 10^{4}} =$ ______________

(b) $\dfrac{8.1 \times 10^{14}}{9.0 \times 10^{-15}} =$ ______________

Go to SELF-EVALUATION I.

SELF-EVALUATION I

Complete the following self-evaluation of your skills in writing numbers in scientific notation and multiplying and dividing such terms. Answers to these questions are on page 273. If one of your answers does not agree with the answer listed, refer to the frame whose number appears in brackets with the answer and rework the problem(s) in that frame.

1. Write the number 7,070,000 in scientific notation.
2. Write the number 2.62×10^{5} in rational form.
3. Write the number 0.0000785 in scientific notation.
4. Calculate the product $(2 \times 10^{6})(3.5 \times 10^{8})$.
5. Calculate the quotient $(1.9 \times 10^{4})/(2.6 \times 10^{5})$ and write the result in proper scientific notation.
6. Calculate the product $(3.6 \times 10^{-1})(4.2 \times 10^{-7})(5.9 \times 10^{8})$ and leave the answer in scientific notation.

Go to frame [21]

---

[21] Analyzing chemical data often requires calculations with powers and roots of numbers. Operations involving numbers written in scientific notation will be considered. This frame illustrates the general case for raising an exponential term to a power. The rule for raising a number to a power is to multiply exponents, as shown in the following examples:

$$(a^{x})^{y} = a^{xy}$$
$$(a^{2})^{3} = a^{(2)(3)} = a^{6}$$

For base 10 terms, the rule is illustrated by the following:

$$(10^{2})^{3} = 10^{6}$$
$$(10^{-2})^{4} = 10^{-8}$$

Using the rule given above, compute the following products:

(a) $(10^2)^5$ = __________ (c) $(10^{1/2})^4$ = __________

(b) $(10^4)^{-2}$ = __________ (d) $(10^{2.5})^2$ = __________

[22] Raising a number written in scientific notation to a power follows from the general rules in frame 21. In general, the product of two numbers raised to a power is merely the product of each term raised to the indicated power. Study the examples below and then make the indicated calculations.

$$(2 \times 10^2)^2 = 2^2 \times (10^2)^2 = 4 \times 10^4$$
$$(3.0 \times 10^3)^3 = 3.0^3 \times (10^3)^3 = 27 \times 10^9 = 2.7 \times 10^{10}$$
$$(2.50 \times 10^4)^2 = 2.50^2 \times (10^4)^2 = 6.25 \times 10^8$$
$$(3.0 \times 10^{-4})^3 = 3.0^3 \times (10^{-4})^3 = 27 \times 10^{-12} = 2.7 \times 10^{-11}$$

Evaluate the following expressions:

(a) $(4.0 \times 10^2)^3$ = __________ (c) $(2.50 \times 10^{-3})^3$ = __________

(b) $(5.0 \times 10^5)^2$ = __________

[23] The rule for extracting the root of a number is essentially the same as that for raising the number to a power. The rule in frame 22 is rewritten for roots as follows: The root of a product involving two or more factors is merely the product of the root of each factor taken collectively. First, consider the extraction of roots for exponential terms. Recall that the square root of a given number is equivalent to the number raised to the power 1/2 and the cube root of a given number is equivalent to the number raised to the power 1/3. Study the following examples:

$$\sqrt{10^2} = (10^2)^{1/2} = 10^{2/2} = 10^1$$
$$\sqrt[3]{10^3} = (10^3)^{1/3} = 10^{3/3} = 10^1$$
$$\sqrt[4]{10^{-24}} = (10^{-24})^{1/4} = 10^{-24/4} = 10^{-6}$$

Use the rule for extracting roots to calculate the values for the following expressions:

(a) $\sqrt{10^{-6}}$ = __________ (c) $\sqrt[3]{10^{-6}}$ = __________

(b) $\sqrt{10^4}$ = __________ (d) $\sqrt[5]{10^{-25}}$ = __________

[24] The exercises in frame 23 involved extracting roots of numbers whose exponents were evenly divisible by the root. For cases in which this is not true, the exponential term must be rewritten as a product of two numbers in which the rational term is greater than 1 and the exponent is evenly divisible by the root. As an example, $\sqrt{1.0 \times 10^3}$ is calculated in the following manner:

$$\sqrt{1.0 \times 10^3} = \sqrt{10. \times 10^2}$$

Now apply the general rule in frame 23:

$$\sqrt{10. \times 10^2} = \sqrt{10.} \times \sqrt{10^2} = (10.)^{1/2} \times (10^2)^{1/2}$$

$$(10.)^{1/2} = 3.2 \quad \text{and} \quad (10^2)^{1/2} = 10^1$$

Therefore, $\sqrt{1.0 \times 10^3} = 3.2 \times 10^1$.

By using the procedure in this example, calculate the value for each of the following expressions. An electronic calculator or a slide rule is essential for the computation work.

(a) $\sqrt{10^5}$ = __________ (c) $\sqrt{10^7}$ = __________

(b) $\sqrt[3]{10^7}$ = __________ (d) $\sqrt[3]{1.0 \times 10^8}$ = __________

[25] The concept developed in frames 23 and 24 can be expanded to include numbers written in scientific notation. The cube root of 44,000 is calculated in the following manner: First, the number is written in exponential form, having the exponent of the exponential term evenly divisible by 3. That is,

$$44{,}000 = 44 \times 10^3$$

and

$$(44{,}000)^{1/3} = (44 \times 10^3)^{1/3} = (44)^{1/3} \times (10^3)^{1/3} = 3.5 \times 10^1$$

Apply the procedure above in calculating the following roots:

(a) $\sqrt{2500}$ = _____ (c) $\sqrt{26{,}200}$ = _____

(b) $\sqrt[3]{1440}$ = _____ (d) $\sqrt[3]{18{,}000}$ = _____

[26] Roots of small numbers are calculated in the same manner. Study the following examples and then calculate the roots for the exercises:

$$(0.000074)^{1/2} = (7.4 \times 10^{-5})^{1/2}$$

$$= (74 \times 10^{-6})^{1/2} = 8.6 \times 10^{-3}$$

$$(0.00000081)^{1/3} = (8.1 \times 10^{-7})^{1/3} = (810 \times 10^{-9})^{1/3}$$

$$= (810)^{1/3} \times (10^{-9})^{1/3} = 9.3 \times 10^{-3}$$

(a) $(0.000045)^{1/2}$ = _____ (c) $(0.00061)^{1/3}$ = _____

(b) $(0.00030)^{1/2}$ = _____ (d) $(9.6 \times 10^{-16})^{1/3}$ = _____

[27] Fractional roots, such as 2/3, 0.40, and 1/5, will be considered in the discussion of logarithms.

Go to SELF-EVALUATION II.

SELF-EVALUATION II

Complete the following self-evaluation of your skills in raising numbers to powers and in extracting roots of numbers. Answers to these questions are on page 274. If one of your answers does not agree with the answer listed, refer to the frame whose number appears in brackets with the answer and rework the problem(s) in that frame.

1. The value of $(10^8)^3$ is ______________.

2. The value of $(2.5 \times 10^4)^2$ is ______________.

3. The cube root of 25,000 is ______________.

4. The square root of 36,800 is ______________.

5. Calculate the cube root of 0.000065.

6. Calculate the square root of 0.000049.

Go to LOGARITHMS and proceed with frame [28].

LOGARITHMS

[28] This section is concerned with arithmetical operations that involve exponential numbers. The study is limited to operations involving base 10 numbers. Answers to questions in this part are on page 274.

The definition of the term *logarithm* of a number is the exponent to which 10 must be raised to equal the number. For example, $100 = 10^2$; that is, the logarithm (abbreviated "log") of 100 is 2. A general interpretation of logs for base 10 numbers is

$$10^x = N$$

where x is the log of the number N. This statement is conventionally rewritten in a form that is more useful for computational work, as shown.

$$\log_{10} N = x$$

The new statement is read in the following manner: "x is the number to which 10 must be raised to equal N." Or alternatively, "x is the base 10 logarithm of N." Using this definition, one can find logs of numbers such as 10, 100, 0.1, and 0.01 as shown below:

$$\log 0.01 = \log 10^{-2} = -2$$
$$\log 0.1 = \log 10^{-1} = -1$$
$$\log 1 = \log 10^{0} = 0$$
$$\log 10 = \log 10^{1} = 1$$
$$\log 100 = \log 10^{2} = 2$$

Find the log of each of the following numbers:

(a) 100,000 = _____ (c) 0.00001 = _____

(b) 10,000 = _____ (d) 0.000001 = _____

[29] As illustrated in frame 28, logs of numbers are whole numbers when the numbers in question are integral powers of 10. But what about all other numbers? For instance, 25 is between 10 and 100. The exponent to which 10 must be raised to equal 25 is between 1 and 2. The log of 25 also is a number between 1 and 2. Perhaps it is obvious at this point that the

log of 25 is 1 plus a decimal fraction. In fact, logs always consist of two parts. Consider the following example:

$$100 = 10^{2.0000}$$

The number 2 is called the *characteristic* and the decimal fraction is called the *mantissa*. Four decimal places are used after the 2 to establish the accuracy of the logarithm. The characteristic is defined as the whole number representing the lower limit of the exponential range within which a number is located. That is, 25 lies between 10^1 and 10^2 and the characteristic of 25 is 1; 3500 lies between 10^3 and 10^4 and the characteristic of 3500 is 3. Determine the characteristic of each of the following numbers:

(a) 175 = _____ (c) 1250 = _____

(b) 7 = _____ (d) 657,000 = _____

[30] Mantissas for numbers have been computed and compiled into tables such as the table of logs inside the back cover of this book. Turn to the log table and locate the following numbers: the number 20 in the column headed by *N* and the number 3010 under the heading 0 to the right of 20. The numbers are to be interpreted in the following manner: the number 20 should not be read "twenty" but rather "two zero," which could represent the numbers, 2, 20, 200, 2000, and so on. To use the table, mentally place a decimal point to the right of the first significant figure of the numbers under the heading *N*, thus assigning values ranging from 1.0 to 9.9 for numbers in the *N* column. As a result of this assignment, all the numbers in the columns headed by 0 through 9 become decimals ranging from .0000 to .9996.

To determine the logarithm of 250, first determine the characteristic to be 2. Next locate 25 in the *N* column, and to the right in the 0 column find 3979. The log of 250 = 2.3979 to four decimal places. Also, $10^{2.3979} = 250$. In the same manner, determine the logarithm for each of the following numbers:

(a) log 3500 = __________ (c) log 450,000 = __________

(b) log 27,000 = __________

[31] The *N*, or number, column of the log table in this book contains two-digit numbers, and to determine logs for numbers with three or more digits, the columns headed by 0 through 9 must be used as shown in the following example. Suppose you wish to find the log of 225. Because 225 lies between 100 and 1000, its characteristic is 2. Now locate 22 in the *N* column; and to the right of 22 in the column headed by 5, find 3522. The log of 225 is 2.3522. The columns 0 to 9 effectively expand the *N* column to a 3-digit column. Now determine logs as indicated below.

(a) log 374 = __________ (c) log 1270 = __________

(b) log 2410 = __________ (d) log 575,000 = __________

[32] At this point you are able to find the log of a number; for logs to be useful you must also be able to find the number that corresponds to a given log. This process is called finding the antilogarithm, and the

procedure is just the reverse of finding the log of a number. As an example, find the number whose log is 2.9782. First, locate the mantissa, 9782, in the table. The mantissa is found to the right of 95 and in the 1 column. This number should be read "nine five one." Second, the magnitude of the number is established by recalling that 0.9782 is the log of 9.51, and the characteristic, 2, indicates that the decimal point is to be shifted two places to the right. That is, 2.9782 is the log of 951. Study the following two examples and then complete the exercises.

Examples:

$\log x = 3.8779$ .8779 corresponds to 7.55

$x = 7550$

$\log x = 6.7042$ .7042 corresponds to 5.06

$x = 5{,}060{,}000$

(a) $\log x = 4.8722$ $x =$ ______   (c) $\log x = 2.4955$ $x =$ ______

(b) $\log x = 3.7745$ $x =$ ______

[33] The table of logs in this book can be used to find logs of three-digit numbers directly, and a fourth place can be found by using the proportional parts (see last three columns of log table). Electronic calculators equipped with log circuits are convenient for working with logs. These calculators usually display logs to eight or nine places. With calculators, antilogs are computed directly by reversing the log function. Consult the instruction manual of your calculator for proper operation. Using proportional parts to determine logs or antilogs is good practice, but not necessary if you have an appropriate calculator. You should work the remaining exercises, with a calculator if you wish, to gain practice with the fundamental logarithm concepts.

The proportional parts table is used in the following manner. In frame 31, you found the log of 225. Now find the log of 225.4. Because 225.4 lies between 225 and 226, the log of 225.4 must lie between log of 225 and log of 226; that is, between 2.3522 and 2.3541. To find the log of 225.4, go to the 4 column in the proportional parts section of the table on the same line as 22. There find the number 8. To the log 225, namely 2.3522, add 8 to the last digit to get 2.3530. As another example, the log of 3765 is found in a similar manner.

$\log 3760 = 3.5752 \qquad \log 3765 = ? \qquad \log 3770 = 3.5763$

From proportional parts section, column 5, find that 6 must be added to get the log of 3765: $\log 3765 = 3.5758$. Find the indicated logs.

(a) log 69340 = __________   (b) log 7469 = __________

[34] The proportional parts table also is used to establish the fourth digit in the antilog procedure. In fact, the log-finding procedure is just reversed. The antilog of 2.6458 is found as follows:

$$\begin{array}{rl} & 2.6458 = \log x \\ & \underline{2.6454 = \log 442} \\ \text{Diff.} & 0.0004 \end{array}$$

Locate the 4 in the proportional parts table in the same line as 442. Note that the 4 is in the 4 column. Therefore, 2.6458 is the log of 442.4. Using the same procedure, find the antilogs as indicated.

(a) $\log x = 1.8989$ (b) $\log x = 2.7791$ (c) $\log x = 3.8429$

Go to SELF-EVALUATION III

SELF-EVALUATION III

Complete the following self-evaluation of your skills in finding logs of numbers and antilogs. Answers to these questions are on page 275. If one of your answers does not agree with the answer listed, refer to the frame whose number appears in brackets with the answer and rework the problem(s) in that frame.

1. The log of 0.00001 is __________.
2. The log of 125 is __________.
3. The antilog of 2.5502 is __________.
4. By using the proportional parts section of the log table, find the antilog of 1.9677.
5. Find the log of 56.78.

Go to frame [35]

---

[35] The common arithmetic operations — multiplication, division, powers, and roots — through the use of logarithms will be considered next. Rules for performing these operations with exponents were treated in the section on exponential numbers. Because logs are exponents, the rules for operations with exponents apply to log operations, except that one more step is involved to determine answers. Multiplication will be discussed first.

Recall that the product of two exponential numbers to the same base is the base raised to the sum of the exponents. Therefore, the log of a product is merely the sum of the logs of each factor. In general,

$$\log_{10} (x)(y) = \log_{10} x + \log_{10} y$$

With numbers,

$$\log (10^4)(10^5) = \log (10^4) + \log (10^5)$$
$$= 4 + 5 = 9$$

The log of the product equals 9; therefore, the product equals 10^9.

For the following indicated products, compute the logs:

(a) $\log (44)(51) =$ __________ (c) $\log (65)(7850) =$ __________

(b) $\log (685)(72) =$ __________ (d) $\log (45)(421)(7850) =$ __________

[36] Products can be computed by calculating the log of a product, then finding the antilog. The table of proportional parts is invaluable at this point. The product of two numbers, such as (75)(85), is computed in the following manner:

$$\log (75)(85) = \log 75 + \log 85 = 1.8751 + 1.9294 = 3.8045$$

Therefore, $(75)(85) = 10^{3.8045}$, or, from the antilog, 6376. If this product is computed by long-hand multiplication, its value is found to be 6375. The apparent discrepancy is due to rounding in the evaluation of logs. However, answers generally will be accurate to the number of allowable significant figures. In this case the product has two significant figures and should be left as 6400. Use the procedure above to compute the following products:

(a) (44)(51) (b) (685)(72) (c) (65)(7850) (d) (45)(421)(7850)

[37] All examples involving logs encountered thus far have dealt with numbers greater than 1. The use of logs is confined to all positive numbers, that is, numbers greater than zero. Now consider finding the log of a number whose value is greater than 0 and less than 1. As an example, the log of 0.5 is found in the following manner: First, rewrite the number in scientific notation as 5×10^{-1}; now compute the log of the product.

$$\log 5 \times 10^{-1} = \log 5 + \log 10^{-1} = 0.6990 + (-1) = -0.301$$

Therefore, $0.5 = 10^{-0.301}$.

Compute the following logs:

(a) log 0.0065 (b) log 0.0043 (c) log 0.000675

[38] Finding the antilog in which the log is negative requires one more step than if the log is positive. Although several methods can be used for this procedure, one method, indicated in the following example, is considered here. Find the number whose log is -2.1871.

Let x be the number in question. Therefore,

$$\log x = -2.1871$$

$$x = 10^{-2.1871}$$

All log values in the table are positive, so the log must be rewritten as the sum of two numbers of which one is an integral power of 10 and the other is greater than 0 but less than 1, their sum equaling the value of the log. For the above case, the log can be written

$$10^{-2.1871} = 10^{3-2.1871} \times 10^{-3} = 10^{0.8129} \times 10^{-3}$$

In other words, what value can be added to -3, the next-higher negative number, to get -2.1871? The value of x is the antilog of 0.8129 times 10^{-3}: 6.50×10^{-3}.

Find the antilogs of the following logs:

(a) -1.6884 (b) -3.6590 (c) -4.8652

[39] Division operations with logs are in a sense opposite to multiplication operations; that is, as the log of the product of two terms is the sum of the logs, so the log of the quotient of two terms is the difference of the logs. In general,

$$\log_{10} \frac{x}{y} = \log_{10} x - \log_{10} y$$

$$\log \frac{10^4}{10^2} = \log 10^4 - \log 10^2 = 4 - 2 = 2$$

$$\log \frac{4}{2} = \log 4 - \log 2 = 0.6021 - 0.3010 = 0.3011$$

Then $4/2 = 10^{0.3011} = 2$.

Applying the procedure above, calculate the quotient for the following exercises:

(a) $\frac{4650}{321}$ (b) $\frac{69,500}{225}$

[40] Now consider a division operation with logs in which the numerator is smaller than the denominator. As an example, the quotient 2/4 is calculated in the following manner:

$$\log \frac{2}{4} = \log 2 - \log 4 = 0.3010 - 0.6021 = -0.3011$$

Then,

$$\frac{2}{4} = 10^{-0.3011} = 10^{0.6989} \times 10^{-1}$$

The antilog of 0.6989 is 5 and the quotient of 2/4 is 5×10^{-1}, or 0.5.

Calculate the following quotients by using logs:

(a) $\frac{425}{575}$ (b) $\frac{1650}{18,250}$ (c) $\frac{(465)(830)}{960}$

[41] Last, consider the use of logs for raising numbers to powers and for extracting roots of numbers. Raising numbers to powers greater than 2 or 3 becomes cumbersome and time-consuming if one does not use a calculator or logs. Logs are exponents and the rules for raising exponential terms to a power or root apply to logs. That is,

$$\log_{10} (N)^x = x \log_{10} N$$

With numbers,

$$\log_{10} (2)^5 = 5 \log_{10} 2 = 5(0.3010) = 1.5050$$

then,

$$(2)^5 = 10^{1.5050} = 32$$

As another example, find the value of $(25)^{2.6}$.

$$\log (25)^{2.6} = 2.6 \log 25 = 2.6(1.3979) = 3.6345$$

$$(25)^{2.6} = 10^{3.6345} = 4310$$

Using logs, raise the following numbers to the powers indicated.

(a) $(1250)^4$ (b) $(21)^{3.4}$ (c) $(145)^{8.67}$

[42] Extracting roots with logs involves exactly the same operations as raising numbers to powers. For example, the cube root of 265 is found in the following manner:

$$\log (265)^{1/3} = \frac{1}{3} \log 265$$

$$= \frac{1}{3}(2.4232) = 0.8077$$

$$\text{antilog of } 0.8077 = 6.42$$

Calculate the indicated roots.

(a) $(475)^{1/5}$ (b) $(1430)^{2/5}$

Go to SELF-EVALUATION IV

SELF-EVALUATION IV

Complete the following self-evaluation of your skills in using logs for multiplication, division, raising numbers to powers, and extracting roots of numbers. Answers to these questions are on page 277. If one of your answers does not agree with the answer listed, refer to the frame whose number appears in brackets with the answer and rework the problem(s) in that frame.

1. Calculate the product of (250)(325)(460).
2. Calculate the quotient of 4850/2320.
3. Calculate the cube root of 4365.
4. Calculate the quotient of 275/865.
5. Evaluate $(117)^4$.

ANSWERS FOR PROGRAMMED UNITS

1. $40 = 4 \times 10 = 4 \times 10^1$

   $400 = 4 \times 100 = 4 \times 10^2$

2. $750 = 7.5 \times 100 = 7.5 \times 10^2$

   $625 = 6.25 \times 100 = 6.25 \times 10^2$

3. (a) $7700 = 7.7 \times 1000 = 7.7 \times 10^3$

   (b) $8250 = 8.25 \times 1000 = 8.25 \times 10^3$

   (c) $6255 = 6.255 \times 1000 = 6.255 \times 10^3$

4. Since 1,000,000,000 contains 9 zeros, the exponent of 10 is 9. In exponential notation 1,000,000,000 is written 1×10^9.

5. (a) $5^0 = 1$

   (b) $100^0 = 1$

   (c) $5^0 \times 10^0 = 1 \times 1 = 1$

   (d) $a^0 \times 10^0 \times b^0 = 1 \times 1 \times 1 = 1$

6. (a) $7655 = 7.655 \times 10^3$

(b) $2{,}225{,}000{,}000 = 2.225 \times 10^9$

(c) $786{,}000 = 7.86 \times 10^5$

7. (a) $2250 = 2.25 \times 10^3$

(b) $60{,}200{,}000{,}000 = 6.02 \times 10^{10}$

(c) $5{,}575{,}000{,}000 = 5.575 \times 10^9$

8. (a) 7×10^{-6}

(b) 4.25×10^{-5}

(c) 6.08×10^{-5}

9. $0.000007 = 7 \times 10^{-6}$

$0.0000425 = 4.25 \times 10^{-5}$

10. No answer required.

11. (a) 3×10^{-3}

(b) 7.5×10^{-3}

(c) 2.354×10^{-4}

(d) 7.2×10^{-6}

12. (a) $23.5 \times 10^4 = 2.35 \times 10^5$

(b) $6.8 \times 10^{-4} = 68 \times 10^{-5}$

13. (a) $10^1 \cdot 10^5 = 10^{1+5} = 10^6$

(b) $10^3 \cdot 10^9 = 10^{3+9} = 10^{12}$

(c) $10^4 \cdot 10^{10} = 10^{4+10} = 10^{14}$

14. (a) $10^{-8} \cdot 10^{-10} = 10^{-8+(-10)} = 10^{-18}$

(b) $10^{-5} \cdot 10^{-3} = 10^{-5+(-3)} = 10^{-8}$

(c) $10^{-5} \cdot 10^{-5} = 10^{-5+(-5)} = 10^{-10}$

15. (a) $10^{-15} \cdot 10^{12} = 10^{-15+12} = 10^{-3}$

(b) $10^{-2} \cdot 10^{-3} \cdot 10^6 = 10^{-2+(-3)+6}$

$= 10^{-5+6} = 10^1$

(c) $10^{14} \cdot 10^{-12} \cdot 10^3$

$= 10^{14+(-12)+3} = 10^5$

16. (a) $(3.2 \times 10^4)(2.0 \times 10^5)$

$= (3.2)(2.0)(10^4)(10^5)$

$= 6.4 \times 10^9$

(b) $(4.0 \times 10^6)(8.0 \times 10^{-5})$

$= (4.0)(8.0)(10^6)(10^{-5})$

$= 32 \times 10^1$ or 3.2×10^2

(c) $(8.0 \times 10^6)(9.1 \times 10^{-2})$

$= (8.0)(9.1)(10^6)(10^{-2})$

$= 7.3 \times 10^5$

(d) $(2.0 \times 10^3)(3.0 \times 10^{-5}) \times$

$(4.0 \times 10^2) = 2.4 \times 10^1$

17. No answer required.

18. (a) $\frac{10^4}{10^1} = 10^{4-1} = 10^3$

(b) $\frac{10^6}{10^2} = 10^{6-2} = 10^4$

(c) $\frac{10^5}{10^{-5}} = 10^{5-(-5)}$

$= 10^{5+5} = 10^{10}$

(d) $\frac{10^{1.2}}{10^{2.4}} = 10^{1.2-2.4} = 10^{-1.2}$

19. (a) $\frac{5.0 \times 10^4}{2.0 \times 10^2} = \frac{5.0}{2.0} \times \frac{10^4}{10^2}$

$= 2.5 \times 10^{4-2}$

$= 2.5 \times 10^2$

(b) $\frac{7.5 \times 10^{-1}}{3.8 \times 10^3} = \frac{7.5}{3.8} \times \frac{10^{-1}}{10^3}$

$= 2.0 \times 10^{-1-3}$

$= 2.0 \times 10^{-4}$

(c) $\frac{9.6 \times 10^{13}}{2.4 \times 10^{-12}} = \frac{9.6}{2.4} \times \frac{10^{13}}{10^{-12}}$

$= 4.0 \times 10^{13-(-12)}$

$= 4.0 \times 10^{25}$

20. (a) $\dfrac{2.2 \times 10^{-2}}{6.6 \times 10^{-4}} = \dfrac{2.2}{6.6} \times \dfrac{10^{-2}}{10^{-4}}$

$= 0.33 \times 10^{2}$;

This quotient should be expressed as 3.3×10^{1}, derived as follows:

$(0.33)(10^{1}) \times \dfrac{10^{2}}{10^{1}} = 3.3 \times 10^{1}$

(b) $\dfrac{8.1 \times 10^{14}}{9.0 \times 10^{-15}} = \dfrac{8.1}{9.0} \times \dfrac{10^{14}}{10^{-15}}$

$= 0.90 \times 10^{29}$

and is expressed as 9.0×10^{28}, derived from

$(0.90)(10^{1}) \times \dfrac{10^{29}}{10^{1}} = 9.0 \times 10^{28}$

(c) $\dfrac{1.42 \times 10^{-3}}{96.2 \times 10^{4}} = \dfrac{1.42}{96.2} \times \dfrac{10^{-3}}{10^{4}}$

$= 0.0148 \times 10^{-7}$

$= 1.48 \times 10^{-9}$

as derived from

$(0.0140)(10^{2}) \times \dfrac{10^{-7}}{10^{2}} = 1.48 \times 10^{-9}$

Self-evaluation I Answers

1. 7.07×10^{6} [1 through 7]
2. 262,000 [6 and 7]
3. 7.85×10^{-5} [8, 9, and 10]
4. 7×10^{14} [16]
5. 7.3×10^{-2} [19] and [20]
6. 8.9×10^{1} [20]

---

21. (a) $(10^{2})^{5} = 10^{(2)(5)} = 10^{10}$

(b) $(10^{4})^{-2} = 10^{-8}$

(c) $(10^{1/2})^{4} = 10^{4/2} = 10^{2}$

(d) $(10^{2.5})^{2} = 10^{(2.5)(2)} = 10^{5}$

22. (a) $(4.0 \times 10^{2})^{3} = (4.0)^{3}(10^{2})^{3}$

$= 64 \times 10^{6} = 6.4 \times 10^{7}$

(b) $(5.0 \times 10^{5})^{2} = (5.0)^{2}(10^{5})^{2}$

$= 25 \times 10^{10} = 2.5 \times 10^{11}$

(c) $(2.50 \times 10^{-3})^{3}$

$= (2.50)^{3}(10^{-3})^{3}$

$= 15.6 \times 10^{-9}$

$= 1.56 \times 10^{-8}$

23. (a) $\sqrt{10^{-6}} = (10^{-6})^{1/2}$

$= 10^{-6/2} = 10^{-3}$

(b) $\sqrt{10^{4}} = 10^{4/2} = 10^{2}$

(c) $\sqrt[3]{10^{-6}} = (10^{-6})^{1/3}$

$= 10^{-6/3} = 10^{-2}$

(d) $\sqrt[5]{10^{-25}} = 10^{-5}$

24. (a) $\sqrt{10^{5}} = (10^{5})^{1/2} = (1 \times 10^{5})^{1/2}$

$= (10 \times 10^{4})^{1/2} = 3 \times 10^{2}$

(b) $\sqrt[3]{10^{7}} = (1 \times 10^{7})^{1/3}$

$= (10 \times 10^{6})^{1/3}$

$= (10)^{1/3} \times (10^{6})^{1/3}$

$= 2 \times 10^{2}$

(c) $\sqrt{10^{7}} = (1 \times 10^{7})^{1/2}$

$= (10 \times 10^{6})^{1/2} = 3 \times 10^{3}$

(d) $\sqrt[3]{1.0 \times 10^{8}} = (1.0 \times 10^{8})^{1/3}$

$= (100 \times 10^{6})^{1/3}$

$= (100)^{1/3} \times (10^{6})^{1/3}$

$= 4.6 \times 10^{2}$

25. (a) $\sqrt{2500} = (25 \times 10^{2})^{1/2}$

$= 5.0 \times 10^{1}$

(b) $\sqrt[3]{1440} = (1.44 \times 10^{3})^{1/3}$

$= (1.44)^{1/3} \times (10^{3})^{1/3}$

$= 1.13 \times 10^{1}$

(c) $\sqrt{26,200} = (2.62 \times 10^{4})^{1/2}$

$= 1.62 \times 10^{2}$

(d) $\sqrt[3]{18,000} = (18 \times 10^{3})^{1/3}$

$= 2.6 \times 10^{1}$

26. (a) $(0.000045)^{1/2}$

$= (45 \times 10^{-6})^{1/2}$

$= 6.7 \times 10^{-3}$

(b) $(0.00030)^{1/2}$

$= (3.0 \times 10^{-4})^{1/2}$

$= 1.7 \times 10^{-2}$

(c) $(0.00061)^{1/3}$

$= (610 \times 10^{-6})^{1/3}$

$= 8.5 \times 10^{-2}$

(d) $(9.6 \times 10^{-16})^{1/3}$

$= (960 \times 10^{-18})^{1/3}$

$= 9.9 \times 10^{-6}$

27. No answer required.

Self-evaluation II Answers

1. $(10^8)^3 = 10^{24}$ [21]
2. $(2.5 \times 10^4)^2 = (2.5)^2 \times (10^4)^2$
   $= 6.2 \times 10^8$ [22]
3. $(25,000)^{1/3} = 2.9 \times 10^1$ [25]
4. $(36,800)^{1/2} = 1.92 \times 10^2$ [25]
5. $(0.000065)^{1/3} = 4.0 \times 10^{-2}$ [26]
6. $(0.000049)^{1/2} = 7.0 \times 10^{-3}$ [26]

---

28. (a) $\log 100,000 = \log 10^5 = 5$

(b) $\log 10,000 = \log 10^4 = 4$

(c) $\log 0.00001 = \log 10^{-5} = -5$

(d) $\log 0.000001 = \log 10^{-6} = -6$

29. (a) $10^2 - 175 - 10^3$ $\underline{2}$

(b) $10^0 - 7 - 10^1$ $\underline{0}$

(c) $10^3 - 1250 - 10^4$ $\underline{3}$

(d) $10^5 - 657,000 - 10^6$ $\underline{5}$

30. (a) $3500 = 3.5 \times 10^3$

The characteristic is 3.
The mantissa is .5441.

log 3500 = 3.5441 or
$10^{3.5441} = 3500$

(b) $27,000 = 2.7 \times 10^4$

Characteristic = 4.
The mantissa is .4314.

log 27,000 = 4.4314 or
$10^{4.4314} = 27,000$

(c) $450,000 = 4.5 \times 10^5$

Characteristic = 5.
The mantissa is .6532.

log 450,000 = 5.6532 or
$10^{5.6532} = 450,000$

31. (a) $374 = 3.74 \times 10^2$

Characteristic = 2.
The mantissa is .5729.

log 374 = 2.5729

(b) $2410 = 2.41 \times 10^3$

Characteristic = 3.
The mantissa is .3820.

log 2410 = 3.3820 or
$10^{3.3820} = 2410$

(c) $1270 = 1.27 \times 10^3$

Characteristic = 3.
The mantissa is .1038.

log 1270 = 3.1038 or
$10^{3.1038} = 1270$

(d) log 575,000 = 5.7597

32. (a) $\log x = 4.8722$ or $x = 10^{4.8722}$

$10^{.8722} = 7.45$ so

$10^{4.8722} = \underline{74500}.$

4 places

(b) $\log x = 3.7745$ or $x = 10^{3.7745}$

$10^{.7745} = 5.95$ so

$10^{3.7745} = \underline{595}0.$

3 places

(c) $\log x = 2,4955$ or $x = 10^{2.4955}$

$10^{.4955} = 3.13$ so

$10^{2.4955} = \underline{31}3.$

2 places

33. (a) log 69340 = 4.8407 + .0002

= 4.8409

(b) log 7469 = 3.8727 + .0005

= 3.8732

34. (a) From the tabled values, 8989 corresponds to 7920 + 3 = 7923. The characteristic is 1; therefore, $x = 79.23$.

(b) x = 6010 + 3; 601.3

(c) x = 6960 + 5; 6965

Self-evaluation III Answers

1. -5 [27]
2. 2.0969 [28]
3. 355.0 [32]
4. 92.83 [34]
5. 1.7542 [33]

---

35. (a) log (44)(51) = log 44 + log 51
log 44 = 1.6435
log 51 = 1.7076
log (44)(51) = 3.3511

(b) log (685)(72) = log 685 + log 72
log 685 = 2.8357
log 72 = 1.8573
log (685)(72) = 4.6930

(c) log (65)(7850)
= log 65 + log 7850
log 65 = 1.8129
log 7850 = 3.8949
log (65)(7850) = 5.7078

(d) log (45)(421)(7850)
= log 45 + log 421
+ log 7850
log 45 = 1.6532
log 421 = 2.6243
log 7850 = 3.8949
log (45)(421)(7850) = 8.1724

36. (a) log (44)(51)
= log 44 + log 51
= 1.6434 + 1.7076
= 3.3510
Then,
$(44)(51) = 10^{3.3510}$
$= 2.2 \times 10^3$

(b) log (685)(72)
= log 685 + log 72
= 2.8357 + 1.8573
= 4.6930
Then,
$(685)(72) = 10^{4.6930}$
$= 4.9 \times 10^4$

(c) log (65)(7850)
= log 65 + log 7850
= 1.8129 + 3.8949
= 5.7078
Then,
$(65)(7850) = 10^{5.7078}$
$= 5.1 \times 10^5$

(d) log (45)(421)(7850)
= log 45 + log 421
+ log 7850
= 1.6532 + 2.6243 + 3.8949
= 8.1724
Then,
(45)(421)(7850)
$= 10^{8.1724}$
$= 1.5 \times 10^8$

37. (a) $\log 0.0065$
$= \log 6.5 \times 10^{-3}$
$= \log 6.5 + \log 10^{-3}$
$= 0.8129 + (-3)$
$= -2.1871$

(b) $\log 0.0043$
$= \log 4.3 \times 10^{-3}$
$= \log 4.3 + \log 10^{-3}$
$= 0.6335 + (-3)$
$= -2.3665$

(c) $\log 0.000675$
$= \log 6.75 \times 10^{-4}$
$= \log 6.75 + \log 10^{-4}$
$= 0.8293 + (-4)$
$= -3.1707$

38. (a) $x = 10^{-1.6884}$
$= 10^{0.3410} \times 10^{-2}$
$= 2.05 \times 10^{-2}$

(b) $x = 10^{-3.6590}$
$= 10^{0.3410} \times 10^{-4}$
$= 2.19 \times 10^{-4}$

(c) $x = 10^{-4.8652}$
$= 10^{0.1348} \times 10^{-5}$
$= 1.36 \times 10^{-5}$

39. (a) $\log \frac{4650}{321} = \log 4650 - \log 321$
$= 3.6675 - 2.5065$
$= 1.1610$

Then,

$\frac{4650}{321} = 10^{1.1610} = 14.5$

(b) $\log \frac{69{,}500}{225}$
$= \log 69{,}500 - \log 225$
$= 4{,}8420 - 2.3522$
$= 2.4898$

Then,

$\frac{69{,}500}{225} = 10^{2.4898} = 309$

40. (a) $\log \frac{425}{575} = \log 425 - \log 575$
$= 2.6284 - 2.7597$
$= -0.1313$

Then,

$\frac{425}{575} = 10^{-0.1313}$
$= 10^{0.8687} \times 10^{-1}$

antilog of $0.8687 = 7.391$

$\frac{425}{575} = 7.39 \times 10^{-1}$

(b) $\log \frac{1650}{18{,}250}$
$= \log 1650 - \log 18{,}250$
$= 3.2175 - 4.2613$
$= -1.0438$

Then,

$\frac{1650}{18{,}250} = 10^{-1.0438}$

antilog $= 9.04 \times 10^{-2}$

(c) $\log \frac{(465)(830)}{960}$
$= \log 465 + \log 830 - \log 960$
$= 2.6675 + 2.9191 - 2.9823$
$= 2.6043$

antilog $= 402$

41. (a) $\log (1250)^4$
$= 4 \log 1250$
$= 4(3.0969) = 12.3876$
$(1250)^4 = 2.44 \times 10^{12}$

(b) $\log (21)^{3.4}$
$= 3.4 \log 21$
$= 3.4(1.3222) = 4.4955$
$(21)^{3.4} = 3.1 \times 10^4$

(c) $(145)^{8.67} = 5.48 \times 10^{18}$

42. (a) $\log (475)^{1/5}$

$$= \frac{1}{5} \log 475$$

$$= \frac{1}{5}(2.6767) = 0.5353$$

antilog of $0.5353 = 3.43$

(b) $\log (1430)^{2/5}$

$$= \frac{2}{5} \log 1430$$

$$= \frac{2}{5}(3.1553) = 1.2621$$

antilog of $1.2621 = 18.3$

Self-evaluating IV Answers

1. 3.7×10^7 [36]
2. 2.09 [37]
3. 16.34 [42]
4. 0.3179 [38 and 39]
5. 1.87×10^8 [41]

APPENDIXES

APPENDIX C

Units and Conversion Factors

Base Units of International System of Units (SI)

Physical Property	Name of Unit	Symbol
Length	Meter	m
Mass	Kilogram	kg
Time	Second	s
Electric current	Ampere	A
Thermodynamic temperature	Kelvin	K
Luminous intensity	Candela	cd
Quantity of substance	Mole	mol

Units of Length

Meter (m) = 39.37 inches (in) = 1.094 yards (yd)	Yard = 0.9144 m (exact)
Centimeter (cm) = 0.01 m	Inch = 2.54 cm (exact)
Millimeter (mm) = 0.001 m	
Kilometer (km) = 1000 m	Mile (U.S.) = 1.60934 km
Angstrom unit (Å) = 10^{-8} cm = 10^{-10} m	

Units of Volume

Liter (L) = 0.001 m^3 = 1000 cm^3	Liquid quart (U.S.) = 0.9463 L
Milliliter (mL) = 0.001 L = 1 cm^3	Dry quart = 1.1012 L
	Cubic foot (U.S.) = 28.316 L

Units and Conversion Factors (continued)

Units of Weight	
Gram (g) = 0.001 kg	Ounce (oz) (avoirdupois) = 28.35 g
Milligram (mg) = 0.001 g	Pound (lb) (avoirdupois) = 0.45359237 kg
Kilogram (kg) = 1000 g	Ton (short) = 2000 lb = 907.185 kg
Ton (metric) = 1000 kg = 2204.62 lb	Ton (long) = 2240 lb = 1.016 metric ton

Units of Energy

4.184 joule (J) = 1 thermochemical calorie (cal) = 4.184×10^7 erg
Erg = 10^{-7} J
Electron-volt (eV) = 1.602189×10^{-12} erg = 23.061 kcal mol^{-1}
Liter atmosphere = 24.217 cal = 101.32 J

Unit of Force

Newton (N) = 1 kg m s^{-2} (force that when applied for 1 second will give to a 1-kilogram mass a speed of 1 meter per second)

Units of Pressure

Torr = 1 mmHg
Atmosphere (atm) = 760 mm Hg = 760 torr = 101,325 N m^{-2} = 101,325 Pa
Pascal (Pa) = kg m^{-1} s^{-2} = N m^{-2}

APPENDIX E

Solubility Products

Substance	K_{sp} at 25°C
Aluminum	
$Al(OH)_3$	1.9×10^{-33}
Barium	
$BaCO_3$	8.1×10^{-9}
$BaC_2O_4 \cdot 2H_2O$	1.1×10^{-7}
$BaSO_4$	1.08×10^{-10}
$BaCrO_4$	2×10^{-10}
BaF_2	1.7×10^{-6}
$Ba(OH)_2 \cdot 8H_2O$	5.0×10^{-3}
$Ba_3(PO_4)_2$	1.3×10^{-29}
$Ba_3(AsO_4)_2$	1.1×10^{-13}
Bismuth	
$BiO(OH)$	1×10^{-12}
$BiOCl$	7×10^{-9}
Bi_2S_3	1.6×10^{-72}
Cadmium	
$Cd(OH)_2$	1.2×10^{-14}
CdS	3.6×10^{-29}
$CdCO_3$	2.5×10^{-14}
Calcium	
$Ca(OH)_2$	7.9×10^{-6}
$CaCO_3$	4.8×10^{-9}
$CaSO_4 \cdot 2H_2O$	2.4×10^{-5}
$CaC_2O_4 \cdot H_2O$	2.27×10^{-9}
$Ca_3(PO_4)_2$	1×10^{-25}
$CaHPO_4$	5×10^{-6}
CaF_2	3.9×10^{-11}
Chromium	
$Cr(OH)_3$	6.7×10^{-31}
Cobalt	
$Co(OH)_2$	2×10^{-16}
$CoS(\alpha)$	5.9×10^{-21}
$CoS(\beta)$	8.7×10^{-23}
$CoCO_3$	1.0×10^{-12}
$Co(OH)_3$	2.5×10^{-43}
Copper	
$CuCl$	1.85×10^{-7}
$CuBr$	5.3×10^{-9}
CuI	5.1×10^{-12}
$CuSCN$	4×10^{-14}
Cu_2S	1.6×10^{-48}
$Cu(OH)_2$	5.6×10^{-20}
CuS	8.7×10^{-36}
$CuCO_3$	1.37×10^{-10}
Iron	
$Fe(OH)_2$	7.9×10^{-15}
$FeCO_3$	2.11×10^{-11}
FeS	1×10^{-19}
$Fe(OH)_3$	1.1×10^{-36}
Lead	
$Pb(OH)_2$	2.8×10^{-16}
PbF_2	3.7×10^{-8}
$PbCl_2$	1.7×10^{-5}
$PbBr_2$	6.3×10^{-6}
PbI_2	8.7×10^{-9}
$PbCO_3$	1.5×10^{-13}
PbS	8.4×10^{-28}
$PbCrO_4$	1.8×10^{-14}
$PbSO_4$	1.8×10^{-8}
$Pb_3(PO_4)_2$	3×10^{-44}
Magnesium	
$Mg(OH)_2$	1.5×10^{-11}

Solubility Products (continued)

Substance	K_{sp} at 25°C	Substance	K_{sp} at 25°C
$MgCO_3 \cdot 3H_2O$	*ca* 1×10^{-5}	AgCl	1.8×10^{-10}
$MgNH_4PO_4$	2.5×10^{-13}	AgBr	3.3×10^{-13}
MgF_2	6.4×10^{-9}	AgI	1.5×10^{-16}
MgC_2O_4	8.6×10^{-5}	AgCN	1.2×10^{-16}
Manganese		AgSCN	1.0×10^{-12}
$Mn(OH)_2$	4.5×10^{-14}	Ag_2S	1.0×10^{-51}
$MnCO_3$	8.8×10^{-11}	Ag_2CO_3	8.2×10^{-12}
MnS	5.6×10^{-16}	Ag_2CrO_4	9×10^{-12}
Mercury		$Ag_4Fe(CN)_6$	1.55×10^{-41}
$Hg_2O \cdot H_2O$	1.6×10^{-23}	Ag_2SO_4	1.18×10^{-5}
Hg_2Cl_2	1.1×10^{-18}	Ag_3PO_4	1.8×10^{-18}
Hg_2Br_2	1.26×10^{-22}	Strontium	
Hg_2I_2	4.5×10^{-29}	$Sr(OH)_2 \cdot 8H_2O$	3.2×10^{-4}
Hg_2CO_3	9×10^{-17}	$SrCO_3$	9.42×10^{-10}
Hg_2SO_4	6.2×10^{-7}	$SrCrO_4$	3.6×10^{-5}
Hg_2S	1×10^{-45}	$SrSO_4$	2.8×10^{-7}
Hg_2CrO_4	2×10^{-9}	$SrC_2O_4 \cdot H_2O$	5.61×10^{-8}
HgS	3×10^{-53}	Thallium	
Nickel		TlCl	1.9×10^{-4}
$Ni(OH)_2$	1.6×10^{-14}	TlSCN	5.8×10^{-4}
$NiCO_3$	1.36×10^{-7}	Tl_2S	1.2×10^{-24}
NiS(α)	3×10^{-21}	$Tl(OH)_3$	1.5×10^{-44}
NiS(β)	1×10^{-26}	Tin	
NiS(γ)	2×10^{-28}	$Sn(OH)_2$	5×10^{-26}
Potassium		SnS	8×10^{-29}
$KClO_4$	1.07×10^{-2}	$Sn(OH)_4$	*ca* 1×10^{-56}
K_2PtCl_6	1.1×10^{-5}	Zinc	
$KHC_4H_4O_6$	3×10^{-4}	$ZnCO_3$	6×10^{-11}
Silver		$Zn(OH)_2$	4.5×10^{-17}
$\frac{1}{2}Ag_2O$ ($Ag^+ + OH^-$)	2×10^{-8}	ZnS	1.1×10^{-21}

APPENDIX F

Formation Constants for Complex Ions

Equilibrium	K_f
$Al^{3+} + 6F^- \rightleftharpoons [AlF_6]^{3-}$	5×10^{23}
$Cd^{2+} + 4NH_3 \rightleftharpoons [Cd(NH_3)_4]^{2+}$	4.0×10^{6}
$Cd^{2+} + 4CN^- \rightleftharpoons [Cd(CN)_4]^{2-}$	1.3×10^{17}
$Co^{2+} + 6NH_3 \rightleftharpoons [Co(NH_3)_6]^{2+}$	8.3×10^{4}
$Co^{3+} + 6NH_3 \rightleftharpoons [Co(NH_3)_6]^{3+}$	4.5×10^{33}
$Cu^+ + 2CN^- \rightleftharpoons [Cu(CN)_2]^-$	1×10^{16}
$Cu^{2+} + 4NH_3 \rightleftharpoons [Cu(NH_3)_4]^{2+}$	1.2×10^{12}
$Fe^{2+} + 6CN^- \rightleftharpoons [Fe(CN)_6]^{4-}$	1×10^{37}
$Fe^{3+} + 6CN^- \rightleftharpoons [Fe(CN)_6]^{3-}$	1×10^{44}
$Fe^{3+} + 6SCN^- \rightleftharpoons [Fe(CNS)_6]^{3-}$	3.2×10^{3}
$Hg^{2+} + 4Cl^- \rightleftharpoons [HgCl_4]^{2-}$	1.2×10^{15}
$Ni^{2+} + 6NH_3 \rightleftharpoons [Ni(NH_3)_6]^{2+}$	1.8×10^{8}
$Ag^+ + 2Cl^- \rightleftharpoons [AgCl_2]^-$	2.5×10^{5}
$Ag^+ + 2CN^- \rightleftharpoons [Ag(CN)_2]^-$	1×10^{20}
$Ag^+ + 2NH_3 \rightleftharpoons [Ag(NH_3)_2]^+$	1.6×10^{7}
$Zn^{2+} + 4CN^- \rightleftharpoons [Zn(CN)_4]^{2-}$	1×10^{19}
$Zn^{2+} + 4OH^- \rightleftharpoons [Zn(OH)_4]^{2-}$	2.9×10^{15}

APPENDIX G

Ionization Constants of Weak Acids

Acid	Formula	K_a at 25°C
Acetic	CH_3CO_2H	1.8×10^{-5}
Arsenic	H_3AsO_4	4.8×10^{-3}
	$H_2AsO_4^-$	1×10^{-7}
	$HAsO_4^{2-}$	1×10^{-13}
Arsenous	H_3AsO_3	5.8×10^{-10}
Boric	H_3BO_3	5.8×10^{-10}
Carbonic	H_2CO_3	4.3×10^{-7}
	HCO_3^-	7×10^{-11}
Cyanic	$HCNO$	3.46×10^{-4}
Formic	HCO_2H	1.8×10^{-4}
Hydrazoic	HN_3	1×10^{-4}
Hydrocyanic	HCN	4×10^{-10}
Hydrofluoric	HF	7.2×10^{-4}
Hydrogen peroxide	H_2O_2	2.4×10^{-12}
Hydrogen selenide	H_2Se	1.7×10^{-4}
	HSe^-	1×10^{-10}
Hydrogen sulfate ion	HSO_4^-	1.2×10^{-2}
Hydrogen sulfide	H_2S	1.0×10^{-7}
	HS^-	1.3×10^{-13}
Hydrogen telluride	H_2Te	2.3×10^{-3}
	HTe^-	1×10^{-5}
Hypobromous	$HBrO$	2×10^{-9}
Hypochlorous	$HClO$	3.5×10^{-8}
Nitrous	HNO_2	4.5×10^{-4}
Oxalic	$H_2C_2O_4$	5.9×10^{-2}
	$HC_2O_4^-$	6.4×10^{-5}
Phosphoric	H_3PO_4	7.5×10^{-3}
	$H_2PO_4^-$	6.3×10^{-8}
	HPO_4^{2-}	3.6×10^{-13}
Phosphorous	H_3PO_3	1.6×10^{-2}
	$H_2PO_3^-$	7×10^{-7}
Sulfurous	H_2SO_3	1.2×10^{-2}
	HSO_3^-	6.2×10^{-8}

APPENDIX H

Ionization Constants of Weak Bases

Base	Ionization Equation	K_b at 25°C
Ammonia	$NH_3 + H_2O \rightleftharpoons NH_4^+ + OH^-$	1.8×10^{-5}
Dimethylamine	$(CH_3)_2NH + H_2O \rightleftharpoons (CH_3)_2NH_2^+ + OH^-$	7.4×10^{-4}
Methylamine	$CH_3NH_2 + H_2O \rightleftharpoons CH_3NH_3^+ + OH^-$	4.4×10^{-4}
Phenylamine (aniline)	$C_6H_5NH_2 + H_2O \rightleftharpoons C_6H_5NH_3^+ + OH^-$	4.6×10^{-10}
Trimethylamine	$(CH_3)_3N + H_2O \rightleftharpoons (CH_3)_3NH^+ + OH^-$	7.4×10^{-5}

APPENDIX I

Standard Electrode (Reduction) Potentials

Half-Reaction	$E°$, V
$Li^+ + e^- \longrightarrow Li$	−3.09
$K^+ + e^- \longrightarrow K$	−2.925
$Rb^+ + e^- \longrightarrow Rb$	−2.925
$Ra^{2+} + 2e^- \longrightarrow Ra$	−2.92
$Ba^{2+} + 2e^- \longrightarrow Ba$	−2.90
$Sr^{2+} + 2e^- \longrightarrow Sr$	−2.89
$Ca^{2+} + 2e^- \longrightarrow Ca$	−2.87
$Na^+ + e^- \longrightarrow Na$	−2.714
$La^{3+} + 3e^- \longrightarrow La$	−2.52
$Ce^{3+} + 3e^- \longrightarrow Ce$	−2.48
$Nd^{3+} + 3e^- \longrightarrow Nd$	−2.44
$Sm^{3+} + 3e^- \longrightarrow Sm$	−2.41
$Gd^{3+} + 3e^- \longrightarrow Gd$	−2.40
$Mg^{2+} + 2e^- \longrightarrow Mg$	−2.37
$Y^{3+} + 3e^- \longrightarrow Y$	−2.37
$Am^{3+} + 3e^- \longrightarrow Am$	−2.32
$Lu^{3+} + 3e^- \longrightarrow Lu$	−2.25
$\frac{1}{2}H_2 + e^- \longrightarrow H^-$	−2.25
$Sc^{3+} + 3e^- \longrightarrow Sc$	−2.08
$[AlF_6]^{3-} + 3e^- \longrightarrow Al + 6F^-$	−2.07
$Pu^{3+} + 3e^- \longrightarrow Pu$	−2.07
$Th^{4+} + 4e^- \longrightarrow Th$	−1.90
$Np^{3+} + 3e^- \longrightarrow Np$	−1.86
$Be^{2+} + 2e^- \longrightarrow Be$	−1.85
$U^{3+} + 3e^- \longrightarrow U$	−1.80
$Hf^{4+} + 4e^- \longrightarrow Hf$	−1.70
$SiO_3^{2-} + 3H_2O + 4e^- \longrightarrow Si + 6OH^-$	−1.70
$Al^{3+} + 3e^- \longrightarrow Al$	−1.66
$Ti^{2+} + 2e^- \longrightarrow Ti$	−1.63
$Zr^{4+} + 4e^- \longrightarrow Zr$	−1.53
$ZnS + 2e^- \longrightarrow Zn + S^{2-}$	−1.44
$Cr(OH)_3 + 3e^- \longrightarrow Cr + 3OH^-$	−1.3
$[Zn(CN)_4]^{2-} + 2e^- \longrightarrow Zn + 4CN^-$	−1.26
$Zn(OH)_2 + 2e^- \longrightarrow Zn + 2OH^-$	−1.245
$[Zn(OH)_4]^{2-} + 2e^- \longrightarrow Zn + 4OH^-$	−1.216
$CdS + 2e^- \longrightarrow Cd + S^{2-}$	−1.21
$[Cr(OH)_4]^- + 3e^- \longrightarrow Cr + 4OH^-$	−1.2
$[SiF_6]^{2-} + 4e^- \longrightarrow Si + 6F^-$	−1.2
$V^{2+} + 2e^- \longrightarrow V$	*ca* −1.18
$Mn^{2+} + 2e^- \longrightarrow Mn$	−1.18
$[Cd(CN)_4]^{2-} + 2e^- \longrightarrow Cd + 4CN^-$	−1.03
$[Zn(NH_3)_4]^{2+} + 2e^- \longrightarrow Zn + 4NH_3$	−1.03
$FeS + 2e^- \longrightarrow Fe + S^{2-}$	−1.01
$PbS + 2e^- \longrightarrow Pb + S^{2-}$	−0.95
$SnS + 2e^- \longrightarrow Sn + S^{2-}$	−0.94
$Cr^{2+} + 2e^- \longrightarrow Cr$	−0.91
$Fe(OH)_2 + 2e^- \longrightarrow Fe + 2OH^-$	−0.877
$SiO_2 + 4H^+ + 4e^- \longrightarrow Si + 2H_2O$	−0.86
$NiS + 2e^- \longrightarrow Ni + S^{2-}$	−0.83
$2H_2O + 2e^- \longrightarrow H_2 + 2OH^-$	−0.828
$Zn^{2+} + 2e^- \longrightarrow Zn$	−0.763
$Cr^{3+} + 3e^- \longrightarrow Cr$	−0.74
$HgS + 2e^- \longrightarrow Hg + S^{2-}$	−0.72
$[Cd(NH_3)_4]^{2+} + 2e^- \longrightarrow Cd + 4NH_3$	−0.597
$Ga^{3+} + 3e^- \longrightarrow Ga$	−0.53
$S + 2e^- \longrightarrow S^{2-}$	−0.48
$[Ni(NH_3)_6]^{2+} + 2e^- \longrightarrow Ni + 6NH_3$	−0.47
$Fe^{2+} + 2e^- \longrightarrow Fe$	−0.440
$[Cu(CN)_2]^- + e^- \longrightarrow Cu + 2CN^-$	−0.43
$Cr^{3+} + e^- \longrightarrow Cr^{2+}$	−0.41
$Cd^{2+} + 2e^- \longrightarrow Cd$	−0.403
$Se + 2H^+ + 2e^- \longrightarrow H_2Se$	−0.40
$[Hg(CN)_4]^{2-} + 2e^- \longrightarrow Hg + 4CN^-$	−0.37
$ClO_4^- + H_2O + 2e^- \longrightarrow ClO_3^- + 2OH^-$	−0.36
$PbSO_4 + 2e^- \longrightarrow Pb + SO_4^{2-}$	−0.356
$In^{3+} + 3e^- \longrightarrow In$	−0.342
$[Ag(CN)_2]^- + e^- \longrightarrow Ag + 2CN^-$	−0.31
$Co^{2+} + 2e^- \longrightarrow Co$	−0.277
$[SnF_6]^{2-} + 4e^- \longrightarrow Sn + 6F^-$	−0.25
$Ni^{2+} + 2e^- \longrightarrow Ni$	−0.250
$Sn^{2+} + 2e^- \longrightarrow Sn$	−0.136
$CrO_4^{2-} + 4H_2O + 3e^- \longrightarrow Cr(OH)_3 + 5OH^-$	−0.13
$Pb^{2+} + 2e^- \longrightarrow Pb$	−0.126
$MnO_2 + 2H_2O + 2e^- \longrightarrow Mn(OH)_2 + 2OH^-$	−0.05
$[HgI_4]^{2-} + 2e^- \longrightarrow Hg + 4I^-$	−0.04
$2H^+ + 2e^- \longrightarrow H_2$	0.00
$NO_3^- + H_2O + 2e^- \longrightarrow NO_2^- + 2OH^-$	+0.01
$[Ag(S_2O_3)_2]^{3-} + e^- \longrightarrow Ag + 2S_2O_3^{2-}$	+0.01
$[Co(NH_3)_6]^{3+} + e^- \longrightarrow [Co(NH_3)_6]^{2+}$	+0.1
$S + 2H^+ + 2e^- \longrightarrow H_2S$	+0.141
$Sn^{4+} + 2e^- \longrightarrow Sn^{2+}$	+0.15
$Cu^{2+} + e^- \longrightarrow Cu^+$	+0.153
$Co(OH)_3 + e^- \longrightarrow Co(OH)_2 + OH^-$	+0.17
$[HgBr_4]^{2-} + 2e^- \longrightarrow Hg + 4Br^-$	+0.21
$AgCl + e^- \longrightarrow Ag + Cl^-$	+0.222
$Hg_2Cl_2 + 2e^- \longrightarrow 2Hg + 2Cl^-$	+0.27
$ClO_3^- + H_2O + 2e^- \longrightarrow ClO_2^- + 2OH^-$	+0.33
$Cu^{2+} + 2e^- \longrightarrow Cu$	+0.337
$[Fe(CN)_6]^{3-} + e^- \longrightarrow [Fe(CN)_6]^{4-}$	+0.36
$[Ag(NH_3)_2]^+ + e^- \longrightarrow Ag + 2NH_3$	+0.373
$O_2 + 2H_2O + 4e^- \longrightarrow 4OH^-$	+0.401
$[RhCl_6]^{3-} + 3e^- \longrightarrow Rh + 6Cl^-$	+0.44
$Ag_2CrO_4 + 2e^- \longrightarrow 2Ag + CrO_4^{2-}$	+0.446
$NiO_2 + 2H_2O + 2e^- \longrightarrow Ni(OH)_2 + 2OH^-$	+0.49
$Cu^+ + e^- \longrightarrow Cu$	+0.521
$TeO_2 + 4H^+ + 4e^- \longrightarrow Te + 2H_2O$	+0.529
$I_2 + 2e^- \longrightarrow 2I^-$	+0.5355
$[PtBr_4]^{2-} + 2e^- \longrightarrow Pt + 4Br^-$	+0.58
$MnO_4^- + 2H_2O + 3e^- \longrightarrow MnO_2 + 4OH^-$	+0.588
$[PdCl_4]^{2-} + 2e^- \longrightarrow Pd + 4Cl^-$	+0.62
$ClO_2^- + H_2O + 2e^- \longrightarrow ClO^- + 2OH^-$	+0.66
$[PtCl_6]^{2-} + 2e^- \longrightarrow [PtCl_4]^{2-} + 2Cl^-$	+0.68
$O_2 + 2H^+ + 2e^- \longrightarrow H_2O_2$	+0.682
$[PtCl_4]^{2-} + 2e^- \longrightarrow Pt + 4Cl^-$	+0.73
$Fe^{3+} + e^- \longrightarrow Fe^{2+}$	+0.771
$Hg_2^{2+} + 2e^- \longrightarrow 2Hg$	+0.789

Standard Electrode (Reduction) Potentials (continued)

Half-Reaction	$E°$, V	Half-Reaction	$E°$, V
$Ag^+ + e^- \longrightarrow Ag$	+0.7991	$Cr_2O_7^{2-} + 14H^+ + 6e^- \longrightarrow 2Cr^{3+} + 7H_2O$	+1.33
$Hg^{2+} + 2e^- \longrightarrow Hg$	+0.854	$Cl_2 + 2e^- \longrightarrow 2Cl^-$	+1.3595
$HO_2^- + H_2O + 2e^- \longrightarrow 3OH^-$	+0.88	$HClO + H^+ + 2e^- \longrightarrow Cl^- + H_2O$	+1.49
$ClO^- + H_2O + 2e^- \longrightarrow Cl^- + 2OH^-$	+0.89	$Au^{3+} + 3e^- \longrightarrow Au$	+1.50
$2Hg^{2+} + 2e^- \longrightarrow Hg_2^{2+}$	+0.920	$MnO_4^- + 8H^+ + 5e^- \longrightarrow Mn^{2+} + 4H_2O$	+1.51
$NO_3^- + 3H^+ + 2e^- \longrightarrow HNO_2 + H_2O$	+0.94	$Ce^{4+} + e^- \longrightarrow Ce^{3+}$	+1.61
$NO_3^- + 4H^+ + 3e^- \longrightarrow NO + H_2O$	+0.96	$HClO + H^+ + e^- \longrightarrow \frac{1}{2}Cl_2 + H_2O$	+1.63
$Pd^{2+} + 2e^- \longrightarrow Pd$	+0.987	$HClO_2 + 2H^+ + 2e^- \longrightarrow HClO + H_2O$	+1.64
$Br_2(l) + 2e^- \longrightarrow 2Br^-$	+1.0652	$Au^+ + e^- \longrightarrow Au$	*ca* +1.68
$ClO_4^- + 2H^+ + 2e^- \longrightarrow ClO_3^- + H_2O$	+1.19	$NiO_2 + 4H^+ + 2e^- \longrightarrow Ni^{2+} + 2H_2O$	+1.68
$Pt^{2+} + 2e^- \longrightarrow Pt$	*ca* +1.2	$PbO_2 + SO_4^{2-} + 4H^+ + 2e^- \longrightarrow PbSO_4 + 2H_2O$	+1.685
$ClO_3^- + 3H^+ + 2e^- \longrightarrow HClO_2 + H_2O$	+1.21	$H_2O_2 + 2H^+ + 2e^- \longrightarrow 2H_2O$	+1.77
$O_2 + 4H^+ + 4e^- \longrightarrow 2H_2O$	+1.23	$Co^{3+} + e^- \longrightarrow Co^{2+}$	+1.82
$MnO_2 + 4H^+ + 2e^- \longrightarrow Mn^{2+} + 2H_2O$	+1.23	$F_2 + 2e^- \longrightarrow 2F^-$	+2.87

APPENDIX J

Standard Molar Enthalpies of Formation, Standard Molar Free Energies of Formation, and Absolute Standard Entropies [298.15 K (25°C), 1 atm]

Substance	$\Delta H^\circ_{f298.15}$, kJ mol^{-1}	$\Delta G^\circ_{f298.15}$, kJ mol^{-1}	$S^\circ_{298.15}$, J K^{-1} mol^{-1}
Aluminum			
Al(*s*)	0	0	28.3
Al(*g*)	326	286	164.4
Al_2O_3(*s*)	−1676	−1582	50.92
AlF_3(*s*)	−1504	−1425	66.44
$AlCl_3$(*s*)	−704.2	−628.9	110.7
$AlCl_3 \cdot 6H_2O$(*s*)	−2692	—	—
Al_2S_3(*s*)	−724	−492.4	—
$Al_2(SO_4)_3$(*s*)	−3440.8	−3100.1	239
Antimony			
Sb(*s*)	0	0	45.69
Sb(*g*)	262	222	180.2
Sb_4O_6(*s*)	−1441	−1268	221
$SbCl_3$(*g*)	−314	−301	337.7
$SbCl_5$(*g*)	−394.3	−334.3	401.8
Sb_2S_3(*s*)	−175	−174	182
$SbCl_3$(*s*)	−382.2	−323.7	184
SbOCl(*s*)	−374	—	—
Arsenic			
As(*s*)	0	0	35
As(*g*)	303	261	174.1
As_4(*g*)	144	92.5	314
As_4O_6(*s*)	−1313.9	−1152.5	214
As_2O_5(*s*)	−924.87	−782.4	105
$AsCl_3$(*g*)	−258.6	−245.9	327.1
As_2S_3(*s*)	−169	−169	164
AsH_3(*g*)	66.44	68.91	222.7
H_3AsO_4(*s*)	−906.3	—	—

Standard Molar Enthalpies of Formation, Standard Molar Free Energies of Formation, and Absolute Standard Entropies [298.15 K (25°C), 1 atm] (continued)

Substance	$\Delta H^\circ_{f298.15}$, kJ mol^{-1}	$\Delta G^\circ_{f298.15}$, kJ mol^{-1}	$S^\circ_{298.15}$, J K^{-1} mol^{-1}
Barium			
$Ba(s)$	0	0	66.9
$Ba(g)$	175.6	144.8	170.3
$BaO(s)$	−558.1	−528.4	70.3
$BaCl_2(s)$	−860.06	−810.9	126
$BaSO_4(s)$	−1465	−1353	132
Beryllium			
$Be(s)$	0	0	9.54
$Be(g)$	320.6	282.8	136.17
$BeO(s)$	−610.9	−581.6	14.1
Bismuth			
$Bi(s)$	0	0	56.74
$Bi(g)$	207	168	186.90
$Bi_2O_3(s)$	−573.88	−493.7	151
$BiCl_3(s)$	−379	−315	177
$Bi_2S_3(s)$	−143	−141	200
Boron			
$B(s)$	0	0	5.86
$B(g)$	562.7	518.8	153.3
$B_2O_3(s)$	−1272.8	−1193.7	53.97
$B_2H_6(g)$	36	86.6	232.0
$B(OH)_3(s)$	−1094.3	−969.01	88.83
$BF_3(g)$	−1137.3	−1120.3	254.0
$BCl_3(g)$	−403.8	−388.7	290.0
$B_3N_3H_6(l)$	−541.0	−392.8	200
$HBO_2(s)$	−794.25	−723.4	40
Bromine			
$Br_2(l)$	0	0	152.23
$Br_2(g)$	30.91	3.142	245.35
$Br(g)$	111.88	82.429	174.91
$BrF_3(g)$	−255.6	−229.5	292.4
$HBr(g)$	−36.4	−53.43	198.59
Cadmium			
$Cd(s)$	0	0	51.76
$Cd(g)$	112.0	77.45	167.64
$CdO(s)$	−258	−228	54.8
$CdCl_2(s)$	−391.5	−344.0	115.3
$CdSO_4(s)$	−933.28	−822.78	123.04
$CdS(s)$	−162	−156	64.9
Calcium			
$Ca(s)$	0	0	41.6
$Ca(g)$	192.6	158.9	154.78
$CaO(s)$	−635.5	−604.2	40
$Ca(OH)_2(s)$	−986.59	−896.76	76.1
$CaSO_4(s)$	−1432.7	−1320.3	107
$CaSO_4 \cdot 2H_2O(s)$	−2021.1	−1795.7	194.0
$CaCO_3(s)$ (calcite)	−1206.9	−1128.8	92.9
$CaSO_3 \cdot 2H_2O(s)$	−1762	−1565	184
Carbon			
$C(s)$ (graphite)	0	0	5.740
$C(s)$ (diamond)	1.897	2.900	2.38
$C(g)$	716.681	671.289	157.987
$CO(g)$	−110.52	−137.15	197.56
$CO_2(g)$	−393.51	−394.36	213.6

Standard Molar Enthalpies of Formation, Standard Molar Free Energies of Formation, and Absolute Standard Entropies [298.15 K (25°C), 1 atm] (continued)

Substance	$\Delta H^\circ_{f_{298.15}}$, kJ mol^{-1}	$\Delta G^\circ_{f_{298.15}}$, kJ mol^{-1}	$S^\circ_{298.15}$, J K^{-1} mol^{-1}
$CH_4(g)$	−74.81	−50.75	186.15
$CH_3OH(l)$	−238.7	−166.4	127
$CH_3OH(g)$	−200.7	−162.0	239.7
$CCl_4(l)$	−135.4	−65.27	216.4
$CCl_4(g)$	−102.9	−60.63	309.7
$CHCl_3(l)$	−134.5	−73.72	202
$CHCl_3(g)$	−103.1	−70.37	295.6
$CS_2(l)$	89.70	65.27	151.3
$CS_2(g)$	117.4	67.15	237.7
$C_2H_2(g)$	226.7	209.2	200.8
$C_2H_4(g)$	52.26	68.12	219.5
$C_2H_6(g)$	−84.68	−32.9	229.5
$CH_3COOH(l)$	−484.5	−390	160
$CH_3COOH(g)$	−432.25	−374	282
$C_2H_5OH(l)$	−277.7	−174.9	161
$C_2H_5OH(g)$	−235.1	−168.6	282.6
$C_3H_8(g)$	−103.85	−23.49	269.9
$C_6H_6(g)$	82.927	129.66	269.2
$C_6H_6(l)$	49.028	124.50	172.8
$CH_2Cl_2(l)$	−121.5	−67.32	178
$CH_2Cl_2(g)$	−92.47	−65.90	270.1
$CH_3Cl(g)$	−80.83	−57.40	234.5
$C_2H_5Cl(l)$	−136.5	−59.41	190.8
$C_2H_5Cl(g)$	−112.2	−60.46	275.9
$C_2N_2(g)$	308.9	297.4	241.8
$HCN(l)$	108.9	124.9	112.8
$HCN(g)$	135	124.7	201.7
Chlorine			
$Cl_2(g)$	0	0	222.96
$Cl(g)$	121.68	105.70	165.09
$ClF(g)$	−54.48	−55.94	217.8
$ClF_3(g)$	−163	−123	281.5
$Cl_2O(g)$	80.3	97.9	266.1
$Cl_2O_7(l)$	238	—	—
$Cl_2O_7(g)$	272	—	—
$HCl(g)$	−92.307	−95.299	186.80
$HClO_4(l)$	−40.6	—	—
Chromium			
$Cr(s)$	0	0	23.8
$Cr(g)$	397	352	174.4
$Cr_2O_3(s)$	−1140	−1058	81.2
$CrO_3(s)$	−589.5	—	—
$(NH_4)_2Cr_2O_7(s)$	−1807	—	—
Cobalt			
$Co(s)$	0	0	30.0
$CoO(s)$	−237.9	−214.2	52.97
$Co_3O_4(s)$	−891.2	−774.0	103
$Co(NO_3)_2(s)$	−420.5	—	—
Copper			
$Cu(s)$	0	0	33.15
$Cu(g)$	338.3	298.5	166.3
$CuO(s)$	−157	−130	42.63
$Cu_2O(s)$	−169	−146	93.14
$CuS(s)$	−53.1	−53.6	66.5

Standard Molar Enthalpies of Formation, Standard Molar Free Energies of Formation, and Absolute Standard Entropies [298.15 K (25°C), 1 atm] (continued)

Substance	$\Delta H^\circ_{f298.15}$, kJ mol^{-1}	$\Delta G^\circ_{f298.15}$, kJ mol^{-1}	$S^\circ_{298.15}$, J K^{-1} mol^{-1}
$Cu_2S(s)$	−79.5	−86.2	121
$CuSO_4(s)$	−771.36	−661.9	109
$Cu(NO_3)_2(s)$	−303	—	—
Fluorine			
$F_2(g)$	0	0	202.7
$F(g)$	78.99	61.92	158.64
$F_2O(g)$	−22	−4.6	247.3
$HF(g)$	−271	−273	173.67
Hydrogen			
$H_2(g)$	0	0	130.57
$H(g)$	217.97	203.26	114.60
$H_2O(l)$	−285.83	−237.18	69.91
$H_2O(g)$	−241.82	−228.59	188.71
$H_2O_2(l)$	−187.8	−120.4	110
$H_2O_2(g)$	−136.3	−105.6	233
$HF(g)$	−271	−273	173.67
$HCl(g)$	−92.307	−95.299	186.80
$HBr(g)$	−36.4	−53.43	198.59
$HI(g)$	26.5	1.7	206.48
$H_2S(g)$	−20.6	−33.6	205.7
$H_2Se(g)$	30	16	218.9
Iodine			
$I_2(s)$	0	0	116.14
$I_2(g)$	62.438	19.36	260.6
$I(g)$	106.84	70.283	180.68
$IF(g)$	95.65	−118.5	236.1
$ICl(g)$	17.8	−5.44	247.44
$IBr(g)$	40.8	3.7	258.66
$IF_7(g)$	−943.9	−818.4	346
$HI(g)$	26.5	1.7	206.48
Iron			
$Fe(s)$	0	0	27.3
$Fe(g)$	416	371	180.38
$Fe_2O_3(s)$	−824.2	−742.2	87.40
$Fe_3O_4(s)$	−1118	−1015	146
$Fe(CO)_5(l)$	−774.0	−705.4	338
$Fe(CO)_5(g)$	−733.9	−697.26	445.2
$FeSeO_3(s)$	−1200	—	—
$FeO(s)$	−272	—	—
$FeAsS(s)$	−42	−50	120
$Fe(OH)_2(s)$	−569.0	−486.6	88
$Fe(OH)_3(s)$	−823.0	−696.6	107
$FeS(s)$	−100	−100	60.29
$Fe_3C(s)$	25	20	105
Lead			
$Pb(s)$	0	0	64.81
$Pb(g)$	195	162	175.26
$PbO(s)$ (yellow)	−217.3	−187.9	68.70
$PbO(s)$ (red)	−219.0	−188.9	66.5
$Pb(OH)_2(s)$	−515.9	—	—
$PbS(s)$	−100	−98.7	91.2
$Pb(NO_3)_2(s)$	−451.9	—	—
$PbO_2(s)$	−277	−217.4	68.6
$PbCl_2(s)$	−359.4	−314.1	136

Standard Molar Enthalpies of Formation, Standard Molar Free Energies of Formation, and Absolute Standard Entropies [298.15 K (25°C), 1 atm] (continued)

Substance	$\Delta H^\circ_{f298.15}$, kJ mol^{-1}	$\Delta G^\circ_{f298.15}$, kJ mol^{-1}	$S^\circ_{298.15}$, J K^{-1} mol^{-1}
Lithium			
$Li(s)$	0	0	28.0
$Li(g)$	155.1	122.1	138.67
$LiH(s)$	−90.42	−69.96	25
$Li(OH)(s)$	−487.23	−443.9	50.2
$LiF(s)$	−612.1	−584.1	35.9
$Li_2CO_3(s)$	−1215.6	−1132.4	90.4
Manganese			
$Mn(s)$	0	0	32.0
$Mn(g)$	281	238	173.6
$MnO(s)$	−385.2	−362.9	59.71
$MnO_2(s)$	−520.03	−465.18	53.05
$Mn_2O_3(s)$	−959.0	−881.2	110
$Mn_3O_4(s)$	−1388	−1283	156
Mercury			
$Hg(l)$	0	0	76.02
$Hg(g)$	61.317	31.85	174.8
$HgO(s)$ (red)	−90.83	−58.555	70.29
$HgO(s)$ (yellow)	−90.46	−57.296	71.1
$HgCl_2(s)$	−224	−179	146
$Hg_2Cl_2(s)$	−265.2	−210.78	192
$HgS(s)$ (red)	−58.16	−50.6	82.4
$HgS(s)$ (black)	−53.6	−47.7	88.3
$HgSO_4(s)$	−707.5	—	—
Nitrogen			
$N_2(g)$	0	0	191.5
$N(g)$	472.704	455.579	153.19
$NO(g)$	90.25	86.57	210.65
$NO_2(g)$	33.2	51.30	239.9
$N_2O(g)$	82.05	104.2	219.7
$N_2O_3(g)$	83.72	139.4	312.2
$N_2O_4(g)$	9.16	97.82	304.2
$N_2O_5(g)$	11	115	356
$NH_3(g)$	−46.11	−16.5	192.3
$N_2H_4(l)$	50.63	149.2	121.2
$N_2H_4(g)$	95.4	159.3	238.4
$NH_4NO_3(s)$	−365.6	−184.0	151.1
$NH_4Cl(s)$	−314.4	−201.5	94.6
$NH_4Br(s)$	−270.8	−175	113
$NH_4I(s)$	−201.4	−113	117
$NH_4NO_2(s)$	−256	—	—
$HNO_3(l)$	−174.1	−80.79	155.6
$HNO_3(g)$	−135.1	−74.77	266.2
Oxygen			
$O_2(g)$	0	0	205.03
$O(g)$	249.17	231.75	160.95
$O_3(g)$	143	163	238.8
Phosphorus			
$P(s)$	0	0	41.1
$P(g)$	58.91	24.5	280.0
$P_4(g)$	314.6	278.3	163.08
$PH_3(g)$	5.4	13	210.1
$PCl_3(g)$	−287	−268	311.7
$PCl_5(g)$	−375	−305	364.5

Standard Molar Enthalpies of Formation, Standard Molar Free Energies of Formation, and Absolute Standard Entropies [298.15 K (25°C), 1 atm] (continued)

Substance	$\Delta H^\circ_{f_{298.15}}$, kJ mol^{-1}	$\Delta G^\circ_{f_{298.15}}$, kJ mol^{-1}	$S^\circ_{298.15}$, J K^{-1} mol^{-1}
$P_4O_6(s)$	−1640	—	—
$P_4O_{10}(s)$	−2984	−2698	228.9
$HPO_3(s)$	−948.5	—	—
$H_3PO_2(s)$	−604.6	—	—
$H_3PO_3(s)$	−964.4	—	—
$H_3PO_4(s)$	−1279	−1119	110.5
$H_3PO_4(l)$	−1267	—	—
$H_4P_2O_7(s)$	−2241	—	—
$POCl_3(l)$	−597.1	−520.9	222.5
$POCl_3(g)$	−558.48	−512.96	325.3
Potassium			
$K(s)$	0	0	63.6
$K(g)$	90.00	61.17	160.23
$KF(s)$	−562.58	−533.12	66.57
$KCl(s)$	−435.868	−408.32	82.68
Silicon			
$Si(s)$	0	0	18.8
$Si(g)$	455.6	411	167.9
$SiO_2(s)$	−910.94	−856.67	41.84
$SiH_4(g)$	34	56.9	204.5
$H_2SiO_3(s)$	−1189	−1092	130
$H_4SiO_4(s)$	−1481	−1333	190
$SiF_4(g)$	−1614.9	−1572.7	282.4
$SiCl_4(l)$	−687.0	−619.90	240
$SiCl_4(g)$	−657.01	−617.01	330.6
$SiC(s)$	−65.3	−62.8	16.6
Silver			
$Ag(s)$	0	0	42.55
$Ag(g)$	284.6	245.7	172.89
$Ag_2O(s)$	−31.0	−11.2	121
$AgCl(s)$	−127.1	−109.8	96.2
$Ag_2S(s)$	−32.6	−40.7	144.0
Sodium			
$Na(s)$	0	0	51.0
$Na(g)$	108.7	78.11	153.62
$Na_2O(s)$	−415.9	−377	72.8
$NaCl(s)$	−411.00	−384.03	72.38
Sulfur			
$S(s)$ (rhombic)	0	0	31.8
$S(g)$	278.80	238.27	167.75
$SO_2(g)$	−296.83	−300.19	248.1
$SO_3(g)$	−395.7	−371.1	256.6
$H_2S(g)$	−20.6	−33.6	205.7
$H_2SO_4(l)$	−813.989	690.101	156.90
$H_2S_2O_7(s)$	−1274	—	—
$SF_4(g)$	−774.9	−731.4	291.9
$SF_6(g)$	−1210	−1105	291.7
$SCl_2(l)$	−50	—	—
$SCl_2(g)$	−20	—	—
$S_2Cl_2(l)$	−59.4	—	—
$S_2Cl_2(g)$	−18	−32	331.4
$SOCl_2(l)$	−246	—	—
$SOCl_2(g)$	−213	−198	309.7

Standard Molar Enthalpies of Formation, Standard Molar Free Energies of Formation, and Absolute Standard Entropies [298.15 K (25°C), 1 atm] (continued)

Substance	$\Delta H^\circ_{f_{298.15}}$, kJ mol^{-1}	$\Delta G^\circ_{f_{298.15}}$, kJ mol^{-1}	$S^\circ_{298.15}$, J K^{-1} mol^{-1}
$SO_2Cl_2(l)$	−394	—	—
$SO_2Cl_2(g)$	−364	−320	311.8
Tin			
$Sn(s)$	0	0	51.55
$Sn(g)$	302	267	168.38
$SnO(s)$	−286	−257	56.5
$SnO_2(s)$	−580.7	−519.7	52.3
$SnCl_4(l)$	−511.2	−440.2	259
$SnCl_4(g)$	−471.5	−432.2	366
Titanium			
$Ti(s)$	0	0	30.6
$Ti(g)$	469.9	425.1	180.19
$TiO_2(s)$	−944.7	−889.5	50.33
$TiCl_4(l)$	−804.2	−737.2	252.3
$TiCl_4(g)$	−763.2	−726.8	354.8
Tungsten			
$W(s)$	0	0	32.6
$W(g)$	849.4	807.1	173.84
$WO_3(s)$	−842.87	−764.08	75.90
Zinc			
$Zn(s)$	0	0	41.6
$Zn(g)$	130.73	95.178	160.87
$ZnO(s)$	−348.3	−318.3	43.64
$ZnCl_2(s)$	−415.1	−369.43	111.5
$ZnS(s)$	−206.0	−201.3	57.7
$ZnSO_4(s)$	−982.8	−874.5	120
$ZnCO_3(s)$	−812.78	−731.57	82.4
Complexes			
$[Co(NH_3)_4(NO_2)_2]NO_3$, *cis*	−898.7	—	—
$[Co(NH_3)_4(NO_2)_2]NO_3$, *trans*	−896.2	—	—
$NH_4[Co(NH_3)_2(NO_2)_4]$	−837.6	—	—
$[Co(NH_3)_6][Co(NH_3)_2(NO_2)_4]_3$	−2733	—	—
$[Co(NH_3)_4Cl_2]Cl$, *cis*	−997.0	—	—
$[Co(NH_3)_4Cl_2]Cl$, *trans*	−999.6	—	—
$[Co(en)_2(NO_2)_2]NO_3$, *cis*	−689.5	—	—
$[Co(en)_2Cl_2]Cl$, *cis*	−681.1	—	—
$[Co(en)_2Cl_2]Cl$, *trans*	−677.4	—	—
$[Co(en)_3](ClO_4)_3$	−762.7	—	—
$[Co(en)_3]Br_2$	−595.8	—	—
$[Co(en)_3]I_2$	−475.3	—	—
$[Co(en)_3]I_3$	−519.2	—	—
$[Co(NH_3)_6](ClO_4)_3$	−1035	−227	636
$[Co(NH_3)_5NO_2](NO_3)_2$	−1089	−418.4	350
$[Co(NH_3)_6](NO_3)_3$	−1282	−530.5	469
$[Co(NH_3)_5Cl]Cl_2$	−1017	−582.8	366
$[Pt(NH_3)_4]Cl_2$	−728.0	—	—
$[Ni(NH_3)_6]Cl_2$	−994.1	—	—
$[Ni(NH_3)_6]Br_2$	−923.8	—	—
$[Ni(NH_3)_6]I_2$	−808.3	—	—

APPENDIX K

Composition of Commercial Acids and Bases

Acid or Base	Specific Gravity	Percentage by Mass	Molarity	Normality
Hydrochloric acid	1.19	38%	12.4	12.4
Nitric acid	1.42	70%	15.8	15.8
Sulfuric acid	1.84	95%	17.8	35.6
Acetic acid	1.05	99%	17.3	17.3
Aqueous ammonia	0.90	28%	14.8	14.8

APPENDIX L

Half-Life Times for Several Radioactive Isotopes

(Symbol in parentheses indicates type of emission; *E.C.* = K-electron capture, *S.F.* = spontaneous fission; y = years, d = days, h = hours, m = minutes, s = seconds.)

Isotope	Half-Life	Emission
$^{14}_{6}C$	5770 y	(β^-)
$^{13}_{7}N$	10.0 m	(β^+)
$^{24}_{11}Na$	15.0 h	(β^-)
$^{32}_{15}P$	14.3 d	(β^-)
$^{40}_{19}K$	1.3×10^9 y	(β^- or *E.C.*)
$^{60}_{27}Co$	5.2 y	(β^-)
$^{87}_{37}Rb$	4.7×10^{10} y	(β^-)
$^{90}_{38}Sr$	28 y	(β^-)
$^{115}_{49}In$	6×10^{14} y	(β^-)
$^{131}_{53}I$	8.05 d	(β^-)
$^{142}_{58}Ce$	5×10^{15} y	(α)
$^{198}_{79}Au$	64.8 h	(β^-)
$^{208}_{81}Tl$	3.1 m	(β^-)
$^{210}_{82}Pb$	21 y	(β^-)
$^{212}_{82}Pb$	10.6 h	(β^-)
$^{214}_{82}Pb$	26.8 m	(β^-)
$^{206}_{83}Bi$	6.3 d	(β^+ or *E.C.*)
$^{210}_{83}Bi$	5.0 d	(β^-)
$^{212}_{83}Bi$	60.5 m	(α or β^-)
$^{207}_{84}Po$	5.7 h	(α, β^+, or *E.C.*)
$^{210}_{84}Po$	138.4 d	(α)
$^{212}_{84}Po$	3×10^{-7} s	(α)
$^{216}_{84}Po$	0.16 s	(α)
$^{218}_{84}Po$	3.0 m	(α or β^-)
$^{215}_{85}At$	10^{-4} s	(α)
$^{218}_{85}At$	1.3 s	(α)
$^{220}_{86}Rn$	54.5 s	(α)
$^{222}_{86}Rn$	3.82 d	(α)
$^{224}_{88}Ra$	3.64 d	(α)
$^{226}_{88}Ra$	1590 y	(α)
$^{228}_{88}Ra$	6.7 y	(β^-)
$^{228}_{89}Ac$	6.13 h	(β^-)
$^{228}_{90}Th$	1.90 y	(α)
$^{232}_{90}Th$	1.39×10^{10} y	(α, β^-, or *S.F.*)
$^{233}_{90}Th$	23 m	(β^-)
$^{234}_{90}Th$	24.1 d	(β^-)
$^{223}_{91}Pa$	27 d	(β^-)
$^{233}_{92}U$	1.62×10^5 y	(α)
$^{234}_{92}U$	2.4×10^5 y	(α or *S.F.*)
$^{235}_{92}U$	7.3×10^8 y	(α or *S.F.*)
$^{238}_{92}U$	4.5×10^9 y	(α or *S.F.*)
$^{239}_{92}U$	23 m	(β^-)
$^{239}_{93}Np$	2.3 d	(β^-)
$^{239}_{94}Pu$	24,360 y	(α or *S.F.*)
$^{240}_{94}Pu$	6.58×10^3 y	(α or *S.F.*)
$^{241}_{94}Pu$	13 y	(α or β^-)
$^{241}_{95}Am$	458 y	(α)
$^{242}_{96}Cm$	163 d	(α or *S.F.*)
$^{243}_{97}Bk$	4.5 h	(α or *E.C.*)
$^{245}_{98}Cf$	350 d	(α or *E.C.*)
$^{253}_{99}Es$	20.0 d	(α or *S.F.*)
$^{254}_{100}Fm$	3.24 h	(*S.F.*)
$^{255}_{100}Fm$	22 h	(α)
$^{256}_{101}Md$	1.5 h	(*E.C.*)
$^{254}_{102}No$	3 s	(α)
$^{257}_{103}Lr$	8 s	(α)
$^{263}_{106}(106)$	0.9 s	(α)

APPENDIX M

Vapor Pressure of Ice and Water at Various Temperatures

Temperature, °C	*Pressure, mmHg*	*Temperature, °C*	*Pressure, mmHg*	*Temperature, °C*	*Pressure, mmHg*
-10	2.1	18	15.5	80	355.1
- 5	3.2	19	16.5	90	525.8
- 2	4.0	20	17.5	95	633.9
- 1	4.3	21	18.7	96	657.6
0	4.6	22	19.8	97	682.1
1	4.9	23	21.1	98	707.3
2	5.3	24	22.4	99	733.2
3	5.7	25	23.8	99.1	735.9
4	6.1	26	25.2	99.2	738.5
5	6.5	27	26.7	99.3	741.2
6	7.0	28	28.3	99.4	743.9
7	7.5	29	30.0	99.5	746.5
8	8.0	30	31.8	99.6	749.2
9	8.6	31	33.7	99.7	751.9
10	9.2	32	35.7	99.8	754.6
11	9.8	33	37.7	99.9	757.3
12	10.5	34	39.9	100.0	760.0
13	11.2	35	42.2	100.1	762.7
14	12.0	40	55.3	100.2	765.5
15	12.8	50	92.5	100.3	768.2
16	13.6	60	149.4	100.5	773.7
17	14.5	70	233.7	101.0	787.5

APPENDIX B

Four-Place Table of Logarithms

No.	0	1	2	3	4	5	6	7	8	9	1	2	3	4	5	6	7	8	9
10	0000	0043	0086	0128	0170	0212	0253	0294	0334	0374	4	8	12	17	21	25	29	33	37
11	0414	0453	0492	0531	0569	0607	0645	0682	0719	0755	4	8	11	15	19	23	26	30	34
12	0792	0828	0864	0899	0934	0969	1004	1038	1072	1106	3	7	10	14	17	21	24	28	31
13	1139	1173	1206	1239	1271	1303	1335	1367	1399	1430	3	6	10	13	16	19	23	26	29
14	1461	1492	1523	1553	1584	1614	1644	1673	1703	1732	3	6	9	12	15	18	21	24	27
15	1761	1790	1818	1847	1875	1903	1931	1959	1987	2014	3	6	8	11	14	17	20	22	25
16	2041	2068	2095	2122	2148	2175	2201	2227	2253	2279	3	5	8	11	13	16	18	21	24
17	2304	2330	2355	2380	2405	2430	2455	2480	2504	2529	2	5	7	10	12	15	17	20	22
18	2553	2577	2601	2625	2648	2672	2695	2718	2742	2765	2	5	7	9	12	14	16	19	21
19	2788	2810	2833	2856	2878	2900	2923	2945	2967	2989	2	4	7	9	11	13	16	18	20
20	3010	3032	3054	3075	3096	3118	3139	3160	3181	3201	2	4	6	8	11	13	15	17	19
21	3222	3243	3263	3284	3304	3324	3345	3365	3385	3404	2	4	6	8	10	12	14	16	18
22	3424	3444	3464	3483	3502	3522	3541	3560	3579	3598	2	4	6	8	10	12	14	15	17
23	3617	3636	3655	3674	3692	3711	3729	3747	3766	3784	2	4	6	7	9	11	13	15	17
24	3802	3820	3838	3856	3874	3892	3909	3927	3945	3962	2	4	5	7	9	11	12	14	16
25	3979	3997	4014	4031	4048	4065	4082	4099	4116	4133	2	3	5	7	9	10	12	14	15
26	4150	4166	4183	4200	4216	4232	4249	4265	4281	4298	2	3	5	7	8	10	11	13	15
27	4314	4330	4346	4362	4378	4393	4409	4425	4440	4456	2	3	5	6	8	9	11	13	14
28	4472	4487	4502	4518	4533	4548	4564	4579	4594	4609	2	3	5	6	8	9	11	12	14
29	4624	4639	4654	4669	4683	4698	4713	4728	4742	4757	1	3	4	6	7	9	10	12	13
30	4771	4786	4800	4814	4829	4843	4857	4871	4886	4900	1	3	4	6	7	9	10	11	13
31	4914	4928	4942	4955	4969	4983	4997	5011	5024	5038	1	3	4	6	7	8	10	11	12
32	5051	5065	5079	5092	5105	5119	5132	5145	5159	5172	1	3	4	5	7	8	9	11	12
33	5185	5198	5211	5224	5237	5250	5263	5276	5289	5302	1	3	4	5	6	8	9	10	12
34	5313	5328	5340	5353	5366	5378	5391	5403	5416	5428	1	3	4	5	6	8	9	10	11
35	5441	5453	5465	5478	5490	5502	5514	5527	5539	5551	1	2	4	5	6	7	9	10	11
36	5563	5575	5587	5599	5611	5623	5635	5647	5658	5670	1	2	4	5	6	7	8	10	11
37	5682	5694	5705	5717	5729	5740	5752	5763	5775	5786	1	2	3	5	6	7	8	9	10
38	5798	5809	5821	5832	5843	5855	5866	5877	5888	5899	1	2	3	5	6	7	8	9	10
39	5911	5922	5933	5944	5955	5966	5977	5988	5999	6010	1	2	3	4	5	7	8	9	10
40	6021	6031	6042	6053	6064	6075	6085	6096	6107	6117	1	2	3	4	5	6	8	9	10
41	6128	6138	6149	6160	6170	6180	6191	6201	6212	6222	1	2	3	4	5	6	7	8	9
42	6232	6243	6253	6263	6274	6284	6294	6304	6314	6325	1	2	3	4	5	6	7	8	9
43	6335	6345	6355	6365	6375	6385	6395	6405	6415	6425	1	2	3	4	5	6	7	8	9
44	6435	6444	6454	6464	6474	6484	6493	6503	6513	6522	1	2	3	4	5	6	7	8	9
45	6532	6542	6551	6561	6571	6580	6590	6599	6609	6618	1	2	3	4	5	6	7	8	9
46	6628	6637	6646	6656	6665	6675	6684	6693	6702	6712	1	2	3	4	5	6	7	7	8
47	6721	6730	6739	6749	6758	6767	6776	6785	6794	6803	1	2	3	4	5	5	6	7	8
48	6812	6821	6830	6839	6848	6857	6866	6875	6884	6893	1	2	3	4	4	5	6	7	8
49	6902	6911	6920	6928	6937	6946	6955	6964	6972	6981	1	2	3	4	4	5	6	7	8
50	6990	6998	7007	7016	7024	7033	7042	7050	7059	7067	1	2	3	3	4	5	6	7	8
51	7076	7084	7093	7101	7110	7118	7126	7135	7143	7152	1	2	3	3	4	5	6	7	8
52	7160	7168	7177	7185	7193	7202	7210	7218	7226	7235	1	2	2	3	4	5	6	7	7
53	7243	7251	7259	7267	7275	7284	7292	7300	7308	7316	1	2	2	3	4	5	6	6	7
54	7324	7332	7340	7348	7356	7364	7372	7380	7388	7396	1	2	2	3	4	5	6	6	7
	0	1	2	3	4	5	6	7	8	9	1	2	3	4	5	6	7	8	9

2 3 4 5 6 7 8 9 0